UNLICENSED MOBILE ACCESS TECHNOLOGY

UNLICENSED MOBILE ACCESS TECHNOLOGY

Protocols, Architecture, Security, Standards and Applications

Edited by
Yan Zhang • Laurence T. Yang • Jianhua Ma

CRC Press
Taylor & Francis Group
Boca Raton   London   New York

CRC Press is an imprint of the
Taylor & Francis Group, an **informa** business
AN AUERBACH BOOK

Auerbach Publications
Taylor & Francis Group
6000 Broken Sound Parkway NW, Suite 300
Boca Raton, FL 33487-2742

© 2009 by Taylor & Francis Group, LLC
Auerbach is an imprint of Taylor & Francis Group, an Informa business

International Standard Book Number-13: 978-1-4200-5537-5 (Hardcover)

Library of Congress Cataloging-in-Publication Data

Zhang, Yan.
    Unlicensed mobile access technology : protocols, architectures, security, standards and applications / edited by Yan Zhang, Laurence T. Yang, Jianhua Ma.
        p. cm. -- (Wireless networks and mobile communications ; 11)
    Includes bibliographical references and index.
    ISBN-13: 978-1-4200-5537-5
    ISBN-10: 1-4200-5537-2
    1. Mobile computing--Congresses. 2. Mobile communication systems--Congresses. I. Yang, Laurence Tianruo. II. Ma, Jianhua. III. Title. IV. Series.

QA76.59.Z43 2008
004.165--dc22   2008008315

Visit the Taylor & Francis Web site at
http://www.taylorandfrancis.com

and the Auerbach Web site at
http://www.auerbach-publications.com

Contents

PART III: STANDARDS AND APPLICATIONS

Preface

This is the first book providing readers a complete cross-reference for unlicensed mobile access (UMA) technology. UMA technology targets to provide seamless access to global system for mobile communication (GSM) and general packet radio service (GPRS) mobile service networks over unlicensed spectrum technologies, including Bluetooth and Wi-Fi (IEEE 802.11), and possibly emerging WiMAX (IEEE 802.16). With a dual-mode enabled mobile terminal, a subscriber is able to roam freely and seamlessly handoff between cellular networks and unlicensed wireless networks. With intelligent horizontal and vertical handoff techniques in UMA, subscribers receive voice and data services continuously, smoothly, and transparently. To achieve these aims, there are a number of challenges. Mobility management is one of the most important issues to address. Vertical and horizontal handoff algorithms shall be intelligently designed to adapt to heterogeneous wireless environments. In addition, guaranteeing quality-of-service (QoS) during movement and handoff is also of great importance to satisfy subscribers' requirements. Furthermore, software-defined radio or cognitive radio is a key enabling technology for the success of UMA.

The book covers basic concepts, advances, and latest standard specifications in UMA technology, and also UMA-relevant technologies Bluetooth, Wi-Fi, and WiMAX. The subject is explored in a variety of scenarios, applications, and standards. The book comprises 17 chapters, topics of which span comprehensively to cover almost all essential issues in UMA. In particular, the discussed topics include system/network architecture, mobility management, vertical handoff, routing, Medium Access Control, scheduling, QoS, congestion control, dynamic channel assignment, and security. The book aims to provide readers with an all-in-one reference containing all aspects of the technical and practical issues in UMA technology.

The chapters in this book are organized into three parts:

- Part I: Architectures
- Part II: Protocols and Security
- Part III: Standards and Applications

Part I introduces the basics, QoS, resource management, mobility management, and security in UMA technology. Part II concentrates on the protocol issues and security challenges in UMA-related technologies, including WirelessPAN, Wi-Fi, and WiMAX. Part III presents the standard specifications and various applications.

This book has the following salient features:

- Provides a comprehensive reference for UMA technology
- Introduces basic concepts, efficient techniques, and future directions
- Explores standardization activities and specifications in UMA and related wireless networks Bluetooth, Wi-Fi, and WiMAX
- Offers illustrative figures that enable easy understanding

The book can serve as a useful reference for students, educators, faculties, telecommunication service providers, research strategists, scientists, researchers, and engineers in the field of wireless networks and mobile communications.

We would like to acknowledge the effort and time invested by all contributors for their excellent work. All of them are extremely professional and cooperative. Our thanks also go to the anonymous chapter reviewers, who have provided invaluable comments and suggestions that helped to significantly improve the whole text. Special thanks go to Richard O'Hanley, Catherine Giacari, and Stephanie Morkert of Taylor & Francis Group for their support, patience, and professionalism during the entire publication process of this book. Last but not least, special thanks should also go to our families and friends for their constant encouragement, patience, and understanding throughout the writing of this book.

Yan Zhang, Laurence T. Yang, and Jianhua Ma

Editors

Dr. Yan Zhang received his PhD from the School of Electrical and Electronics Engineering, Nanyang Technological University, Singapore. Since August 2006, he has been working with the Simula Research Laboratory, Norway (http://www.simula.no/). He is associate editor of *Security and Communication Networks* (Wiley), and he is on the editorial boards of the *International Journal of Network Security*, *Transactions on Internet and Information Systems*, *International Journal of Autonomous and Adaptive Communications Systems*, and the *International Journal of Smart Home*. He is the editor for the Auerbach Wireless Networks and Mobile Communications series. Dr. Zhang has served as guest coeditor for a few journals and selected papers. He has coedited numerous books, including, *Resource, Mobility and Security Management in Wireless Networks and Mobile Communications*; *Wireless Mesh Networking: Architectures, Protocols and Standards*; *Millimeter-Wave Technology in Wireless PAN, LAN and MAN*; *Distributed Antenna Systems: Open Architecture for Future Wireless Communications*; *Security in Wireless Mesh Networks*.

He has served as the workshop general cochair for COGCOM 2008, WITS-08, and CONET 2008, and has organized and cochaired numerous conferences since 2006. He has been a member of technical program committees for numerous international conferences including ICC, PIMRC, CCNC, AINA, GLOBECOM, and ISWCS. He received the best paper award and outstanding service award in the IEEE 21st International Conference on Advanced Information Networking and Applications. His research interests include resource, mobility, spectrum, energy, and security management in wireless networks and mobile computing. He is a member of IEEE and IEEE ComSoc.

Dr. Laurence T. Yang is a professor of computer science at St. Francis Xavier University, Antigonish, Nova Scotia, Canada. His research includes high-performance computing and networking, embedded systems, ubiquitous/pervasive computing, and intelligence.

He has published around 280 papers in refereed journals, conference proceedings, and book chapters in these areas. He has been involved in more than 100 conferences and workshops as a program/general conference chair and in more than 200 conferences and workshops as a program committee member. He has served as a chair, vice-chair, or cochair on a variety of IEEE Technical Committees and Task Forces.

In addition, he is the editor-in-chief of 10 international journals and a few book series. He is also an editor for 20 international journals. He has edited or contributed to 30 books and has won numerous best paper awards from the IEEE.

Dr. Jianhua Ma is a professor at the Faculty of Computer and Information Sciences, Hosei University, Japan, since 2000. He has had 15 years teaching/research experience at National University of Defense Technology, Xidian University, and the University of Aizu. From 1983 to 2003, his research focused on applications of wireless and mobile Web communications, e-learning, graphics rendering, Internet audio and video, and more. Since 2003 he has devoted his time to "smart worlds" and ubiquitous computing.

Dr. Ma is the coeditor-in-chief of three international journals and is the assistant editor-in-chief of the *International Journal of Pervasive Computing and Communications*. He is on the editorial board of *IJCPOL*, *IJDET*, *IJWMC*, and *IJSH*, and has edited more than 10 journal special issues as a guest editor. He has served as chair and committee member in many conferences/workshops.

Dr. Ma received many annual excellent paper awards from the Chinese Information Theory Society, Electronics Society, and the Association of Hunan Science and Technology. He received the best paper award at the IEEE International Conference on Information Society in the 21st Century (2000), and the highly commended paper award from the IEEE International Conference on Advanced Information Networking and Applications (2004). He received an appreciation certificate from the IEEE Computer Society for the years 2004–2007.

Contributors

Baher Abdulhai
Department of Civil Engineering
University of Toronto
Toronto, Ontario, Canada

Omar Ashagi
School of Computer Science and Informatics
University College Dublin
Dublin, Ireland

Adam Bachorek
Distributed Computer Systems Lab (DISCO)
University of Kaiserslautern
Kaiserslautern, Germany

Manish Kumar Batsa
Department of Electronics and Computer
   Engineering
Indian Institute of Technology
Roorkee, Uttarakhand, India

Nicolas Bihannic
CORE Networks Department
Orange Labs
Lannion, France

Bego Blanco
Department of Computer Languages
   and Systems
University of the Basque Country
Bilbao, Spain

Lin Cai
Department of Electrical and Computer
   Engineering
University of Victoria
Vancouver, British Columbia, Canada

Henry C.B. Chan
Department of Computing
Hong Kong Polytechnic University
Kowloon, Hong Kong

Lawrence Wong Wai Choong
Department of Electrical and Computer
   Engineering
National University of Singapore
Kent Ridge, Singapore

Mohamed El-Darieby
Software Systems Engineering
University of Regina
Regina, Saskatchewan, Canada

Armando Ferro
Department of Electronics
   and Telecommunications
University of the Basque Country
Bilbao, Spain

Ajay Gupta
Department of Computer Science
Western Michigan University
Kalamazoo, Michigan

Usman Javaid
CORE Networks Department
Orange Labs
Lannion, France

Jose Luis Jodra
Department of Electronics
   and Telecommunications
University of the Basque Country
Bilbao, Spain

Zill-E-Huma Kamal
Department of Computer Science
Western Michigan University
Kalamazoo, Michigan

Muhi A.I. Khair
Department of Computer Science
University of Manitoba
Winnipeg, Manitoba, Canada

S.P.T. Krishnan
Cryptography and Security Department
Institute for Infocomm Research
Singapore

Victor C.M. Leung
Department of Electrical and Computer
   Engineering
University of British Columbia
Vancouver, British Columbia, Canada

Fidel Liberal
Department of Electronics
   and Telecommunications
University of the Basque Country
Bilbao, Spain

Leszek Lilien
Department of Computer Science
Western Michigan University
Kalamazoo, Michigan

Kuang-Hao Liu
Department of Electrical and Computer
   Engineering
University of Waterloo
Waterloo, Ontario, Canada

Ivan Martinovic
Distributed Computer Systems Lab (DISCO)
University of Kaiserslautern
Kaiserslautern, Germany

Djamal-Eddine Meddour
CORE Networks Department
Orange Labs
Lannion, France

Jelena Mišić
Department of Computer Science
University of Manitoba
Winnipeg, Manitoba, Canada

Vojislav B. Mišić
Department of Computer Science
University of Manitoba
Winnipeg, Manitoba, Canada

Yasser Morgan
Software Systems Engineering
University of Regina
Regina, Saskatchewan, Canada

Hassnaa Moustafa
France Telecom R&D (Orange Labs)
Issy Les Moulineaux, France

Liam Murphy
School of Computer Science and Informatics
University College Dublin
Dublin, Ireland

Seán Murphy
School of Computer Science and Informatics
University College Dublin
Dublin, Ireland

Samuel Pierre
Department of Computer Engineering
Ecole Polytechnique de Montreal
Montreal, Quebec, Canada

Veselin Rakocevic
School of Engineering and Mathematical
   Sciences
City University
London, United Kingdom

Tinku Rasheed
Pervaise Group
Create-Net Research Center
Trento, Italy

Haidar Safa
Department of Computer Science
American University of Beirut
Beirut, Lebanon

Jens B. Schmitt
Distributed Computer Systems Lab (DISCO)
University of Kaiserslautern
Kaiserslautern, Germany

Xuemin (Sherman) Shen
Department of Electrical and Computer
    Engineering
University of Waterloo
Waterloo, Ontario, Canada

Enrique Stevens-Navarro
Department of Electrical and Computer
    Engineering
University of British Columbia
Vancouver, British Columbia, Canada

Bharadwaj Veeravalli
Department of Electrical and Computer
    Engineering
National University of Singapore
Kent Ridge, Singapore

Mohamed K. Watfa
Department of Computer Science
American University of Beirut
Beirut, Lebanon

Vincent W.S. Wong
Department of Electrical and Computer
    Engineering
University of British Columbia
Vancouver, British Columbia, Canada

Daqing Xu
Department of Information and Computing
    Science
Changsha University
Changsha, China

Zijiang Yang
Department of Computer Science
Western Michigan University
Kalamazoo, Michigan

Frank A. Zdarsky
Distributed Computer Systems Lab (DISCO)
University of Kaiserslautern
Kaiserslautern, Germany

Jie Zhang
Department of Electrical and Computer
    Engineering
University of British Columbia
Vancouver, British Columbia, Canada

Li Jun Zhang
Department of Computer Engineering
Ecole Polytechnique de Montreal
Montreal, Quebec, Canada

Yan Zhang
Simula Research Laboratory
Fornebu, Norway

ARCHITECTURES

I

Chapter 1

UMA Technology: Architecture, Applications, and Security Means

Hassnaa Moustafa

CONTENTS

Unlicensed mobile access (UMA) technology was born from the requirements of mobile integrated operators to deliver high-performance, low-cost, mobile voice and data services to subscribers at home and the office. With UMA, mobile operators can leverage the cost and performance advantages of Internet Protocol (IP) access technologies of fixed networks (DSL, cable, Wi-Fi, etc.) to deliver good-quality, low-cost, mobile voice and data services in locations where subscribers spend most of their time (home and office). Another trend in UMA technology is to extend the technology beyond homes and offices, precisely to hot spot areas.

Operators and service providers are seizing opportunities in fixed–mobile convergence (FMC) presented in UMA to expand their service offerings and to explore new business models and next generation technology for new revenue streams. Moreover, home and office users benefit from attractive pricing in addition to the advantage of always using the same terminal everywhere (inside home/office and outside) while reducing financial (pricing) and technological (radio signals being at home) burdens. This is in turn advantageous for operators and service providers in terms of attracting more clients. This growing interconnection among heterogeneous and diverse network systems presents just one of the many Achilles' heels of security issues facing operators and service providers. In fact, pure UMA security is crucial because the advent of dual-mode phones based on UMA technologies makes an operator's core infrastructure vulnerable to attacks from infected devices, while the subscribers may face service abuse such as stealth attacks and voice spam. In stealth attacks, the attacker could disconnect the network (e.g., by causing partitions or isolating nodes) to degrade its performance or could eventually modify routing information to hijack traffic. Moreover, UMA networks (UMANs) have numerous unique vulnerabilities at the application layer. These vulnerabilities can be exploited to launch a variety of attacks including floods, fuzzing, and stealth attacks. Consequently, reliability and performance in UMA is a major concern, with security being the key concern. Building a secure foundation is key for protecting future investment returns for operators and service providers, and a new level of security requirements should exist.

This chapter gives an overview of the architecture and services of UMA, discussing the different threats in UMA technology and presenting some security requirements for operators and service providers. The security solutions defined in the UMA standard are also presented giving an idea on how operators and service providers can build a secure foundation based on people, policy, and technology. Finally, the security implications of UMA for global system for mobile communications (GSM) security are illustrated especially focusing on the impact of open terminal platforms, where a number of countermeasures for mitigating risks are given.

1.1 UMA: Brief History and Evolution

Currently, the definition of standards allowing for transparent handover of the user connection between different radio technologies (vertical handover) is an area of intense activity. A number of standards in this domain have been approved or are under development, for example, the IEEE

802.21 standard. This is especially true for 802.11 and cellular technologies, aiming to exploit the rapid deployment of broadband and the use of wireless LANs (WLANs) within homes, offices, and hot spots. A concrete example is providing a high bandwidth and low-cost wireless access network, which is integrated into an operator cellular core network, enabling roaming between access networks with seamless continuity of service. In this context, the UMAC (Unlicensed Mobile Access Consortium) was formed by leading companies within the wireless industry to promote UMA technology and to develop its specifications. The UMAC worked with the 3GPP (Third Generation Partnership Project), which was established in 1998 through a collaboration agreement between different telecommunication standards bodies, to develop formal standards for UMA. The initial set of UMA specifications was published in September 2004, which details the use of the same device over a licensed radio spectrum connection (GSM) when users are outside the UMA coverage and using an unlicensed radio spectrum (Bluetooth or Wi-Fi) when being inside the UMA coverage. 3GPP defined UMA as a part of 3GPP release 6 (3GPP TS 43.318) under the name of GAN (generic access network).

UMA defines a parallel radio access network (RAN) known as the UMAN that interfaces with the mobile cellular core network using existing GSM-defined standard interfaces. This solution uses the IP tunneling technique to transparently extend mobile voice, data, and IP multimedia subsystem (IMS) services to mobile users through enabling service delivery to mobile phones over any WLAN Access Point (including Wi-Fi and Bluetooth). For seamless integration between existing mobile networks and unlicensed spectrum networks, a UMA-enabled handset is defined with dual-mode operation capable of connecting within both networks.

1.1.1 UMA Architecture

UMA technology allows mobile subscribers to seamlessly roam between mobile and home wireless networks or WLAN hot spots. As subscribers move between networks, they continue to receive mobile voice and data services in a consistent manner. In fact, subscribers within buildings (indoors) can obtain good-quality voice due to improved signal strength. Thanks to UMA, mobile users can take advantage of potentially faster data services through avoiding the bandwidth constraints of the GSM. Figure 1.1 illustrates the general UMA concept.

As illustrated in Figure 1.2 [1], connection to the fixed network occurs automatically when a mobile subscriber with a UMA-enabled dual-mode mobile handset moves within range of an unlicensed wireless network to which the handset is allowed to connect. Upon connecting, the handset contacts the UMA network controller (UNC) over the broadband IP access network to be authenticated and authorized to access GSM voice and GPRS data services via the unlicensed wireless network. If approved, the subscriber's current location information stored in the core network is updated, and from this point on, all mobile voice and data traffic is routed to the handset via the UMAN rather than the cellular RAN.

1.1.2 UMA Services

UMA technology delivers a number of key service advantages [2,3]. With UMA, mobile operators can allow millions of subscribers to securely access the mobile core service network over an IP access network (including the Internet). Because UMA is an IP layer solution that does not impact the physical radio access layer, different wireless technologies such as Wi-Fi, Bluetooth, or even next generation wireless IP technologies such as worldwide interoperability for microwave

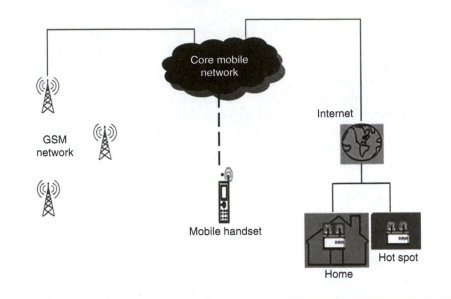

Figure 1.1   The UMA concept.

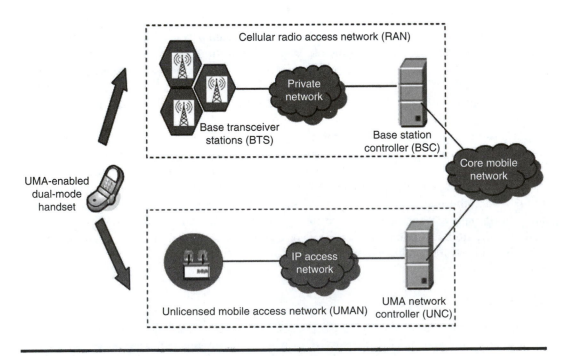

Figure 1.2   UMA architecture.

access (WiMAX). Consequently, services can be provided at different environments such as home, office, hot spot, coffee shop, campus, and airport.

Through UMA, all services available over GSM networks are available over IP access networks in a transparent manner. The following are some examples:

- Seamless mobility between cellular and IP access networks allows for providing true voice calls and data sessions continuity.
- Mobile users are able to make use of existing as well as new data services for entertainment, business, and education in a seamless manner. Also, advanced data services can be obtained thanks to the higher data rates compared to cellular networks.
- Always-on services such as IM, SMS, and MMS sessions do not have to end when the user goes home.
- Bandwidth-intensive mobile services such as mobile games and MP3 downloads do not have to end when the user goes home.
- Future high-value multimedia services over IMS such as push-to-talk, Voice-over-IP (VoIP), and IP video are also available.

1.1.3 Benefits of UMA for Mobile Operators and Service Providers Benefit

Over recent years, a number of market trends and industry developments have combined to make a practical business proposition for UMA. Mobile operators and service providers can thus exploit the rollout of broadband data connections and WLANs to offer a single user device for both cellular and fixed-line connectivity. UMA technology can allow mobile operators and service providers to maximize their revenue potential and improve subscriber retention by increased use of mobile phones. The following benefits for mobile operators, service providers, as well as clients could hence be achieved:

- Optimizing the use of GSM radio network resources by using an alternative lower-cost and higher-bandwidth access network.
- Reducing capital and operational expenditure on radio networks by using an alternative low-cost access network.
- Providing advanced and consistent services over both fixed and mobile networks.
- Offering bundled fixed and mobile services, making the mobile handset the customer's only phone, thereby increasing their share of the customer's total expenditure.
- Greatly increasing the use of mobile voice and data services in locations where usage was discouraged due to cost or network coverage.
- Delivering enhanced reach as well as improved voice quality.
- Bringing increased usage and allowing new services to be offered, thanks to delivering broadband data rates to the handsets.
- Because operators have a lower cost to deliver the service, they will be able in the near future to achieve higher margins and offer more aggressive pricing to their subscribers.
- Clients (users) have the advantage of using the same terminal everywhere (inside home and outside).
- Clients (users) benefit from economical (special pricing) and technical advantages (radio coverage at their homes and offices).

1.2 UMA Threat Analysis

Although UMA technology enables operators to easily expand their coverage and introduce new mobile data services, such services will not be widely adopted if there is a threat to their availability or integrity. Thus, the security and the availability of services are important in driving the success of new service offerings. Consequently, network operators and service providers could not launch UMA technology without knowing how to secure it. Also, the latter would not permit poor security to spoil their business. This section gives an analysis of the possible threats and types of attack in UMANs. In addition, some important security requirements are illustrated.

1.2.1 Different UMA Threats and Possible Attacks

Nowadays, riding on the momentum of FMC new services are being rolled out by service providers in an unprecedented manner. Consequently, failures in security implications can threaten gains in revenue and brand recognition for any new service offering.

It is observed from the security risk assessments of several leading service providers' networks [4] that the core operational infrastructure of these networks could be easily accessed and compromised. Facing this fact are two types of risks. First, the critical infrastructure of service providers is at risk of significant damage by attackers. Also, security incidents can negatively impact a service provider's reputation, leading directly to brand damage and loss of revenues. Second, entrepreneurs served by mobile service providers could be highly concerned with security, and they would slow their investment in mobile technology until the security issue is addressed.

In fact, the introduction of the UNC into the GSM/GPRS core also exposes the network to new security threats. Consequently, a number of threats could result, due to these main reasons:

- Opening traditional GSM/GPRS RAN to a public IP world increases the attacks against the network, especially man-in-the-middle attacks and denial-of-service (DoS) attacks, which could highly impact the services' access.
- The fact that UNC is publicly reachable threatens the network's functionality and hence the clients access to the offered services.
- Known security concerns also exist in WLAN, for example, eavesdropping.

As a result, UMA technology is vulnerable to two main types of threats: (1) UMA subscriber threats and (2) UMA subscriber service threats.

In UMA subscriber threats, a malicious subscriber can act as an intruder with a cloned or stolen handset and data terminal. Also, the Internet allows a number of attacks against subscribers. As a consequence, some possible attacks arise taking the following forms:

- Malicious exploitation causing system shutdown or connection disturbance
- Intrusion attacks that can lead to unauthorized access (of a nonlegitimate subscriber) as well as unauthorized installation (through a UMA subscriber or the Internet), thus damaging the whole communication
- DoS attacks from UMA subscribers or from the Internet
- Man-in-the-middle attacks from the Internet that can lead to traffic redirection or even data manipulation
- Stealth attacks and voice spam

On the other hand, UMA subscriber service threats are mainly similar to GSM/GPRS threats as well as some UMA-specific threats. The following are some possible attacks that can take place:

- DoS attacks from GSM/GPRS access network to UMA network and subscribers
- DoS attacks from Gi side public interface via UNC or to UMA subscribers

Some of these security challenges can be mitigated by technical solutions. For example, adding an additional security gateway (SGW) may address some of the potential malicious attacks. However, to appropriately address most of the security challenges, service providers need to think beyond technology and add a policy process into the overall solution. Section 1.2.2 highlights a number of requirements in UMA security.

1.2.2 UMA Security Requirements

Because UMA opens the mobile packet core to the public Internet, network-based security is thus a critical component in UMA deployment. In this context, the 3GPP specification for the UMA requires subscriber security and employs the SGW to provide subscriber-facing security. The following protocols are required to achieve this (more details on UMA security specification are given in Section 1.2.3):

- Internet Key Exchange v2 (IKEv2) with Extensible Authentication Protocol-Subscriber Identity Module (EAP-SIM) for registration, authentication, and integrity verification of mobile users
- IPSec encryption to ensure traffic privacy

It is observed that UNC is the core element in UMA technology, performing the same function as a base station controller (BSC) in a GSM/GPRS network. In this context, 3GPP had defined a standard interface specification on the UNC to address basic security requirements. These include unlicensed interface security, Up interface security, authentication and GSM/GPRS ciphering, and data application security (e.g., HTTPS). This standard-based security only provides part of the solution, providing a base level of security that some service providers may find to be acceptable. However, they do not address all security dimensions within the UMA operational environment supported by people, process, and technology. Service providers thus need to consider some security implications of adding the UNC into their network to protect their investment, brand image, and new revenue generating services. Indeed, service providers require cost-effective solutions that not only meet the required standards for securing subscriber connections but also provide comprehensive network-based security, massive scalability, and carrier-class reliability [5].

Finally, one should notice that most service providers implementing UMA already have some level of security architecture; UMA hence needs to be integrated into the existing security architecture and the operational security environment. In fact, a complete security solution first requires an in-depth investigation of the corresponding service provider/organization's goals, assets, and associated threats, then a security policy should be carefully determined together with the technologies to be integrated with a given set of technical solutions and the service provider's current environment. Consequently, one can notice that service providers should review existing security process, policy, and technology as a part of UMA implementation to truly understand potential security pitfalls.

1.2.3 Security Countermeasures in UMA

Security pitfalls are found to be mostly common among network operators and service providers, which can threaten UMA technology. In this context, the following countermeasures are useful and are simple to be deployed [4]:

- Increasing service providers' comprehensive perimeter of security measures
- Enhancing security patching and update processes
- Changing password policies that are seldom followed or updated
- Preventing control of network management equipment by unauthorized users
- Assuring nonvisibility of cellular subscribers to other subscribers and the Internet
- Maintaining confidentiality and integrity of sensitive information (for instance, information related to subscribers' profiles)
- Protecting identities and information communicated by subscribers
- Preventing attacks that deny the availability of services
- Preventing fraudulent use of services

1.3 UMA Security Solutions

UMA opens the mobile packet core to the public Internet for the first time through VoIP endpoints, creating security threats to calls and identity privacy. Indeed, UMA addresses the security challenge by incorporating a SGW to secure and aggregate end-user traffic. This gateway must be highly scalable to support millions of subscriber endpoints simultaneously. Although, the gateway must also provide network-based security to maintain the performance and reliability subscribers expect, this is not required in the UMA standards specifications. This section presents the UMA security specified in the UMA standards and presents some proprietary solutions, addressing some issues on gateway reliability and scalability within the UMA architecture.

1.3.1 Standard Security Solutions

UMA addresses the security challenge of opening the mobile packet core to the public Internet through incorporating a highly scalable SGW to secure and aggregate end-user traffic. The SGW provides subscribers confidentiality and data integrity by encapsulating the call and signaling data in secure IPSec tunnels. As shown in Figure 1.3, the SGW is positioned at the access edge of the core network and authenticates/registers users on the network every time the handset roams into a WLAN, regardless of whether or not a call is placed. Once authentication is established, the IPSec connection between the mobile station and the SGW remains active to ensure that the handset could immediately place and receive calls. The gateway must be highly scalable to support millions of subscribers simultaneously. The gateway must also provide network-based security to maintain the performance and the reliability that subscribers expect. The details of the UMA security process [6,7] are explained in the following subsections. These mechanisms aim at protecting the communication between the handset and the UNC; however, security of the GSM/GPRS core network reuses the existing GSM security mechanisms, which are reviewed below.

1.3.1.1 Protecting UMA

A UMA dual-mode handset supports GSM and 802.11 radio (this could also be Bluetooth) technologies and seamlessly routes calls over either a GSM RAN or a broadband access network. Upon

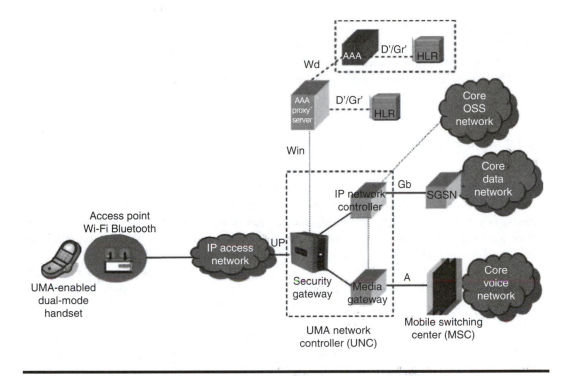

Figure 1.3 Handset establishment of an IPSec tunnel to the SGW.

entering a Wi-Fi hot spot, the handset establishes an IPSec tunnel through a public IP access network to the UNC, allowing the handset to place and receive calls using the UMA architecture. The tunnel is terminated at the UNC, which appears to the mobile core network as a BSC. This allows seamless handover between the in-building and cellular networks in the same way as between cells in an existing GSM network. In effect, subscribers have their own GSM micro-cell within their homes.

The UNC includes a SGW, which plays a number of key roles in the UMA architecture as follows:

1. authenticating the mobile user based on the subscriber profile, location, and activity status information stored in the Home Location Register (HLR),
2. decrypting the incoming traffic before forwarding it to the appropriate UMA application server, and
3. routing voice calls through a media gateway to the core voice network and circuit, and packet-based data services through an IP network controller toward the data core. This is illustrated in Figure 1.3.

The 3GPP has adopted a formal specification for the UMA architecture that places a high priority on subscriber security. The specification calls for the SGW to provide subscriber-facing security. In fact, the mobile user acts as the initiator, while the SGW acts as the responder. In this context, the following protocols are employed [8]:

- IKEv2 with EAP-SIM for authentication of mobile users with SIM only
- IKEv2 with Extensible Authentication Protocol Authentication and Key Agreement (EAP-AKA) for authentication of mobile users with USIM
- IPSec encryption to ensure privacy and data integrity for VoIP traffic

1.3.1.2 User Authentication

As previously mentioned, the UMA specification uses IKEv2 to perform mutual authentication of each mobile user and the core network and to establish and maintain security associations between handsets and SGWs. The advanced encryption standard (AES) encrypts this data traffic to ensure security. When the handset attempts to place a call from a Wi-Fi hot spot, the gateway must first authenticate the end station. The gateway then registers the user with the HLR as a roaming user that the mobile network can reach through the UMA media gateway (rather than a traditional GSM basestation). Once authenticated and registered, the dual-mode handset can place and receive calls through the Wi-Fi infrastructure as long as its UMA security tunnel remains active. It is important to notice that this process is transparent to end users, enabling them to seamlessly roam between GSM and Wi-Fi networks.

Because the mobile user and the SGW use EAP-SIM or EAP-AKA for mutual authentication, IKEv2 mandates that this is used in conjunction with a public key signature-based authentication of the SGW to the mobile user [8].

For integrating the UMA authentication process with existing authentication and billing infrastructure of mobile operators, the sign-on/sign-off process is applied for billing. The Wm interface is used to communicate with the Authentication, Authorization, and Accounting (AAA) server. In fact, UMA deployments initially used the Remote Authentication Dial-In User Service (RADIUS) protocol to authenticate users. However, it is expected that UMANs will migrate to DIAMETER as carriers begin to adopt this maturing technology.

1.3.1.3 Data Encryption

The UMA architecture uses IKEv2 to dynamically establish secure IPSec tunnels between a handset and a SGW. IPSec is useful in ensuring the security and integrity of wireless traffic traveling through one or more alternative provider networks. Four IKEv2-based cryptographic suites are defined in the UMA specification [7], enabling the SGW to encrypt the data (using one of these suites). Each suite includes

1. Encryption scheme to ensure data confidentiality based on either AES or 3DES.
2. Pseudorandom function (PRF) mechanism generating random numbers to be used in deriving new private/public key pairs and session keys. These random numbers are also used for padding.
3. Data integrity mechanism to encrypt and decrypt data using digital signatures with the SHA-1 or Message Digest 5 (MD5) algorithms, or AES-Cipher Block Chaining-Message Authentication Code (AES-CBC-MAC) in conjunction with a Diffie–Hellman public key.

For IPSec Encapsulating Security Payload (ESP), UMA specification defines four profiles for cryptographic algorithms that may be used between the mobile station and the SGW [7]. These are mainly based on AES and 3DES with a specific combination of confidentiality algorithm and an integrity algorithm. The four defined profiles should be supported by the SGW, while at least one profile should be supported by the mobile station.

As it is highly likely that almost all of the traffic initiating from WLAN access points will undergo port and address translation before reaching the public Internet, Network Address Translation-Traversal (NAT-T) is a key part of UMA specification, where the UMA specification assumes that all encryption will be done in an IPSec tunnel mode by a SGW that has built-in support for NAT-T.

1.3.1.4 Mobile Packet Core Protection

To provide another protection layer for core wireless network servers and gateways, the SGW should implement sophisticated quality of service (QoS) mechanisms that can limit the aggregate traffic bandwidth destined for this critical infrastructure. The gateway should also be able to use application and destination classification techniques to prioritize and mark traffic, ensuring that the packet core treats all services appropriately and predictably.

Traffic policing and traffic shaping are also useful for controlling the flow and volume of traffic streaming from the end station to the packet core. With potentially millions of subscriber endpoints requiring simultaneous authentication and access to the network, operators need a massively scalable SGW to ensure network availability and a reliable user experience.

1.3.1.5 GSM Security Mechanisms

In fact, deploying UMA technology does not require any change in the GSM/GPRS core network. Consequently, UMA reuses the existing GSM security mechanisms in this case. The important security features in GSM and GPRS systems are subscriber authentication, protecting transfer over the radio interface, temporary identities usage, and equipment identities usage [9].

- Subscriber authentication: This process is based on a permanent subscriber-specific secret key, which is stored at the authentication center and in the user's SIM which is a tamper-resistant smart card.
- Protecting the transfer over the radio interface: This process is mainly based on encryption, applying a stream cipher keyed by a secret session key. The secret session key is generated during the authentication procedure.
- Temporary identities usage: This process aims at protecting subscriber location privacy by limiting the number of occasions when the permanent identity of the subscriber, the international mobile subscriber identity (IMSI), needs to be sent over the air unencrypted.
- Equipment identities usage: This process aims at preventing the use of stolen phones or phones with severe malfunctions.

1.3.2 Security Gateways: Proprietary Solutions

As service providers interconnect 3G networks to UMA networks, security stands among the highest priorities during deployment. In this context, the UMA SGW is included in the specification of UMA technology. The SGW is designed to extend into UMANs, and in authentication, integrity, and security functions that already exist in wireless networks today.

To capitalize on the UMA market opportunity and ensure service availability, there is a need for a purpose-built SGW that also provides robust QoS, massive scalability, and network-facing security. In fact, a critical consideration for UMA deployment that is not considered by UMA specification is the design of security functionalities protecting core-facing servers and gateways. Consequently, the SGW should implement robust and stateful firewalls and DoS protection mechanisms for safeguarding network-facing servers.

The following subsections present examples of some existing deployment solutions for SGWs.

1.3.2.1 nCite Security Gateway

The nCite SGW [10] is deployed by Netrake and supports a pre-IMS market opportunity with a projected addressable market of 100 million VoIP endpoints expected by 2010. Indeed, the VoIP endpoint requires a security association to the core network to preserve privacy, integrity, and assurance. The nCite SGW provides this security association for both signaling and media using IPSec encryption. It also provides a simple migration strategy to IMS-based mobile and fixed networks and services, thus protecting IP services as they are delivered across any mobile or fixed access networks. This latter point allows meeting UMA requirements for evolution to IMS networks, coping with the rapid multimode convergence in both wireless carriers and broadband ISPs.

The nCite SGW offers multilayer security management for wireless carriers, ensuring IP communications integrity between user endpoints, network elements, and carrier networks. These include (1) network layer security association through employing IPSec, (2) application layer security association based on using Transport Layer Security (TLS), and (3) session layer security association mainly through SBC (session border controller).

Finally, a key feature in the nCite SGW is to enable carriers to transform their networks in a managed way without compromising network security, integrity, or performance. This ability allows carriers to migrate their networks to IMS in a managed environment. For example, carriers may deploy a UMAN to extend reach to residential subscribers or hot spots using IPSec, then upgrade software to SIP-based IMS.

1.3.2.2 Reef Point UMA Security Gateway

Reef Point is a multi-service massively scalable UMA SGW aiming to enable service providers to support much higher subscriber densities for their converged UMA and IMS services. Each Reef Point SGW can establish up to half a million secure IPSec connections, significantly reducing capital and operating costs.

The Reef Point SGW [6] provides comprehensive multi-service security for UMANs. This is achieved through concurrently providing robust threat defence including stateful firewalls with DoS attack prevention, intrusion detection services, custom firewall filtering, dynamic virtual routing with network address translation, and session limiting to protect against external threats. Consequently, session scalability does not cause problems to the performance of other security features.

There are two important features of this SGW. One is the purpose-built, carrier-class design, where uncompromising high-performance and reliability is provided through Reef Point's patented Flow Application Streaming Technology and optimum mix of custom ASICs/FPGAs and network processors. This unique design assures that the complete set of security services are delivered with uncompromising performance, even in the most stringent network deployments. The other is the robust wireless Standards Support, where it supports IKEv2 and EAP-SIM to provide scalable mutual authentication, encryption, and data integrity safeguards for signaling, voice, and data.

1.3.2.3 VPN-1 MASS Security Gateway

As carriers deploy next generation networks for FMC, they face the problem of merging their data and voice networks without compromising security. Check Point VPN-1 MASS (Multi-Access Security Solution) SGW [11] provides scalable secure access for next generation carrier networks.

Check Point VPN-1 MASS delivers the foundation of secure FMC for carriers, enabling them to deliver advanced communications products to their customers without compromising the network's security. With support for 3G Wireless Interworking (3G I-WLAN), UMA, and traditional remote access VPNs, VPN-1 MASS can be scaled up to provide remote access for up to 100,000 secure voice channels and massive amounts of data connections. The expected benefits of deploying the check point SGW are enabling additional access services for carrier customers, ensuring security of carrier networks against attack, and integrating with UMA and I-WLAN networks with ease.

1.4 Implications of UMA for GSM Security

GSM security in its own is seen as successful and reliable. In fact, subscribers do not get charged for calls they did not make, eavesdropping is sufficiently difficult, and security is mostly invisible to users, and does not depend on the user always making the right choices (unlike on the Internet, for instance).

Indeed, a main reason for successful security has been the use of closed platforms that prevent the end user from tampering with GSM protocol stacks. Although it is possible to build phones that do not have such restrictions, this is difficult due to, for example, legislation and technical complexity. Nowadays, with the emergence of UMA technology, access to GSM services is carried out over WLAN or Bluetooth. Consequently, this challenges the assumption of closed platforms. Section 1.4.1 presents the impact of open platforms and the possible resulting threats.

1.4.1 Impact of Open Platforms

The open platform is discussed in Ref. [12] and mainly concerns the existence of an open mobile terminal that can communicate in a GSM/GPRS network by running the UMA protocol stack on top of readily available hardware and operating systems. As a consequence, a number of threats could result.

In fact, the use of open platforms makes it easier to insert malicious software into the terminals of innocent users. Attackers could, for instance, distribute a virus or a Trojan horse that communicates directly with the SIM card, and thus can hijack the victim's identity and subscription. The consequence of this attack in GSM/GPRS networks could be so harmful that the victim pays for calls made by the attacker.

Another possible attack arises from using Bluetooth technology, where the Bluetooth SIM access profile [13] allows other Bluetooth devices to access the phone's SIM card. As a result, there is no need to compromise the dual-mode handset, but rather the virus or Trojan horse can access the handset from the victim's PC if, for instance, the laptop is paired with the handset. When the Bluetooth feature is often used, the SIM card does not require entering a PIN code when powered on and does not require explicit authorization every time the Bluetooth connection is used.

DoS attacks through resource exhaustion are also possible after a successful client authentication procedure. The GSM/GPRS authentication procedure verifies that the supplicant has a valid service subscription, but a successful authentication procedure does not imply that the user behind the device, or the device itself will not try to compromise the network. It is very difficult to trace, especially, an attacker using a prepaid subscription. An attacker could also be masquerading using a victim's compromised device, which in turn authenticates itself to the network transparently from the victim.

Furthermore, there is a risk of eavesdropping even if UMA requires traffic between the mobile terminal and UNC to be protected using IPSec. In fact, this will prevent users who have the capability to tamper with their terminal protocol stacks from eavesdropping other users' communication. However, it should be noted that the UMA specifications [7] state that it is possible to use NULL encryption for the IPSec tunnel, for example, in cases where high trust exists between the UMA operator and the access network provider. This exception is, however, based on a dangerous assumption.

Trust between a UMA operator and an access network provider does not imply by any means that subscribers in an access network of a provider trust each other. For example, communications from a subscriber connected through a WLAN link that uses weak security mechanisms are subject to eavesdropping from an attacker who resides within range of the WLAN link. To support the consumer's legacy WLAN equipment, the UMA specifications do not make any normative requirements on the security capabilities of the WLAN equipment. In addition, because operators do not necessarily have control on the subscribers' WLAN equipment, it is difficult to ensure that the recommended policies are conformed to. Consequently, operators should be very cautious when opting to use null encryption for the IPSec tunnel, thereby assuming confidentiality is accounted for at the lower layers.

1.4.2 Countermeasures for Mitigating Threats in Open Platforms

We notice that the potential attacks against GSM core networks caused by open platforms of the UMS technology are mainly two categories: internal or external attacks. In internal attacks, an attacker modifies his or her own terminal for one's own interest (to send malicious inputs). On the other hand, in external attacks, an attacker compromises a victim's terminal through a virus or a Trojan horse.

The following are some countermeasures against these attacks [12]:

- Protecting nonmalicious users' terminals
- Technical prevention of unapproved terminals
- Legal prevention of unapproved terminals
- Detecting and disabling misbehaving terminals
- Increasing core network resistance to attacks

1.5 Conclusion and Outlook

With UMA, operators can extend high-quality mobile services to provide increased revenue opportunities. Many service providers view UMA as a critical first step toward merging voice and multimedia services over an IMS architecture. Indeed, security is critical for successful UMA deployments and is one of the highest priorities as service providers interconnect 3G wireless networks to UMANs. This priority is reflected in the UMA technology industry group's inclusion of a UMA SGW in its specification. The SGW is designed to extend into UMANs, the authentication, integrity, and security functions that already are integral to wireless networks today.

With UMA being a relatively new technology, many of the specific threats are not yet well understood. This poses new challenges to GSM service providers as they increasingly need to enhance their service portfolio. As a consequence, it is imperative that service providers look beyond the base UMA security requirements defined by the standards but address security within the scope of their entire network.

One can notice that UMA security follows a subscriber-facing security model. However, network-facing protection is also an important point that merits consideration. One method could be through implementing firewall policies by operators to ensure that only Transmission Control Protocol (TCP) traffic or TCP traffic on specific ports is allowed to access the IP network controllers (INCs). This could also ensure that only User Datagram Protocol (UDP) traffic flows to the media gateway.

DoS attack is one of the most difficult attacks that cannot be prevented by simple packet filtering. For the effective prevention of DoS attacks, SGW should implement a stateful firewall controlling the traversing network connections. Consequently, this enables to prevent mal-formed, malicious, or suspicious packets from impacting core wireless infrastructure. Furthermore, single-station session limiting is also an important feature for protecting against DoS attacks that use session flooding to drain available UMA server resources and block legitimate ends from accessing UMA services.

Acknowledgment

Many thanks go to Fabio Costa (France Telecom R&D [Orange Labs]) for his useful comments on this chapter that helped to elaborate this work.

REFERENCES

1. UMA Technology, http://www.umatechnology.org/overview/index.htm.
2. Unlicensed Mobile Access (UMA) User Perspective (Stage 1), R1.0.0 (2004-09-01), Technical Specification. http://www.umatechnology.org/specifications/index.htm.
3. UMA Today, http://www.umatoday.com/.
4. UMA Security—Beyond technology, white paper, June 2006.
5. Securing the UMA network, white paper, ReefPoint Systems, 2005.
6. D. Racca, Security eases migration from UMA to IMS, white paper, Reef Point, 2006.
7. Unlicensed Mobile Access (UMA) Architecture (Stage 2), R1.0.4 (2005-5-2), Technical Specification. http://www.umatechnology.org/specifications/index.htm.
8. Unlicensed Mobile Access (UMA) Protocols (Stage 3), R1.0.4 (2005-5-2), Technical Specification. http://www.umatechnology.org/specifications/index.htm.
9. V. Niemi and K. Nyberg, *UMTS Security*, John Wiley & Sons, England, November 2003.
10. Security gateway, white paper, Netrake, 2005. http://www.audiocodes.com/objects/sbc/Netrake_SGWP.pdf.
11. Pure Security: VPN1 MASS scalable secure access for next generation carrier networks, white paper, Check point, 2007. http://www.checkpoint.com/products/downloads/vpn1_mass_datasheet.pdf.
12. S. Grech and P. Eronen, Implications of Unlicensed Mobile Access (UMA) for GSM Security, IEEE First International Conference on Security and Privacy for Emerging Areas in Communications Networks (SECURECOMM'05), Athens, Greece, September 2005.
13. Bluetooth SIG. SIM Access Profile Interoperability Specification 0.95, August 2002.

Chapter 2

UMA and Related Technologies: The Road Ahead

Usman Javaid, Nicolas Bihannic, Tinku Rasheed, and Djamal-Eddine Meddour

CONTENTS

2.1 Introduction

The proliferation of fixed and mobile access technologies, communication devices, and networks have highly enlarged the choice for network operators and service providers to offer a wide variety of services. This context is illustrated with access networks such as universal mobile telecommunication systems (UMTS), WiMAX, Wi-Fi, DVB, etc., and heterogeneous mobile terminals (e.g., smartphone, Personal Digital Assistant (PDA)), supporting an enhanced integration of applications. Currently, with the massive interest in the deployment of these mobile broadband wireless technologies, which offer complementary coverage and services and are targeting overlapping user chains and markets, the integration and convergence of these networks and technologies become not only possible but also a necessity to provide several value-added services to consumers at an affordable price. In the past few months, seamless convergence has been more than ever at the center of fixed and mobile operators' throughout the world, and it is considered to be the next big step in the evolution of telecommunication networks. For instance, wireless technologies such as Wi-Fi, WiMAX, etc. offer high data rates at low cost but do not guarantee seamless coverage especially with high mobility. Bluetooth technology supports low data rates compared to hot spot technologies, but on the contrary saves on the power consumption required for wireless access. In contrast, cellular networks such as GSM/GPRS and UMTS provide wide area coverage and support high mobility at higher cost but do not offer higher data rates.

In such a diverse environment, the concept of being always connected becomes always best connected [1,2]. This refers to being connected in the best possible way by exploiting the heterogeneity offered by access networks to experience a large variety of network services, particularly in the event of user mobility (accessing to services with various terminals). Moreover, end user devices are being increasingly equipped with multiple interfaces capacitating access to different wireless networks subject to network availability, device characteristics, and the applications used, all of which introduces the need for network interoperability in this heterogeneous environment. Also, the tremendous growth of connected wireless devices has augmented the endless competition for scarce wireless resources and has significantly exposed the challenges in heterogeneous network resource management.

In this increasingly heterogeneous networking architecture, seamless integration and convergence can be achieved in different ways by integrating technologies at different levels and ensuring different kinds of mobility between access networks [3]. From basic commercial convergence (unified billing for fixed/mobile/Internet), passing by service convergence (unified set of service without mobility management), to network convergence (transparent service delivery when changing access, with service continuity—seamless mobility), different strategies can be adopted by operators, depending on the services they want to deliver to their customers. In this chapter, we aim at discussing different technologies that offer seamless handover and access to mobile, voice, video, and data services. We provide a comprehensive survey of each of these technologies, by highlighting their design goals, architecture, and protocols. A comparison study illustrating their differences, advantages, and limitations is also presented. We conclude the chapter by presenting our personal views about the evolution of the discussed technologies toward the true seamless and ubiquitous convergence of these heterogeneous networks.

2.2 Seamless Convergence: Long-Term Vision

Beyond 3G (B3G) networks are composed of several MBWA (mobile broadband wireless access) technologies with heterogeneous elements at different levels ranging from services to user terminals.

This multidimensional heterogeneity is an obstacle for the massive deployment of B3G networks and also for their economic and social viability. Convergence is the key to offer a unified solution where all these heterogeneous components are coupled together to give a homogeneous outlook. Although the research community is interested in the integration and interworking of B3G heterogeneous elements at various scales, we present in this section the different flavors of convergence that can be performed in a service provider network.

Seamless network architectures can roughly be classified as those ensuring either the mobility of the users accessing services with various terminals or the mobility of the terminals accessing services across heterogeneous access networks. Each type of convergence is coupled with services that can be supported with different levels of integration and have different impacts on the network. In this regard, four types of convergence are proposed: convergence in the home network, in the access network, in the core network, and at the application server level.

2.2.1   Home Network Convergence

Home network convergence can be defined as the capability to break the silo approach where a terminal is dedicated to the use of a given service. Home network convergence allows the following benefits:

- User can access different types of services from a same terminal. An example is the ability to handle a call with an handset and be able to display on this handset content retrieved from another device in the home network.
- Service is available on more than one handset. For instance, the video-on-demand (VoD) service is not only displayed onto the user's television screen but can also be viewed on a personal computer or mobile handset.
- Diverse communication technologies are expected within the home network sphere. Most prominent among these are Wi-Fi, PLT, and Giga-Ethernet. Hence, the home network is a convergence arena where the devices are able to interoperate together and also with the service platforms. It also is a strong opportunity for the operator to leverage from these local exchanges to enrich already existing services.

Home network convergence is mainly built around the introduction of a home gateway equipment, embedding features like IP routing and application controls. This convergence mainly concerns fixed operators.

2.2.2   Access Network Convergence

Access network convergence can either be achieved with convergence at the transport layer or convergence at the service-oriented layer. The first one is mainly driven by the reduction in OPerational EXpenditure (OPEX) costs. An example is to aggregate mobile access nodes into a backhaul network shared with a fixed access network, or more generally the use of a shared infrastructure for heterogeneous access solutions. The convergence at the service-oriented layer allows access to a same service irrespective of the access network infrastructure.

2.2.3   Core Network Convergence

Core network convergence addresses convergence in the core network and is typically associated with the definition of a common framework able to handle any service invocation irrespective of the access network. An intuitive example is the specification of the IP Multimedia Subsystem (IMS), specified in the 3GPP and endorsed by Telecoms & Internet converged Services & Protocols for Advanced Network (TISPAN) [4] for the fixed broadband access network.

The IMS is a core network infrastructure to control user sessions for the following services:

- Conversational services with multimedia components like voice, video, etc.
- Real-time data-oriented services like instant messaging, presence, etc.
- Audiovisual services, in the scope of specifications for TISPAN in release 2.

The following are some of the benefits expected by an operator deploying IMS infrastructure:

- The aim for a fixed and mobile operator to use a common functional infrastructure is to control services to be as much access network agnostic as possible. A strong advantage on TTM (time-to-market) performance is also expected with an efficient integration of services once the IMS infrastructure is deployed.
- Enhanced mechanisms to reserve bandwidth on the user data path as negotiated during the session establishment between end user handsets. This allows the operator to better control resources, especially significant in the mobile domain for PS services.
- Service triggering toward application servers in accordance with the user service profiles.
- Solution for Public Switched Telephone Network (PSTN) renewal and expectations on OPEX/CAPEX (CAPital EXpenditure) reductions.

The implementation of an IMS infrastructure has numerous impacts on the services offered by the operator such as new capabilities on the terminal to support the Session Initiation Protocol (SIP) profile, updates of mobile gateways (like GGSN) to support a new interface for resource control, new application servers upon IMS either to handle SIP-based service logic or to interwork with legacy service platforms (like CAMEL (Customized Application for the Mobile network Enhanced Logic)-based for some mobile services), and finally on IS (information systems) for service provisioning.

2.2.4 Application Server Level Convergence

The convergence at the service platform level (also named application server level) can take two directions. First, it can be coupled with the introduction of a common control infrastructure like IMS to address heterogeneous networks (fixed or mobile) from the same service platform with the capability to offer differentiated quality of service (QoS) to end users. This case can largely meet carrier grade strategies. Second, it can be considered as a stand-alone convergence, as discussed below.

First, this stand-alone approach allows service providers to benefit from a generalized IP connectivity of terminals (fixed and mobile) to offer their services. This model is based on the Internet model with best-effort QoS different from IMS that allows the operator to set policy on QoS and potentially to charge services accordingly. This convergence proposed by the service providers (who do not generally own the network) can also be in a decentralized way, also called peer-to-peer (P2P). In this P2P model, the user accesses his or her services (voice, IM, or content sharing) due to the IP connectivity of all terminals in the community. This model can also be coupled with some centralized functions to extend the connectivity to a non-IP environment like PSTN.

This convergence at the application server level is not limited to the interests of service providers as carrier grade can also benefit from this type of convergence to enhance user experience. A possible example is streaming services on a mobile network with the use of application metrics to enrich the user experience: the application servers can directly upgrade or downgrade the streamed content based on the metrics report without triggering enhanced QoS mechanisms for a new network resource reservation as supported by the IMS, for example.

2.3 Existing Solutions toward Seamless Convergence

The past few years have seen tremendous growth in the number of wireless hot spots based on WiFi. Today, widely traveling laptop users access the Internet through WLANs at a variety of places and environment including their homes, offices, and public places. WLANs have emerged as a promising networking platform, which offers high data rates to mobile users at a very low network deployment cost. Anyone can simply plug a WLAN access point to the Internet and make it available to wireless users to enjoy connectivity. Normally, WLAN-based hot spots are deployed in areas with high user density and high bandwidth demands (e.g., in a town centre). In contrast, the BS (base station) in UMTS offers a larger cell with inter-BS links; the UMTS network provides nearly ubiquitous worldwide coverage. The integration between UMTS and WLAN networks extend the existing radio access technologies (RATs) and provide an economical solution to off-load a part for traffic from licensed to unlicensed spectrum technologies. Moreover, UMTS/WLAN integration provides an interesting blend, where the user can leverage the global coverage of UMTS and high data-rate support of WLAN.

To this end, SCCAN (seamless converged communications access networks) proposed to define converged services and mobility management in the private network (IP PBX). UMA (unlicensed mobile access) and I-WLAN (Interworking-WLAN) offer a more generalized approach, where convergence is managed by the access network infrastructure. While the integration of heterogeneous network entities are supported in the core network by architectures like IMS [4] and protocols like MIP (Mobile IP) [5,6], convergence may also be supported at the application level, managed by service platforms. On the other hand, the MIH (media independent handover) entity of IEEE 802.21 [7] is a flexible framework that does not intend to provide a stand-alone solution for FMC (fixed–mobile convergence), but rather assists the intertechnology handover decision and seamless interoperability in coordination with other mechanisms.

In this section, we aim at discussing in detail these complementary technologies by highlighting their design goals, architectures, and protocols.

2.3.1 Seamless Converged Communication across Networks

The SCCAN* is an industry-led standard governed by Motorola, Avaya, and Proxim. SCCAN supports an emerging open specification for technologies that enables seamless converged communications. By incorporating the most popular SIP of the IETF as a control protocol, SCCAN's specifications aim at the convergence of Wi-Fi technology with cellular networks for voice, video, and data services. SCCAN provides an enterprise solution that offers seamless interoperability between the Wi-Fi enabled enterprise network and cellularwide area networks.

SCCAN splits the functionality among dual-mode (Wi-Fi/cellular) handsets, mobility-enabled IP private branch exchanges (PBX), and WLAN gateways, as shown in Figure 2.1. When entering the office premises, the user's session switches from the cellular network to the Wi-Fi network. To assure this functionality in the core network, PBX has an SS7 (Signaling System 7) link to the

---

* SCCAN Forum (www.sccan.org).

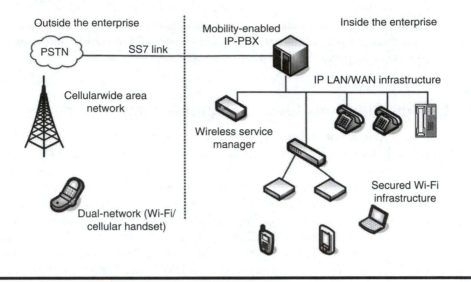

Figure 2.1 SCCAN enterprise solution architecture.

wireless carrier so that the location registration and call control can be performed upon session switching. SCCAN may present some advantages to set up customized business offers. However, from a deployment perspective, such a type of solution presents significant constraints to interconnect with the mobile infrastructure.

2.3.2 Unlicensed Mobile Access

UMA technology is designed to enable fixed–mobile convergence in an access network. It is currently endorsed by the 3GPP [8] under the name of GAN (generic access network). The terminology of GAN remains lesser known than UMA terminology and the latter continues to be used as a marketing term (in the rest of the chapter, we use the terms UMA and GAN interchangeably). The GAN architecture and functional components are shown in Figure 2.2. A major feature of GAN is to offer call continuity from a GAN-capable terminal between a local area network (UWB or 802.11) terminating at a fixed access and GSM infrastructure. Data services are also supported but are limited in throughput because interconnection to the PSCN (packet-switched core network) is

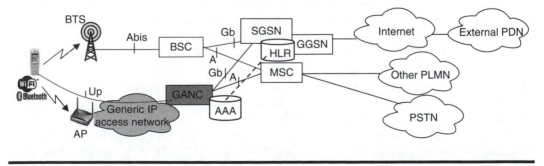

Figure 2.2 GAN architecture and functional components.

performed using the Gb interface. An outstanding evolution of GAN is to enrich user experience for data services as with the use of 3G Radio Resource Protocol and the support of interfaces. More precisely, UMA is today an available technology already deployed by certain operators like Orange with its Unik* offer.

In the GAN architecture, an IPSec (IP Security) tunnel is established on the Up Interface between the GAN terminal and the GANC (GAN controller). This flow tunneling is a strong security requirement that allows conveying both signaling flows and user data flows (GSM/GPRS signaling and user plane flows are piggy backed into GAN-specific protocols and the IPSec tunnel) over an access network (named generic IP access network) that is not supposed to be under the control of the mobile operator. The newly defined GANC entity reuses already 3GPP-defined interfaces namely Gb and A interfaces to interconnect to the PSCN and circuit-switched (CS) core network, respectively. Note that the AAA server is used to authenticate the GAN terminal when it sets up the secure tunnel. The following scheme in Figure 2.2 presents the architecture of GAN and its positioning versus the GSM/GPRS architecture.

2.3.3 Interworking-WLAN

3GPP is developing interworking solutions between 3G and WLAN networks under the auspices of Interworking-WLAN (I-WLAN), aiming to realize UMTS/WLAN integration [9,10]. (I-WLAN is a 3GPP standard that intends to define the Interworking architecture between a WLAN access network and a 3GPP core network).

2.3.3.1 I-WLAN Architecture

In the 3GPP Release 6 specifications that aim at providing access to mobile operator services from a WLAN Access Network, I-WLAN introduces three principal components for 3G/WLAN convergence: a WAG (wireless access gateway), a PDG (packet data gateway), and an AAA Server as shown in Figure 2.3. The UE (user equipment) is typically dual-mode capable: inside WLAN coverage, it is capable of connecting to the WLAN AN (access network) using Wi-Fi or Bluetooth before attachment to the I-WLAN infrastructure and when outside WLAN coverage, it can connect to a UMTS operator network. Data from UEs through ANs is aggregated at WAG, which is further connected to PDG. During roaming, the visited WAG is also able to route packets toward the home domain of the operator to which the user has subscribed The PDG in the I-WLAN architecture works as a gateway toward either the external packet data networks (PDNs) or the operator service infrastructure, as shown in Figure 2.3. PDG also interacts with the AAA server to perform service-level authorization, authentication, and accounting.

When entering into the coverage area of WLAN AN, the UE triggers its attachment procedure with the I-WLAN infrastructure and thus an IPSec tunnel is established between the UE and the PDG. Packet switched (PS) domain signaling and user plane data are carried into this secure tunnel over a Wu interface.

---

* Orange Unik (http://unik.orange.fr).

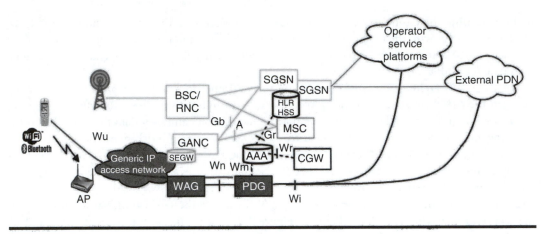

Figure 2.3   I-WLAN R6 architecture and functional components.

2.3.3.2   I-WLAN Protocols

The protocol stack between a WLAN UE and a PDG is shown in Figure 2.4. The protocols introduced in the I-WLAN stick are discussed below:

1. Remote IP Layer: The remote IP layer is used by the WLAN UE to communicate with the external PDN. The PDG routes the remote IP packets without modifying them.
2. Tunneling Layer: The tunneling layer consists of a tunneling header (IPSec), which allows end-to-end tunneling between a WLAN UE and a PDG. The tunneling layer is indeed used to encapsulate remote IP layer packets. The tunneling header contains the information further required by the PDG and the UE to decrypt the IP packets.
3. Transport IP Layer: The transport IP layer is used by the intermediate entities/networks and WLAN AN to transport the remote IP layer packets encapsulated into the IPSec tunnel.

2.3.3.3   I-WLAN Evolution: Release 7

I-WLAN Release 7 (R7) started in January 2005 with the aim of defining an evolved UMTS architecture. On the core network side, a new work item called system architecture evolution (SAE) was

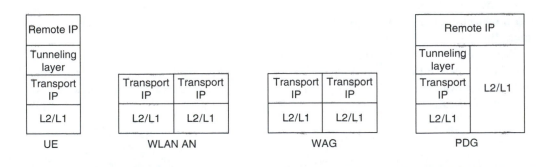

Figure 2.4   I-WLAN R6 protocol stack.

defined. In this evolved UMTS architecture, it is expected that IP-based services will be provided through various access technologies. A mechanism to support seamless mobility between heterogeneous access networks is needed for future network evolution. To this end, I-WLAN is included in the SAE to ensure a smooth migration path from the R6 I-WLAN work to a generic multi-access solution. Apart from seamless mobility across heterogeneous RATs, I-WLAN R7 also supports access to IMS and private networks from I-WLAN and LoCation Service (LCS) to enlarge the scope of location services deployed for GSM/UMTS. Some enhancements to support QoS on the WLAN access network are also in the scope of studies.

2.3.4 Media Independent Handover: IEEE 802.21

The IEEE 802.21 working group (WG) has decided MIH to offer seamless convergence across heterogeneous networks [7]. The MIH defines a framework to support information exchange that aids mobility decisions, as well as a set of functional components to execute these decisions. The MIH shields link-layer heterogeneity and provides a unified interface to upper-layer applications to support transparent service continuity. The handover scenarios considered in 802.21 WG include wired as well as wireless technologies—the complete IEEE 802 group technologies and 3GPP and 3GPP2 access network standards.

The MIH framework provides methods and procedures to gather useful information from the mobile terminal and the network infrastructure to facilitate handover between heterogeneous access networks. The MIH provides network discovery procedures that help the mobile terminal to determine networks that are available in its current neighborhood. The mobile terminal selects the most appropriate network with the help of gathered information such as link type and quality, application class, network policy, user profile, and power constraints.

2.3.4.1 MIH Architecture

The MIHF (media independent handover function) lies at the heart of the MIH architecture and performs an intermediary or a unified interface between the lower-layer heterogeneous access networks and higher-layer components, as shown in Figure 2.5. MIH provides generic access technology independent primitives called service access points (SAPs). SAPs are APIs (application programming interfaces) through which the MIHF can communicate with the upper- and lower-layer entities.

The MIHF facilitates three services namely media independent event service (MIES), media independent command service (MICS), and media independent information service (MIIS), which are responsible for signaling state changes at lower layers, control by higher layers, and provision of information regarding neighboring networks and their capabilities, respectively, as highlighted in Figure 2.5.

2.3.4.2 MIH Functional Components

In the following section, we describe MIH functional components such as MIES, MICS, and MIIS in greater detail.

Media independent event service. MIES provides timely lower-layer events (triggers) to the upper layers to optimize handover performance. The events can be local, that is, that have taken place within a mobile client or remote or sent by the network component. The event model works

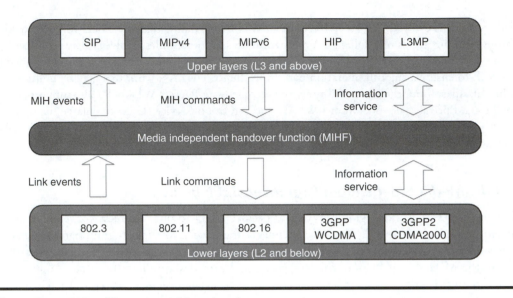

Figure 2.5   MIH architecture and functional components.

according to the notification/subscription principle. Because events are advisory and not mandatory, registration to a specific event is needed for an entity to be notified whenever such an event occurs.

MIES may be broadly classified into two categories: link events and MIH events. Both link and MIH events typically traverse from a lower layer to a higher layer. Link events are defined as events that originate from event source entities below the MIH function and typically terminate at the MIH function. Within the MIHF, link events may be further propagated, with or without additional processing, to upper-layer entities that have registered for the specific event. Events that are propagated by the MIH to the upper layers are defined as MIH events. Some of the common events include Link Going Down, L2 Handover Imminent, Link Parameters Change, etc. On the reception of a certain event, the upper layer makes use of the command service to react to the change in the network state.

Media independent command service. MICS refers to commands sent from higher layers to lower layers in the MIH framework. MIH commands are use to subscribe to certain information from the lower layers such as gathering information about the status of connected links, as well as execute higher-layer mobility and connectivity decisions at the lower layers. Similar to MIH events, commands can also be local and remote. Analogous to MIES, MICS can also be divided into two categories: MIH commands and link commands. Both types of commands follow the same principle as explained for MIES. Some of the common commands include MIH Poll, MIH Scan, MIH Configure, MIH Switch, etc.

Media independent information service. MIIS is used by a mobile node or a network entity to discover and obtain homogeneous and heterogeneous network information. The purpose of information service is to acquire a global view of the heterogeneous networks to facilitate seamless handover across those networks. For instance, when a mobile node is about to move out of the coverage of the current network, it queries the network (MIIS) about the available neighboring networks to optimize the handover process. MIIS provides access to both static and dynamic information. Static information may include names and providers of the neighboring networks. Examples of dynamic information include channel information, MAC address, and security information. MIIS stores the information in a standardized format such as ASN.1 or XML. A common higher-layer mechanism such as MIIS to offer information about the neighboring networks of heterogeneous access technologies alleviates the need for a specific access-dependent discovery method.

2.4 Limits and Potential of Seamless Convergence Solutions

The objective of this section is to provide a comprehensive comparison between seamless convergence techniques by illustrating the advantages and limitations of each solution.

SCCAN limits its scope to the enterprise market where the solution ensures the seamless interoperability between Wi-Fi enabled enterprise networks and cellularwide area networks. SCCAN also imposes strong investment constraints and introduces several security constraints. The proposed architecture works almost in the switched mode (PBX with SS7 signaling), and the evolution of the solution toward Voice-over-IP (VoIP) is a must and should be investigated.

UMA presents a good and evolved solution at short- and mid-terms for voice services over a CS core network. Further, the marriage of UMA and IP-networks has been already investigated and seems to be a viable solution. This does not prevent us from highlighting its many limitations, for example, an important overhead due to the complex protocol stack at the terminal (with the upper layers of a GSM/GPRS protocol stacking over a new UMA protocol stack for CS and PS domains, respectively). UMA also suffers from poor performance in the packet mode. Also, no QoS control mechanisms on the home LAN or on the access network is advocated within the current release of UMA. Finally, UMA also requires the use of a new terminal for the end user to access FMC services built into this technology.

The I-WLAN is an alternative solution to the UMA aiming to integrate UMTS core networks and WLAN. It is designed for packet-based services and is targeting a more long-term evolution than UMA. Indeed, enhancements with 3GPP R7 specifications on the I-WLAN are intended to support mobility and are released with the first stable SAE specification. Other advantages of the I-WLAN R7 are the QoS add-ons within the WLAN access network through the use of DiffServ mechanisms. The weak point for I-WLAN is the lack of support for circuit-based services. Consequently, it requires the operator to couple I-WLAN deployment with an IMS infrastructure with a VCC (voice call continuity) enabler to support voice services and the user mobility into/from legacy CS network.

The MIH is an interesting solution to facilitate seamless interoperability for a long-term network architecture evolution. MIH provides mechanisms to improve mobility policy for services supported over both PS and CS networks. The mobility processing could be triggered and enforced by either the network or the terminal; the interaction with the service platform is considered as an implementation issue. The major point to highlight here is that the mobility procedure remains under the control of the network operator.

MIH, as a generalized framework for mobility, can benefit from the mobility policy in a 3GPP architecture context. A potential architecture for the integration of MIH with the current network architecture is shown in Figure 2.6. Indeed, integration of some MIH functions can provide complementary information to those sent by the mobile terminal for user mobility processing. Particularly, the 3GPP standardization committee has put forth several requirements related to non-3GPP mobility support and they recognize that the network-based mobility scheme needs some form of media independent mobility signaling to coordinate the access change between networks. In this perspective, some MIH functions like those related to generate MIH events could be inserted into some access points composing the convergence network of the operator. As a significant example, the home gateway could inform the mobility manager (in charge of the mobility

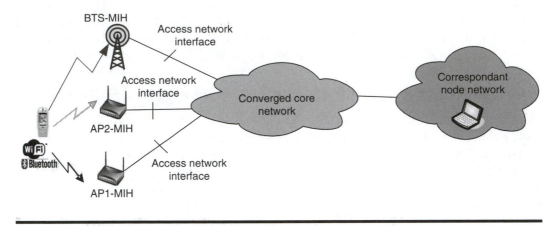

Figure 2.6 MIH potential integration with the current network architecture.

policy) with complementary information not available at the terminal side but necessary for optimal mobility processing, for instance, the available bandwidth on the radio interface used by the considered mobile. Actually, the evolution of the home LAN architecture leads to new constraints, and a global view on the state of resources on the home LAN is needed for better mobility management.

Numerous issues are still to be addressed to integrate MIH capabilities with existing 3GPP mechanisms. A first issue is on methods to convey reports of MIH events from the access points up to the mobility manager (in charge of the mobility policy). In the context of previous 3GPP technologies (GAN and I-WLAN), strong security constraints are set to interconnect with the mobile infrastructure. Two potential orientations can be investigated: the access through the SEGW (secure gateway) or a direct interconnection with the mobility manager. The first proposal is close to the security requirement set for any terminal with an IPSec tunnel establishment coupled with terminal authentication. The second one is more applicable in a context where there is trust between the mobile operator and the access network operator. A second issue is on correlation of reports of MIH events with reports generated by a mobile terminal. This issue is associated with the location management to correlate MIH event reports from a given access point with the reports of terminals attached to this same access point.

A comprehensive comparison of various technologies (assumed to be combined with IMS and Mobile-IP) discussed in this chapter is provided in Table 2.1. From the discussion on the potential advantages and limits of enabling technologies for seamless convergence, the architectural scenarios that could be potentially adopted for short-, mid-, and long-term network evolutions are presented in Table 2.2. This provides a quick overview of future architectures and technologies to be deployed in the network, and how a network operator could progressively evolve its network, providing the customer with a seamless mobility experience.

2.5   Conclusion

In this chapter, we presented an overall view of network convergence coupled with the services and their levels of integration and the different impacts on the network architecture. A comprehensive survey of different standardized seamless convergence solutions were presented with an in-depth discussion and comparison of their potential utilization and disadvantages. The main intention

Table 2.1 Comparison of Technologies for Seamless Convergence

	GAN/UMA	I-WLAN R6	I-WLAN R7	802.21/MIH	SCCAN
Radio Access Technology	Any (mainly WLAN) + 3GPP	Any (mainly WLAN) + 3GPP	Any (mainly WLAN) + 3GPP	802.xx + 3GPP	Enterprise WLAN + 3GPP
Standard status	Finalized	Finalized	Finalized	Almost finalized	Finalized
Convergence type	Fixed–mobile	Fixed–mobile	Fixed–mobile	Fixed–mobile + mobile–mobile	Fixed–mobile
Mobility management intra-AN	+	+	++	++	+
Mobility management inter-AN	++	+	++	++	++
Mobility management inter-CN	+	+	++	NA	NA
Handover support	+	+	+	++	+
Mobile subscriber ease of use	+	+	+	++	+
Security	++	++	++	+	–
Network controlled handover	++	++	++	++	++
Subscriber controlled handover	NA	NA	NA	++	NA
Evolving technical solution	–	+	++	++	–
Billing	–	++	++	NA	–
Impact on terminals	– –	– –	– –	–	– –

Note: ++, Strong advantage; +, advantage; –, drawback; – –, strong drawback; NA, not applicable.

Table 2.2 Recommendations for Seamless Convergence Architectures at Different Periods

	Commercial Launch (BTFusion, Unik)	Mid-Term Solution	Long-Term Solution
Voice	UMA	GAN R6	I-WLAN R7 +
Data	UMA	I-WLAN R6 + MIP	IMS + MIP + MIH
Convergence type	Fixed–mobile	Fixed–mobile	Fixed–mobile + mobile–mobile

Note: ■ Seamless mobility ■ No mobility ■ Mobility partially handled.

of the study was to assess different mid- and long-term architectural scenarios for convergence of heterogeneous access network mechanisms for seamless terminal mobility between different access networks ensuring continuity of service even for the most stringent types of applications.

Seamless convergence of heterogeneous access networks is essential in today's telecommunication systems. It allows operators to provide services without worrying about the user's location, access technology, or device, apart from avoiding the problems of maintaining multiple networks that obviously creates interoperability issues and complicates maintenance, support, and upgrading. Fixed–mobile convergence provides the opportunity for fixed-only operators to defend their business against mobile substitution and also enables integrated operators to avoid developing separate facilities for each type of network. In contrast, the seamless convergence of heterogeneous access technologies enables operators to offer enhanced services and an ultimate user experience. To complete the convergence puzzle, if today's approaches are mainly focused on integrating WLANs with UMA and I-WLAN as short- and mid-term solutions, the distant focus might be to integrate cellular and noncellular technologies. MIH coupled with mobility protocols at the higher layers (MIPv6, e.g.) and in coordination with IMS and I-WLAN R7 could be the future of seamless convergence enabled network architectures.

REFERENCES

1. U. Javaid, T. Rasheed, and D. E. Meddour, Cooperation in 4G systems—A service-oriented perspective, Workshop on Wireless Mesh and Sensor Networks Co-located with IEEE Vehicular Technology Conference (VTC), Dublin, Ireland, April 2007.
2. J. McNair and F. Zhu, Vertical Handoffs in Fourth-Generation Multinetwork Environments, *IEEE Wireless Communications*, 11(3): 8–15, June 2004.
3. European Commission ICT Report for Brussels Round Table, Telecoms in Europe 2015, February 2007.
4. G. Camarillo, *The 3G IP Multimedia Subsystem (IMS): Merging the Internet and the Cellular Worlds* 2nd Ed. John Wiley & Sons, February 2006.
5. I. F. Akyildiz, X. Jiang, and S. Mohanty, A survey of mobility management in next-generation all-IP-based wireless systems, *IEEE Wireless Communication*, 11(4): 16–28, August 2004.
6. IETF RFC 3775: Mobility Support in IPv6.

7. L. Vollero and F. Cacace, Managing mobility and adaptation in upcoming 802.21 enabled devices, Proceedings of the 4th International Workshop on Wireless Mobile Applications and Services on WLAN Hotspots (collocated with MOBICOM 2006), Los Angeles, CA, September 2006.

8. S. Parkvall, Long-Term 3G Evolution—Radio Access, Ericsson Research Report.

9. 3GPP Technical Specification: Group Services and Systems Aspects-Requirements on 3GPP System to WLAN Interworking (Release 6): 3GPP TS 22.234, 2005–2006.

10. 3GPP Technical Specification: Group Services and Systems Aspects-Requirements on 3GPP System to WLAN Interworking (Release 7): 3GPP TS 22.234, 2006.

Chapter 3

Quality of Service Management in UMA

Veselin Rakocevic

CONTENTS

Quality of service (QoS) in a network refers to the properties of the network that directly contribute to the degree of satisfaction that users perceive. Typically, QoS is deployed through traffic differentiation and by using appropriate resource management techniques to prioritize the treatment of different types of traffic. The recent emergence of heterogeneous, next-generation networks created new challenges for QoS management. In networks supporting unlicensed mobile access (UMA) technology, the problem of QoS management has not been investigated in great detail. This chapter analyzes issues that are important for successful QoS management in UMA. Main challenges are highlighted, and guidelines for the design of a policy-based management framework to implement, monitor, and control QoS provision are given.

3.1 Introduction

During the recent decade, mobile network operators have been able to provide voice and multimedia services to a huge number of end users. The penetration of mobile phones in the developed world is very close to the saturation point and new markets in the developing world are emerging. Complex wide-area, cellular networks have been designed to provide simple and good quality services.

However, for the operators to be able to increase the network bandwidth sufficiently to create the environment for the crucial step from the voice-oriented network to the multiservice multimedia network of the future, technical challenges remain. One of the main challenges in this respect is the need for higher bandwidth and end-to-end management of the QoS.

The last decade has seen a number of wireless access network technologies emerge alongside cellular networks. Examples of these new technologies include wireless local area networks (WLANs, better known as wireless fidelity, Wi-Fi), wireless metropolitan area networks wireless metropolitan area networks (WMANs) under the worldwide interoperability for microwave access (WiMAX) umbrella, and a range of wireless private area networks (WPANs), with Bluetooth technology the best known example. Mobile operators are working toward converging all these technologies into an integrated network capable of optimally satisfying user's needs, especially in terms of bandwidth. The unlicensed mobile access (UMA) technology has been developed for this purpose. UMA enables the integration of any fixed or wireless access network with the core GSM/GPRS (Global System Mobile/General Packet Radio Service) network. The principal concept of UMA is to enable any access network to be seen by the mobile core network as a standard GSM/GPRS access network. Using UMA, any access network can be used to access the Universal Mobile Telecommunication Mobile (UMTS) core network, too, in the same way any access GSM/GPRS network can. In this way, UMA enables network operators to easily expand coverage and introduce new mobile data services to existing GSM/GPRS customers. One of the main drivers for the deployment of UMA is the potential to improve indoor coverage. UMA has been developed as a joint standardization effort of major mobile network operators.

In the recent 3GPP (Third Generation Partnership Project) standardization documents and in the technical literature, UMA is referred to as a generic access network (GAN) [1]. Throughout this chapter we will be using both names for the UMA/GAN technology, to maintain consistency with the title of the book and the common use of the term in the literature.

The UMA standard specifies the network entities required to provide seamless connectivity between the GSM/GPRS core network and any access network. The standard defines procedures and functions needed to support seamless handover and roaming between UMA and GSM and also between UMA and UMTS to provide access to services in the mobile core network.

The importance of QoS in the development of multiservice networks has been highlighted in numerous analyses. This chapter contributes to the debate by analyzing the issues that exist in QoS management in UMA/GAN networks. The chapter shows that, for the majority of providers deploying UMA, QoS is not seen as the main issue, mostly because of the limited application of UMA and because of that fact that UMA connects two network systems that will rarely be operated by the same network operator, making it difficult to enforce QoS procedures.

This chapter is organized as follows. QoS is defined first as a general network problem. Existing solutions for QoS management in the fixed Internet and in wireless LANs and cellular networks are analyzed then, followed by a brief presentation of the policy-based QoS management framework, which is emerging as the dominant concept for QoS management in multiservice networks. After this, UMA/GAN and the limited existing QoS support for UMA/GAN networks are presented. The chapter then analyzes the design issues and potential problems in the development of a policy-based QoS management architecture for UMA/GAN, and also analyzes the integration of such a system in the existing architectural solutions for end-to-end QoS in multiservice networks.

3.2 Quality of Service

3.2.1 Overview

The QoS of a network refers to the properties of the network that directly contribute to the degree of satisfaction that users perceive. The perceived QoS depends on the type of application the user is running. There are many examples for this: the same network can have a poor quality if a user wants to hear an audio signal, but can be sufficient to download a text file relatively quicker.

In the business world, the QoS determines whether a normal voice conversation is possible, whether a video conference is of sufficient quality, or if a multimedia application improves productivity for the staff. At home, it determines whether the savings offered by an inexpensive voice service are worthwhile, or if there are complaints about the quality of a video-on-demand movie. Around the world, discriminating businesses and residential users demand higher QoS from the network.

If we consider video traffic as a global Internet service, when video display devices play back frames at rates of between 25 and 30 frames per second using the image persistence provided by display systems like the cathode ray tube, the human eye–brain perceives continuous motion. When losses or errors disrupt a few frames in succession, the human eye–brain detects a discontinuity. On the other hand, looking at the perception of an audio signal, the human ear–brain combination is less sensitive to short dropouts in received speech, being able to accept loss rates ranging from 0.5 percent to 10 percent depending upon the type of voice coding employed. This level of loss may cause an infrequent clicking noise, or a loss of a syllable. The perception of a delay also depends on the type of application. One-way communication like video broadcast (television) or an audio signal (radio) can accept relatively long absolute delays. However, delay impedes two-way, interactive communication if the round-trip latency exceeds 300 ms.

Higher-level protocols can also have an influence to the end user's perception of the network performance. Many audio and video protocols tolerate errors in the received information to a certain degree, but they are highly sensitive to delay variation. Streaming audio and video protocols employ a limited playback buffer to account for delay variation. If too little or too much data arrives when the application is playing back the audio or video, then the application either starves for data or overflows the playback buffer. Some data protocols respond to delay and loss through retransmission strategies to provide guaranteed delivery.

Selecting precise estimates for QoS parameters like loss, delay, and delay variation is not an easy task. Part of the difficulty arises from the subjective nature of perceived quality. A commonly employed approach groups applications with similar QoS requirements into broad generic classes and then specifies the QoS parameters for these classes. The idea is to enable the network to guarantee a specified QoS for traffic that conforms to a precisely defined set of parameters.

The emergence of mobile networks brought fresh challenges to the QoS provision. The randomness of the errors in wireless links increased the probability of lost packets and loss of synchronization for real-time traffic. The emergence of third-generation packet-switched (PS) mobile networks requires a new approach to QoS definition and management. This chapter analyzes QoS in mobile networks, especially for the PS services. The chapter also analyzes the currently prevalent option for QoS management in multiservice network, the policy-based network management (PBNM) paradigm.

3.2.2 Internet QoS

The flexibility of the Internet Protocol (IP) has enabled the huge success of the Internet as a complex network providing delivery service to numerous sophisticated applications. QoS has always been a major issue of debate in the Internet community. Up to now, even though several standards for Internet QoS exist, a majority of providers manage QoS by simple overprovisioning, hoping that this alone would be enough to satisfy the majority of the users.

For the analysis given in this chapter, it is necessary to present the two main QoS architectures developed for the Internet, the Integrated Services (IntServ) Architecture [2] and the Differentiated Services (DiffServ) Architecture [3]. The IntServ Architecture is based on a per-flow reservation of QoS. A special protocol, the Resource Reservation Protocol (RSVP), has been standardized to carry reservation messages through the network. The IntServ Architecture is generally considered to be inefficient because it is not scalable and provides a complex per-flow QoS management. Today, the RSVP is frequently used as the QoS reservation protocol for IP-based networks. The Differentiated Services (DiffServ) Architecture is a much more promising concept, offering a framework within which providers can offer each customer a service differentiation, where a "service" is defined as the overall treatment of a defined subset of customer traffic within the network.

In the DiffServ (DS) Architecture, the network consists of DS domains, contiguous sets of nodes that operate with a common set of service provisioning policies. Customers request a specific performance level on a packet-by-packet basis, by marking the DS field of each IP packet. The DS field is a TOS field of the IPv4 header or a Traffic class field of the IPv6 header. The first 6 bits of the DS field form a DS codepoint. Packets are classified in one of a small number of aggregated flows, based on the setting of bits in the DS codepoint. An aggregated flow is a number of flows that share forwarding state and a single resource reservation along a sequence of routers.

A DS codepoint value specifies the per-hop behavior (PHB) that packets receive on nodes along their path in the DS domain. PHBs are defined to permit a reasonably granular means of allocating buffer and bandwidth resources at each node among competing traffic streams. Therefore, QoS in the DiffServ Architecture can be deployed without any end-to-end signaling. At the DS boundary nodes (DS nodes that connect a DS domain to a node in another DS domain, or in a domain that is not DS-capable), the classification and conditioning of the incoming traffic is done. The packet classification policy identifies the subset of traffic that may receive a differentiated service by being conditioned or mapped to one or more behavior aggregates (by DS codepoint remarking) within the DS domain. Traffic conditioning performs metering, shaping, policing, or remarking to ensure that the traffic entering the DS domain conforms to the rules specified.

Differentiated Services is significantly different from Integrated Services. Because service is allocated in the granularity of a class, the amount of state information is proportional to the number of classes, not proportional to the number of flows. Differentiated Services therefore provide a scalable QoS solution to Internet service provider (ISP) networks. Also, sophisticated classification, authentication, marking, policing, and shaping operations are only needed at the boundary.

ISP interior routers need only to implement a classification scheme and some simple scheduling algorithm. In the DiffServ Architecture, because there is no signaled or per-flow control, performance guarantees rely on accurate dimensioning and the use of policers at the edge of the network to ensure that users remain with their agreed profiles.

The Internet QoS has been analyzed in numerous standards and research documents. For the analysis presented here, the key point is the traffic classification and packet marking concepts introduced by the Differentiated Services Architecture. The concept that the traffic needs to be classified in a limited number of traffic classes, and that each packet should carry a small tag identifying the traffic class is an important and highly scalable concept that is widely accepted as the optimal concept for traffic differentiation. We see in the remainder of this chapter that this is a concept accepted by the 3G QoS architecture standardization.

3.2.3 Quality of Service in Cellular Networks

Wireless cellular networks have been developed in the last decades and have been a huge success in providing means of communications to millions of users. The main application of these networks has always been and is still voice. With this in mind, it is clear that the foundation of any QoS scheme, algorithm, or management framework needs to satisfy the requirements of a large number of voice users. This makes the QoS requirements very different from those on the Internet. The GSM/GPRS network, although providing basic data services to the user, did not have full QoS support for data applications [4]. The implementation of traffic classification and QoS in GPRS is constrained by the fact that GPRS can differentiate QoS only on the basis of the IP address of a mobile station, but not on the basis of individual IP flows. In GPRS, a specific QoS profile (part of the Packet Data Protocol (PDP) context profile) is assigned to every subscriber upon attachment to the network. Further to this, the GPRS core network uses IP tunnels, which makes the applicability of IP QoS schemes troublesome.

On the other hand, the recent emergence of the third-generation mobile network (through the standardization of the UMTS network, e.g.) has made the QoS problem in cellular networks more interesting. The UMTS network contains two major network domains, the circuit switched (CS) for the provision of voice applications and the packet switched (PS) for the provision of multimedia and Internet-based applications. Network operators recognize the potential of multimedia applications and also the need for a sophisticated QoS management framework to support multimedia applications.

UMTS by its definition must be able to provide multimedia applications to end users while keeping the voice quality at the same level as its predecessor, GSM. UMTS achieves QoS management through the establishment of so-called bearer services, which are established between UMTS entities at different layers. To provide service differentiation, a UMTS network supports different bearer services that correspond to differentiation similar to the one deployed in IP network. The UMTS standard recognizes four distinct traffic classes: conversational traffic, streaming traffic, interactive traffic, and background traffic. To communicate the QoS management information to the external IP network, a special IP bearer service manager is needed. This is an important point for the design of the QoS management framework for the UMA network, as this chapter describes in more detail later.

The conversational traffic class is the only one where the traffic characteristics are strictly determined by human perception. In the streaming traffic class, delay variation will be limited to keep it at a manageable level. The limits for delay variation are set by the application. Longer delays and greater delay variation are more acceptable than in the conversational class. The interactive class describes classical data communication and is characterized by the request–response pattern of the end user. Round-trip delay (RTD) is one of the key attributes. Low error rate is important. Finally, the background traffic is delivery-time insensitive. More information about UMTS traffic classification is given in Ref. [5].

The behavior of traffic classes is defined with QoS attributes. The 3GPP document [5] defines a number of parameters: traffic class, maximum bit rate, guaranteed bit rate, indication of delivery order, maximum SDU size, SDU format information, SDU error ratio, residual bit error ratio, indication of delivery of erroneous SDUs, transfer delay, traffic-handling priority, allocation/retention priority, source statistics descriptor, and signaling indication.

In a typical UMTS scenario, the user equipment (UE) identifies the QoS requirement for a particular application. The UMTS standard is not precise on the method and protocol to be used for this identification. RSVP can be used, although the currently adopted method for QoS requirement identification is the use of the Session Description Protocol (SDP) carried over the Session Initiation Protocol (SIP). The SDP/SIP use has been encouraged by the development of the IP multimedia system (IMS) service management framework. IMS is seen as the dominant solution for complex service management in 3G networks. Although some authors see UMA and IMS as competing technologies, the scope of IMS is much wider than the scope of UMA. While UMA is usually used to extend the cell coverage of the cellular system, IMS is seen as a global management framework encompassing service provision, QoS, charging, and network management. UMA does not directly support SDP/SIP, although it can be used to forward any application information to the core network, therefore including the SDP/SIP information.

The process of connection management in UMTS networks is very important. In the PS domain of the UMTS network, connection management is usually referred to as session management. During this process, the PDP context is activated following the QoS negotiation. Different PDPs can be used, with IP the preferred option. The PDP context is a data structure present on both the serving GPRS support node (SGSN) and the gateway GPRS support node (GGSN), which contain the subscriber's session information when the subscriber has an active session.

In general, when a mobile station wants to use GPRS, it must first attach and then activate a PDP context. This allocates a PDP context data structure in the SGSN that the subscriber is currently visiting and the GGSN serving the subscribers access point (AP). The session management process is important for QoS management because within this process the QoS is being negotiated and the QoS class is identified.

QoS architecture design guidelines for 3G networks are given in detail in Refs. [5,6]. These references describe the network and service requirements for QoS architecture design and QoS management. For the analysis presented in this chapter, of special importance are the guidelines for end-to-end QoS management for cases when end-to-end traffic uses UMTS and also other networks other than UMTS to deliver the service. Interaction between UMTS QoS management and the external network is needed in this case.

3.2.4 Quality of Service in Wireless LANs

The success of the wireless LAN technology based on the IEEE 802.11 standard (the so-called Wi-Fi) has been tremendous, mostly due to the simplicity and flexibility of this networking technology. The 802.11 networks can be established anywhere. They operate in a distributed or, more frequently, centralized network topology with a special station—AP—controlling the communication between the end stations and the outside network. The 802.11 standard defines the bottom two layers of the protocol stack, including the medium access control, which is controlled using a Carrier Sense Multiple Access with Collision Avoidance (CSMA/CA).

The basic 802.11 standard offered no specific support for QoS. A good survey of open QoS issues for 802.11-based LANs is given in Ref. [7]. However, a number of recent modifications to the standard do offer some QoS support. The QoS enforcement in Wi-Fi networks is based on the concept of prioritization between participating end stations.

The 802.11p standard offers the MAC frame priority indication, and 802.11q offers virtual LAN tagging. QoS extensions defined in the 802.11e draft will provide fuller support for demanding applications such as voice or video in addition to providing prioritized access on a per-user or per-application basis. Future enhancements such as 802.11r will support faster handoffs, improving application performance for mobile devices.

3.2.5   Policy-Based QoS Management

Management of QoS in modern multiservice networks is a complex task. In general, QoS management refers to the activities of QoS specification, QoS negotiation, and monitoring and control of network resources to meet end-to-end user and application requirements, business objectives, and resource availability.

The problem becomes even more complicated in the heterogeneous network environment, where more than one network is responsible for traffic forwarding and QoS provision. A universal and well-defined method of communicating QoS requirements and QoS management is needed. In recent years, a dominant management concept for complex network operation management has been the PBNM. The PBNM framework defines the space for definition of policies—sets of rules and actions that need to be performed by network entities to achieve certain performance goals.

Internet Engineering Task Force (IETF) has defined a policy framework [8] with two main architectural elements: the policy decision point (PDP) and the policy enforcement point (PEP) (Figure 3.1). The PDP is the element that makes the policy decision. The PDP can make the decision on its own or may consult a remote policy repository (PR). The PDP makes a policy decision on the basis of the information it receives from the end user station, for example. This information can be the QoS requirement sent via a RSVP message. The PEP is the element where the policy is actually enforced. For example, in a Wi-Fi network, the PEP would be located at a wireless router, which connects the Wi-Fi with the external Internet. The PDF would, in most of the cases, be a separate entity or it would be located somewhere in the network of the ISP.

The policy-based QoS management is based on a set of policy rules for the enforcement of QoS in different networks. The detailed QoS request is typically specified in a service level agreement (SLA) between the user and the network. In most cases, the policy rules describe the amount of

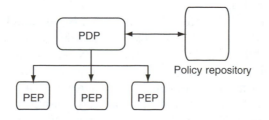

Figure 3.1   PBNM architecture.

network resources required to provide a certain QoS level defined in the SLA. Policy rules can be exchanged between different network domains, and the policy rule translation presents a major issue.

Well defined policies and policy enforcement rules are very important for successful cross-network deployment of QoS management. In the context of networks deploying UMA/GAN technology, the use of PBNM in QoS management is a natural choice, because sessions/connections use more than one network technology to forward the traffic.

Finally, it is important to note that the policy framework employed in the UMTS should, as far as possible, conform to the IETF Internet policy standards. The IETF policy framework may be used for policy decision, authorization, and control of the IP level functionality, at both user and network levels. However, there should be separation between the scope and roles of the UMTS policy mechanisms and the IP policy framework. This is to facilitate separate evolution of these functions.

3.2.6   QoS Management for Heterogeneous Mobile Networks

There is a substantial body of work on QoS management in heterogeneous mobile network environments. Deploying a scalable management framework to cover different network structures is a complicated task. More work is needed, especially in terms of an deployment and testing to ensure efficient and scalable QoS provision.

A majority of the existing solutions are based on PBNM. Architectures are presented that relay policy information, negotiate requirements, and enforce policies on various network technologies. A good example of this is the work done by Chakravorty et al. [9]. They present a complex system for QoS management over a network link that includes a generic IP network and the UMTS network. The analysis is based on different methods of relaying the information between different network points. In more detail, they assume that the mobile station determines the QoS requirements. Crucially, this information is mapped to the PDP context parameters. The main element for the processing of a QoS management model is the GGSN. When the GGSN learns about QoS requirements, it configures the IP classifier and provides interworking between the PDP context and the backbone IP network. The existence of an authorization token is assumed, where the token would include authentication information, packet handling action, and event generation information. Further, they point out that in the case the UE supports RSVP, RSVP can be used instead of PDP. They assume a system consisting of a number of functions responsible for subscription, classification, mapping, and policing.

A substantial body of work deals with the implementation of IMS to manage and provide QoS. The IMS Architecture enables the establishment and management of multimedia sessions on broadband mobile networks, most importantly on UMTS. The IMS architecture is complex, consisting of a number of hardware and software entities and is managed by a range of protocols. For a detailed description of IMS refer to Ref. [10].

Zhuang et al. [11] present a detailed analysis of a policy-based multidomain QoS management architecture in different UMTS–WLAN interworking scenarios. Their architecture is based on the existence of the master PDF, which belongs to the UMTS network operator. For the creation of the multidomain policy-based management framework it is important for them to have a policy-based QoS architecture for the WLANs. A PDF can enforce the policies at the wireless router of the WLAN directly. The PDF can be an independent entity controlling the WLAN only, or it can be an integrated part of the PDF in the UMTS PS domain. When the WLAN is not under the direct management of the operator—which is a general case of interest in this chapter—the wireless PDF (WPDF) is a peer of the MPDF. Network-level (end-to-end) policy rules still can be deployed, though. Zhuang et al. recognize two types of policy rules—static policies that are easy to deploy and dynamic policies that require negotiation of policies between the WPDF and the MPDF. Zhuang et al. propose the use of Common Open Policy Service (COPS) Protocol for this purpose.

In another work, Choi et al. [12] present the so-called IP-triggered resource allocation strategy, where they present how DiffServ/IntServ QoS parameters or classifications can be mapped to two lower layers of the mobile network. Chen [13] proposes an adaptive cross-layer scheme that adapts the resource management procedure to adapt to the dynamic mobile communication environment, application layer, and also error mechanisms.

3.3 Generic Access Network

This section analyzes briefly the UMA/GAN Architecture. We assume that the architecture will be analyzed in more detail in other chapters and only present the basic discussion here.

3.3.1 GAN Architecture Overview

GAN is a 3GPP standard covering the scenario of a new mobile station wanting to access UMTS network services using an access network other than UTRAN/GERAN (GSM with EDGE Radio Access Network/UMTS Terrestrial Radio Access Network). The GAN is a continuation of the standardization effort UMA. The main elements of the GAN architecture are the GAN controller (GANC) and the up interface between the GANC and the mobile station. The GANC communicates to the UMTS core network using standard A and Gb interfaces. The security gateway (SEGW) is also a part of the GANC, communicating using the Wm interface with the AAA server in the UMTS core network (Figure 3.2).

The new mobile station initially sets up a secure tunnel over the up interface with the GANC security gateway. The mobile station needs to be authorized using AAA services and then the GANC sends back the information about its identity. The mobile station then sets up a signaling connection over the Up and registers with GANC. If the GANC accepts the registration, service and system information is then provided to the mobile station.

When needed, the GANC and the mobile station can set up a CS domain and PS domain user plane bearers over the GAN.

Figure 3.3 shows the protocol mapping done in GANC. For the analysis provided in this chapter, of particular importance are the Generic Access Circuit Switched Resources (GA-CSR) and Generic Access Packet Switched Resources (GA-PSR) protocols. The GA-CSR Protocol is one that provides

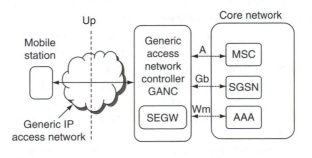

Figure 3.2   GAN architecture.

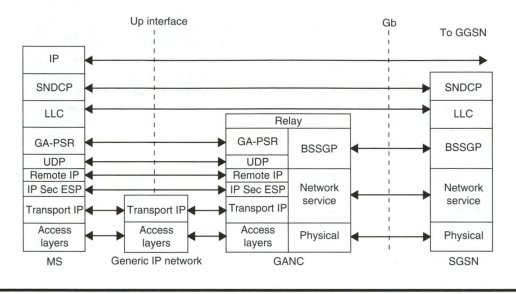

Figure 3.3   Up PS Domain User Plane Protocol Architecture. (From Generic Access to the A/Gb Interface; Stage 2 (Release 7), 3GPP TS 43.318, November 2006.)

a resource management layer for communication with the CS domain of the core network. This protocol is equivalent to the GSM-RR Protocol. The protocol provides the following functions [1]: set up of bearer for CS traffic between the MS and GANC; handover support between GERAN and GAN; and additional functions such as GPRS suspension, paging, ciphering configuration, and classmark change. In a similar fashion, the GA-PSR Protocol provides service for the PS domain of the core network: delivery of GPRS signaling, paging, flow control, GPRS transport channel management; PS handover support between GERAN/UTRAN mode and GAN mode; and finally, transport of GPRS user plane data.

3.4   QoS Management in UMA

It has already been stated in Section 3.1 that UMA has been primarily developed to expand the coverage of cellular networks, especially indoors in remote parts of the house or office. UMA is also seen as a way for the network operators to make sure they are ready to exploit the popularity of Voice-over-IP (VoIP). Because of this primary motivation for UMA, a big question has been raised about the need for QoS support in networks deploying UMA.

A majority of the authors agree that UMA does not support end-to-end QoS management. Also, they are quick to point out that it is unlikely to expect UMA to provide this support, as UMA bridges the mobile core network, which is under the management of the mobile network operator, and the wireless access network, which is in most of the cases a Wi-Fi network that can belong to anyone, and does not have to be under the direct management of the mobile operator. As soon as a part of the network is out of the direct control of the network operator, it is impossible for the operator to guarantee QoS.

Therefore, the main objective of this section is to focus on two issues. First, to explain the elementary level of QoS support UMA has, primarily through a defined process of recognition of the DS QoS architecture and the ability to read DS codepoints and perform appropriate actions.

Second, this section describes the overlay policy-based management system, which will be capable of providing a framework within which the mobile operator can eventually guarantee QoS. As in any other heterogeneous network environment, in networks deploying UMA it is possible to provide end-to-end service as long as there are means to distribute the information to all networks that form the communication path.

In general, the translation of policies and QoS management functions between network domains is necessary. Because UMA is used to connect different network domains, the translation of these policies at the GAN controller point is a natural design decision. The standard defines two alternatives available to an external IP user station at the PDP context activation point: the basic GPRS connectivity service using local bearer policy decisions and the enhanced GPRS connectivity service where network-level policies are applied. The standard defines four methods to achieve interworking:

- Signaling along the flow path
- Packet marking or labeling along the flow
- Interaction between policy control and resource management entities
- SLA enforced by border routers between networks

It is interesting to note here that these methods closely follow the concepts defined in Internet QoS architectures, with the first method relevant to QoS reservation using RSVP, and the second one to packet marking and traffic conditioning typical for the Differentiated Services Architecture.

3.4.1 Support for QoS

The GAN standard [5] identifies basic QoS procedures, mainly to do with the classification and identification of packets using DS codepoint marking. The main recommendation for the mobile station and the SEGW is to observe existing DS codepoints in packets and to react to any observed changes. For a tunnel mode security association (SA), there is typically an outer IP header that specifies the IPsec processing destination, and the inner IP header that specifies the ultimate destination for the packet. We can then recognize two separate set of activities. These activities are interesting as they can be observed as the process that can be controlled by appropriate policies. The activities are

1. When the SEGW receives an IP packet from the GANC, it has to make sure to map the DS codepoints from the inner IP header to the outer IP header when creating the IPsec tunnel toward the mobile station MS. Packet flow identifier can be used when more than one flow is using the same tunnel. Different DS codepoints can then be used for different flows.
2. When the MS receives IP packets from the SEGW, it will observe the DS codepoints in inner IP headers and, if different from the previously used DS codepoints, will save the received DiffServ codepoint (DSCP) and use it for future packets for the same traffic channel. This will be performed regardless of whether the packet carries Transmission Control Protocol/User Datagram Protocol (TCP/UDP) data, or is part of a GA-CSR traffic channel carrying CS data, or is a GA-PSR message containing signaling or SMS. When creating the outer IP header, the MS will use the DSCP value from the inner IP header.

This limited QoS support enables the transparent use of the Differentiated Services concept. The key issue, however, is the management of processing methods in the access network and the core network. In other words, the key issue is whether the DSCP marking can be used to trigger QoS actions, especially in the IP access network. Further, a major issue is the process of QoS monitoring. This needs to be managed and organized by the PBNM. Section 3.4.2 describes the whole process in more detail.

3.4.2 QoS Architecture

We have seen in the previous sections of this chapter that achieving and guaranteeing QoS is a complex task for any network. For the UMA, the problem is additionally complex as UMA by its definition needs to bridge two very different network entities, the core mobile network and the general access IP-based network. The access network in a majority of cases is Wi-Fi, which almost always provides much more bandwidth than a GPRS/UMTS network in the same area. Therefore, it is unlikely that the access network will be the QoS bottleneck in terms of bandwidth.

UMA has never been designed to provide a comprehensive session QoS management. The recent development of IMS gives means to provide complex QoS and security management for complex applications. With this in mind, in this section we point out the main issues in QoS management and provision for QoS and highlight the implementation and evaluation methodology that should be used to finalize the QoS management architecture. In this chapter, we use QoS scenarios identified in Ref. [6].

Figure 3.4 shows the simplified end-to-end QoS scenario for a mobile station (UE) running a communication session with a remote Internet host. We use this topology to present a general case, even though for the QoS management in UMA it is completely irrelevant whether the receiving end is in an external IP network or within the UMTS network domain. QoS management assumes that application-layer QoS negotiation has taken place (this can be done in a number of ways, the current dominant method is using SDP/SIP). This negotiation results in the creation of a logical IP bearer service. We also assume here that the UE is capable of managing the IP bearer service. Physically, the traffic will have to pass through different network domains. The traffic will first use an access IP network to reach the GANC, where it will be forwarded using the mobile network protocol stack to the GGSN. The GGSN is the gateway node in the mobile core network, responsible for communication to the external IP network. From the GGSN to the remote receiving node the traffic will again be transported through the IP network.

To provide end-to-end QoS management for the given communication scenario, IP bearer service management needs to be transparent on all links. In the IP links, Internet QoS architecture (DiffServ) will be used, and in the mobile network, QoS will be controlled using the PDP context and mobile network QoS solutions. The key issue in providing QoS in such a complex infrastructure is QoS negotiation, translation, and monitoring. Figure 3.5 shows the communication phases

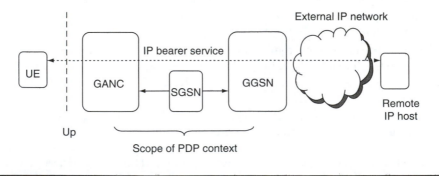

Figure 3.4 End-to-end QoS architecture.

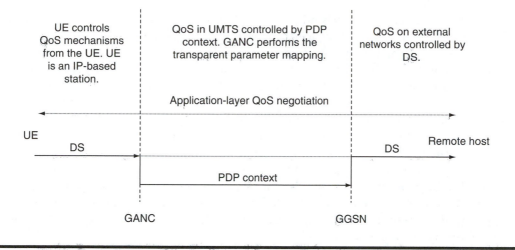

Figure 3.5 Separation of contexts.

and how QoS is being controlled. The key points in this communication are the translations that happen at GANC and again at GGSN. The GANC needs to be able to map the IP layer service requirements to the PDP context parameters. In turn, the GGSN DiffServ edge function needs to be able to set the DSCP that corresponds to the DiffServ arrangement for the external network and also to the received PDP context parameters. This PDP context will be signaled between different UMTS bearer service managers.

It is clear that such an architecture is highly dynamic and that it needs to be adaptive. This makes a good case for the deployment of a PBNM system to monitor and update rules for the QoS translation at edge points, especially at GANC and GGSN. The main tasks of such a PBNM system will include

1. Distribution of policies to control the mapping of IP service requirements to PDP context parameters and distribution of policies to monitor end-to-end QoS. In this respect, it is important to point out the need for an efficient policy negotiation between different networks. Policy negotiation and translation present a major issue [14]. It is interesting to note that, for example, the research work that exists in policy coordination for IMS-based management systems [15,16] largely ignores the issue of policy coordination.
2. Enforcement and monitoring of distributed policies. In terms of enforcement, Section 3.4.1 described briefly the QoS support currently provided in UMA. The mapping process, in combination with the access network ability to support traffic prioritization and QoS, can give the basis for QoS provision. Additional work is needed to identify the details of QoS policy enforcement in different access networks.

3.5 Conclusion and Open Issues

This chapter presented the problem of QoS and analyzed how QoS is currently supported in UMA and what needs to be done to enhance the QoS support. QoS management in UMA networks, although important and required, cannot be considered as the highest priority for the developers and users of UMA technology. However, the establishment of a policy-based QoS management infrastructure in existing UMA networks is desirable, and can prove to be an efficient and scalable method of integrating UMA implementations to a next-generation network that will be fully capable of managing complex multimedia applications.

In other words, the implementation of a QoS management system in UMA can help existing UMA systems to become a more efficient element of the architecture for next-generation mobile networks. The IMS system, which has been only mentioned throughout this chapter, will be able to fully support end-to-end QoS, and UMA can prove to be an important element of this architecture.

A number of technical challenges remain that prevent efficient implementation of a QoS management system in UMA. First, the GAN controller must be enhanced with a PDP functionality and capability to process policy rules from the mobile core network and enforce these on the access network. This functionality also needs to be able to utilize the QoS support that exists in the access network. Work on the development of this PDP functionality and on the fine-tuning for different access networks is needed. A policy coordination framework is required to successfully integrate GANC QoS support with other networks.

Other issues remain unsolved in UMA. Currently, UMA works only if a single device is able to connect to the cellular network and to the access network (i.e., Wi-Fi). Work is needed on vertical handover and identity management systems that will enable seamless handover between two separate devices. Handover between Wi-Fi APs is another open issue for the UMA QoS support that needs to be resolved. Finally, considering the ubiquitous acceptance of the SDP/SIP for QoS negotiation and requirement definition, efficient end-to-end QoS mechanism needs to be able to communicate SIP information among network entities.

In summary, QoS is a complex issue and its full provision is always seen as a trade-off between implementation and maintenance cost and end user requirement. QoS support has been standardized in both the global Internet and in cellular wide-area mobile networks. Work on integrating the standardization guidelines for these two complex networks is ongoing, and provides numerous challenges for the coming years.

REFERENCES

1. Generic Access to the A/Gb Interface; Stage 2 (Release 7), 3GPP TS 43.318, November 2006.
2. R. Braden, D. Clark, and S. Shenker, Integrated services in the Internet architecture: An overview, IETF Request for Comments 1633, January 1994.
3. S. Blake, D. Black, M. Carlson, E. Davies, Z. Wang, and W. Weiss, An architecture for differentiated services, IETF Request for Comments 2475, December 1998.
4. G. Priggouris, S. Hadjiefthymiades, and L. Merekos, Supporting IP QoS in the general packet radio service, *IEEE Network*, September/October 2000.
5. Quality of Service (QoS) Concept and Architecture (Release 7), 3GPP TS 23.107, pp. 18–29, June 2007.
6. End-to-End Quality of Service (QoS) Concept and Architecture (Release 7), 3GPP TS 23.207, June 2007.
7. Q. Ni, L. Romdhani, T. Turletti, and I. Aad, QoS Issues and enhancements for IEEE802.11 wireless LANs, INRIA Technical Report No. 4612, November 2002.
8. F. Yavatkar, D. Pendarakis, and R. Guerin, A framework for policy-based admission control, IETF Request For Comments 2753, January 2000.
9. R. Chakravorty, I. Pratt, and J. Crowcroft, A framework for dynamic SLA-based QoS control for UMTS, *IEEE Wireless Communications*, 10(5), pp. 30–37, October 2003.
10. M. Poikselka, A. Niemi, H. Khartabil, and G. Mayer, *The IMS: IP Multimedia Concepts and Services*, Wiley 2006.

11. W. Zhuang, Y.-S. Gan, K.-J. Loh, and K.-C. Chua, Policy-based QoS management architecture in an integrated UMTS and WLAN environment, *IEEE Communications Magazine*, 41(11), pp. 118–125, November 2003.
12. Y.-J. Choi and K. B. Lee, All-IP 4G network architecture for efficient mobility and resource management, *IEEE Wireless Communications*, 14(2), pp. 42–46, April 2007.
13. Z. Chen, A customizable QoS strategy for convergent heterogeneous wireless communications, *IEEE Wireless Communications*, 14(2), pp. 20–27, April 2007.
14. M. Charalambides, P. Flegkas, G. Pavlou, A. K. Bandana, E. C. Lupu, A. Russo, N. Dulav, M. Sloman, and J. Rubio-Loyola, Policy conflict analysis for quality of service management, Proceedings of 6th IEEE International Workshop on Policies for Distributed Systems and Networks, 2005.
15. F. C. De Gouveia and T. Magedanz, A framework to improve QoS and mobility management for multimedia applications in the IMS, Proceedings of 7th IEEE International Symposium on Multimedia ISM'05, 2005.
16. J. Y. Kim, J. H. Hahm, Y. S. Kim, and J. K. Choi, Policy-based QoS Control Architecture Model using API for streaming services, Proceedings of International Conference on Networking, 2006.

Chapter 4

Radio Resource Management in IEEE 802.11-Based UMA Networks

Frank A. Zdarsky and Ivan Martinovic

CONTENTS

Unlicensed mobile access (UMA) networks increasingly depend on radio resource management techniques to satisfy the rising quality of service (QoS) expectations of their users. The task of radio resource management is to monitor a wireless network's performance and control its allocation of radio resources to meet a predefined performance goal, such as maximizing the minimum fair share of transmission opportunities for each node in the network. In this respect, it differs from the planning of a wireless network, which only covers the parameters that have to be chosen before a network's deployment and remain static during its operation.

This chapter starts with a discussion of why radio resource management is much more challenging in UMA networks than in mobile cellular networks. Furthermore, it contends that, in the light of ever-increasing UMA network densities, radio resource management mechanisms will have to coordinate themselves across network boundaries to be able to provide applications with a high and stable service quality in the future. It continues with a description of the "turning knobs" with which radio resource management controls a wireless network's performance and a discussion of their limits, dependencies, and pitfalls in the context of IEEE 802.11-based networks, the most popular current UMA technology. It also presents selected algorithms for each control parameter. Following a taxonomy of radio resource management architectures, two example systems based on different architectural decisions are described. The chapter then goes on to survey the support for radio resource management in current and upcoming IEEE 802.11 standards, before concluding with a summary and outlook on open research issues.

4.1 Challenges for Radio Resource Management in UMA Networks

UMA networks, in particular those based on IEEE 802.11 technology, have become almost ubiquitous. Numerous small-to-medium sized companies have built their business model around this technology and provide services such as wireless LAN-based video surveillance, facility management, fleet tracking, advertising, etc.

Furthermore, an increasing percentage of all broadband access routers installed in private homes has built-in IEEE 802.11 support that allows comfortable, untethered Internet access everywhere. Thus, the number of deployed wireless LANs is still growing rapidly.

The tremendous success of UMA may be explained in large part by the low deployment and maintenance cost due to the use of unlicensed frequency bands. This allows anyone to set up a private wireless network without high licensing costs. Another important success factor is that, in particular, the IEEE 802.11 technology has been designed from the start to be simple, which resulted in very low cost and consumer-friendly devices, but also to be robust, so that IEEE 802.11-based networks are easy to set up and show reasonably good performance even without an extensive site survey and planning process.

At the beginning, wireless LANs were mainly used for applications that were a little demanding on the quality of the wireless connectivity, such as getting the latest news from the Web, keeping contact via e-mail or text-based chats, occasional file transfers, etc. Increasingly, however, the usage model for wireless LANs is shifting toward real-time and interactive applications

Voice-over-IP (VoIP) video streaming, online games), which are highly sensitive to strong, non-deterministic variations in their connection's throughput, delay, and loss characteristics.

In mobile cellular networks, much effort has been invested to guarantee a high QoS through a combination of careful network planning and radio resource management:

- Network planning concerns all parameters that need to be decided upon before the deployment and operation of a wireless network and remain quasi-static during operation. Examples of these static parameters are the number of base stations, their placement, the antenna configuration (omnidirectional or sectorized), the antenna elevation and orientation, and sometimes also the static assignments of operating channels and transmit powers.
- Radio resource management, in contrast, has the task to monitor a network's performance during its operation and to control the performance by adjusting the dynamic parameters of the network, such as the dynamic channel assignment, transmit power selection, transmit rate selection, assignments of user to base stations (load-balancing), etc. It is thus able to adapt the network to a changing radio environment and may, to a certain extent, even compensate suboptimal decisions during planning and unforeseen changes in the service demand.

Both network planning and radio resource management are well studied in the context of mobile cellular networks. Furthermore, a large base of design principles, best practices, methods, and tools exists to support these processes. It therefore seems appealing to transfer this collective experience to wireless LANs as well. Indeed, early efforts at increasing the service quality and reliability of wireless LANs were very much oriented to those in mobile cellular networks.

However, wireless LANs, like all UMA networks, are very different from mobile cellular networks in that they cannot assume to be the exclusive user of the frequency bands they operate in and thus have to be able to coexist with other wireless networks occupying the same frequency band. More specifically, they need to have mechanisms that let them contend with networks for radio resources. These mechanisms should be distributed in nature (as a central coordinator between networks of different operators would be impractical) and should ensure a somewhat fair sharing of radio resources across networks.

As a consequence, UMA networks usually employ randomized shared medium access schemes, which require a much higher effort to provide at least some form of soft QoS guarantees. Some of these access schemes, like IEEE 802.11's Carrier Sense Multiple Access with Collision Avoidance (CSMA/CA), also have their own peculiarities and display subtle and sometimes nonobvious side effects, such as capture and fairness problems, which radio resource management needs to account for and which further add to the complexity.

A much greater challenge for radio resource management in UMA networks arises as a direct consequence of their success though. As the density of deployed UMA networks increases, so does the intensity with which networks of different operators need to compete for resources. This does not only reduce the long-term share of bandwidth that is available to each network, but also makes throughputs and delays much more variable and unpredictable.

One would assume that critically high densities are still somewhat far down the road. However, in hot spot areas like city centers, large office buildings, and student dormitories, access point densities occasionally already reached levels at which mutual interference seriously degrades the performance—sometimes even to the point of starvation. This is substantiated by a study of wireless LAN deployments in six major U.S. cities, which reports, for example, for Boston that over 20 percent of all deployed access points had more than six other access points within their transmission range, and some of them even had more than 80 [1].

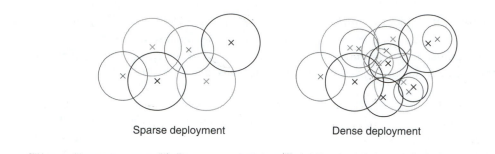

Figure 4.1   Sparse versus dense deployment models.

Until recently, however, research on radio resource management for UMA networks has focused only on isolated wireless networks and networks with few neighbors. Thus, relatively little is known about the properties and behavior of wireless network aggregates in chaotic and dense deployment scenarios.

■ Chaotic deployment denotes a scenario of multiple wireless networks that are both unplanned and unmanaged [1]. They are unplanned in the sense that, when viewed as an aggregate, the positions of access points are highly irregular, unlike to what may be observed in a well-planned cellular network. Furthermore, many of these access points are unmanaged and often operate with factory default settings, a common case for access points in private homes, which tend to crowd on the most common channel 6.

■ Dense deployment refers to a scenario in which access points from multiple wireless networks are positioned so close to each other that it is not feasible for the single network to find a configuration which allows an interference- and contention-free operation [2]. Figure 4.1 illustrates the difference between sparse and dense deployments. Note that even in a single wireless network access points may be so densely deployed that an interference-free operation is not possible, because even at the lowest useful transmit powers, interference ranges may be quite large.

In a chaotic and dense deployment scenario, the radio resource management of a single network is no longer able to provide a sufficiently high QoS for real-time and interactive applications on its own, as disruptions in the network's service are mainly caused by other networks that are not under its control. Instead, there need to be mechanisms to effectively mitigate the intensity and variability of interference between neighboring networks. These mechanisms require some form of coordination between the radio resource management of neighboring networks. It is still an open issue how this coordination may be done both efficiently and in a globally optimal way and whether the coordination should be implicit or explicit.

4.2   Radio Resource Management Tuning Parameters

Radio resource management may have various performance objectives. For instance, its objective could be to maximize the minimum fair share of transmission time for all senders while meeting certain latency and energy budget constraints.*

---

* Although a mobile device's energy supply is an important resource that affects (and is affected by) the device's ability to communicate wirelessly over an extended period, the focus of this chapter is on the QoS aspects of performance. In practice, radio resource management has to incorporate energy constraints in its decisions, of course.

This section describes the most important parameters that radio resource management may adjust to achieve its performance objectives. It also discusses the interactions between these parameters and points out the peculiarities that an IEEE 802.11-specific resource management strategy has to consider in this context. Furthermore, it provides examples of some of the fundamental or popular algorithms that are used to select parameters in a given scenario.

4.2.1 Transmit Power Selection

To be successfully decoded, a transmission needs to reach the receiver at a certain minimum signal strength relative to the sum of the interference from other sources and the background noise (signal to interference and noise ratio, signal-to-noise-plus-interference ratio [SINR]). The higher the transmit rate, the better the SINR has to be for decoding to succeed with a given probability. After accounting for antenna gains, path loss between transmitter and receiver, and link margin (a margin of safety to counter shadowing and fading effects), the necessary minimum transmit power may be calculated.

In principle, using a higher transmit power is beneficial as this increases the maximum transmission range at a given transmit rate or alternatively allows a higher transmit rate at the same distance from the sender. On the downside, higher transmit powers require more energy, which drains the batteries of mobile devices more quickly and also leads to higher interference with the transmissions of other wireless nodes. Figure 4.2 illustrates the problem that the distance at which a transmitter interferes with nearby nodes or even triggers their carrier sense indication is typically much higher than the maximum distance at which a high-rate transmission can still be received successfully.

When selecting the transmit power for a transmission, radio resource management needs to make a trade-off between these aspects, but also has to obey the limits imposed by frequency regulations and the IEEE 802.11 standard. In Europe, the maximum transmit power in the entire 2.4 GHz band is restricted to 100 mW (Equivalent Isotropically Radiated Power EIRP). The 5 GHz band is divided into ranges 5150–5350 MHz for indoor use at maximum 200 mW (EIRP), and 5470–5725 MHz for indoor and outdoor use at maximum 1 W, but only when transmit power control (TPC) and dynamic frequency selection (DFS) mechanisms are used (see Section 4.4.1). Other regulatory domains impose similar restrictions.

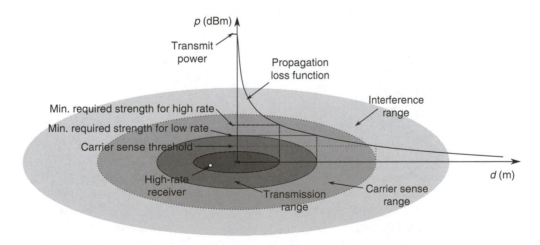

Figure 4.2 Transmission, carrier sense, and interference ranges.

Many IEEE 802.11 radios still do not employ adaptive TPC mechanisms and instead simply transmit all frames at the highest power allowed in the respective regulatory domain, unless manually set to a lower level. If TPC is implemented, it may operate at different granularities: per cell, per node, or per frame. Per cell means the access point transmits all frames at the same power level, just strong enough to provide its most exigent associated station with a signal level above its link margin. With per node control, frames transmitted to a node are sent at a power level just sufficient to meet the receiving node's link margin. Per frame control is similar, but its control loop works on shorter timescales. Finally, TPC may further distinguish between data and control or management frames, for example, to implement a cell breathing technique (see Section 4.2.4).

One would think that the optimal strategy for radio resource management would simply be to always transmit frames at the lowest possible power under the constraint set by the receiver's link margin to minimize interference between nodes. Yet, in CSMA/CA networks, such a strategy may lead to highly unfair allocations of transmit opportunities across nodes. The reasons for this are physical layer capture and asymmetric links.

■ Physical layer capture describes the phenomenon that in a "collision" of two frames, the one with a much stronger signal may still be decoded successfully at the receiver. This can even be the case if the stronger frame is sent later than the other frame, as long as it arrives while the receiver is still attempting to synchronize on the Physical Layer Convergence Protocol (PLCP) preamble of the earlier frame [3]. As only the inferior node then has to increase its exponential backoff window, it is penalized even further.

■ Link asymmetry occurs if the difference between the transmit powers of two nearby nodes is so high that one node can sense the other, but not vice versa. Consider the scenarios in Figure 4.3, in which two access points A and B try to transmit frames to their stations C and D, respectively. In case (a), both A and B use the same transmit power, are able to sense each other's transmissions, and thus defer to each other. This leads to a roughly equal performance degradation for both access points. In case (b), A employs TPC and has therefore reduced its transmit power. A will still defer when it senses transmissions from B, but not vice versa, because A is now hidden from B. Again, the inferior node suffers a higher performance degradation, possibly even to the point of starvation.

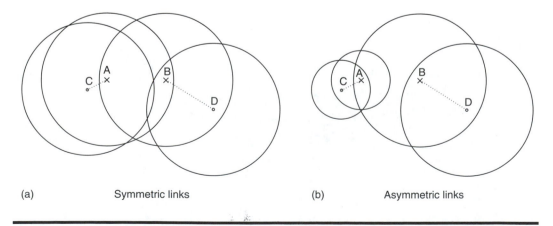

(a)   Symmetric links   (b)   Asymmetric links

Figure 4.3   Scenarios of symmetric and asymmetric links between two nodes A and B.

One solution to the second problem would be to protect the transmission with a Request-to-Send/Clear-To-Send (RTS/CTS) frame exchange sent at a common transmit power (to make the link symmetrical); yet this adds further overhead, increases interference for neighboring nodes, and does not help with physical layer capture. A better approach, described in the next section, is to combine the selection of transmit powers with that of the carrier sense threshold (CST).

4.2.2 Carrier Sense Threshold Selection

In IEEE 802.11, physical carrier sense is performed by the clear channel assessment (CCA) function. This function has various methods (modes) to determine whether the channel is currently occupied by another station or is idle. The most important modes are

- CCA mode 1 (energy detection threshold) (EDT) that determines whether the average energy on the channel is above a certain threshold. It makes no difference whether the energy stems from a single source or from the superposition of multiple sources and whether the source is a IEEE 802.11 node or a microwave.*
- CCA modes 2 + 4 (carrier sense) signal the channel as busy only when they detect the valid signal of a DSSS (Direct-Sequence Spread Spectrum) and HR PHY (High-Rate PHY(sical layer)), respectively. Whether the signal can be detected depends on the sensitivity of the hardware, which determines the lowest possible CST.
- CCA modes 3 + 5 are a combination of the aforementioned modes, which indicate a busy channel only when detecting a valid signal above the EDT.

In Ref. [4], several situations are identified, in which carrier sensing unnecessarily wastes transmit opportunities:

- In exposed station scenarios, in which two transmitting nodes are within each other's carrier sense range, but their respective receivers are separated enough so that the transmissions do not collide at the receivers.
- In scenarios in which the sum of interference is high enough to trigger a transmitter's carrier sense indication, but the sources of interference are too far away to actually disturb the signal at the intended receiver.
- When the physical layer capture effect described in the previous section may be exploited in a useful way, that is, if two transmissions do collide at the receiver, but the useful signal is much stronger than the interfering signal and is therefore successfully received.

The authors therefore suggest switching off carrier sensing when one of the above situations is detected.[†]

A comparison of this list with the one in the previous section suggests that CST selection and transmit power selection are strongly related. In fact, Jason et al. [5] have studied the possibility of increasing spatial reuse in CSMA/CA networks and have argued that the product of transmit power and CST should be constant. The rationale is that a node using a higher transmit power than the average should also be more sensitive to the transmissions of senders that transmit at lower transmit powers and vice versa.

---

* Note that in the literature the term carrier sense threshold is sometimes wrongly used in a context that actually describes the energy detection threshold.
[†] Although their observations were made in the context of wireless sensor networks using B-MAC for medium access, their results should nevertheless be applicable to CSMA/CA-based networks as well.

In Ref. [6], which studies starvation effects caused by asymmetric links, this constant product rule is analytically confirmed. The authors also show how to choose this constant to minimize the potential delay experienced by stations in a network under the assumptions of per cell TPC and downlink traffic only. They show that higher transmit powers should be allocated to cells with a higher number of stations or stations with poor channel conditions.

4.2.3 Transmit Rate Selection

Most modern wireless communication systems support multiple modulation and channel coding schemes, which allows them to dynamically adapt their transmit rates to varying channel conditions. Generally, when the channel quality gets worse and the SINR decreases, it becomes more difficult for the receiver to correctly decode the received signal, which exponentially increases the probability of bit errors occurring during decoding.

If the SINR drops below a certain level, the overhead created by lost frames and retransmissions may become so high that it is more efficient to switch to a lower rate that uses a more robust modulation, a higher redundancy, or both. This is especially true if retransmissions are very costly in a protocol, like in CSMA/CA, where lost frames are only detected after an Acknowledgment (ACK) timeout and the exponential increase of the backoff window introduces additional delays. Thus, it is necessary to find a good trade-off between transmission rates and bit error rates (BER).

The IEEE 802.11 wireless LAN standards also support the selection between various data rates. The rates supported both by IEEE 802.11b and .11g are 1 Mb/s (DBPSK modulated) and 2 Mb/s (DQPSK), and 5.5 Mb/s and 11 Mb/s (CCK). IEEE 802.11a and .11g support rates of 6, 9, 12, 18, 24, 36, 48, and 54 Mb/s by using orthogonal frequency division multiplexing (OFDM) with Binary Phase-Shift Keying (BPSK), Quadrature Phase-Shift Key (QPSK), 16-QAM, or 64-QAM modulations on the subcarriers and 1/3, 2/3, or 3/4 redundancy through punctured convolutional codes.

The rate that should be used to transmit a given frame is only partially regulated by the standard. For example, for reasons of robustness and backward compatibility, it is mandatory to transmit the PLCP preamble and header at the lowest bit rates (1 or 2 Mb/s) and to switch to a possibly higher bit rate only for the medium access control (MAC) frame itself. Furthermore, some frames like ACKs and beacons need to be received by all stations in a basic service set (BSS), and therefore must be sent at no more than the maximum data rate supported by all stations in the BSS.

Other than this, the standard does not give any guidelines on the specific rate control algorithm to be used. Therefore, a plethora of vendor- and operating system-specific rate control algorithms exists. These may be found freely mixed on different nodes of the same wireless LAN.

The principle of transmit rate adaptation is similar to that used in mobile cellular networks. It includes the following three aspects:

1. There needs to be a metric for the current channel quality from the receiver's perspective (e.g., the SINR, the signal strength, the BER, or the frame error rate) and a function that predicts the channel quality for the next transmission based on one or more previous observations of the channel quality.
2. A further function then has to select the rate for the next transmission based on the channel quality prediction.
3. Finally, there has to be a feedback mechanism, as the channel quality is determined at the receiver, but the sender ultimately needs to know the selected rate.

Note that the prediction of the channel quality and the rate selection may actually be performed by the sender or by the receiver.

Again, in wireless LANs, rate control is more difficult, because medium access is half duplex and introduces random delays, so the time between feedback and transmission may be long. Furthermore, there is no direct way to distinguish between frame collisions and losses due to frame corruption when a transmission attempt fails. Finally, as mentioned before, retransmissions cause a relatively large protocol overhead, so a conservative rate control may actually provide higher throughputs.

Feedback on the channel quality may be communicated explicitly by the receiver, for example, by embedding the quality metric into frames returned to the sender, or inferred by the sender from a history of previous transmission failures, usually based on the absence of acknowledgment frames. The advantage of explicit feedback is that it allows a more direct signaling of the current channel condition and is also timelier, so that prediction accuracies are potentially much higher. On the other hand, it requires changes to the IEEE 802.11 Protocol (e.g., adding fields to RTS/CTS frames or using reserved fields of the PLCP header), which are not backward compatible or not easily deployable without hardware upgrades. Furthermore, explicit feedback is only useful if supported by all nodes in a BSS, which is a very limiting assumption in practice.

The earliest representatives of these two fundamental feedback approaches are auto rate fallback (ARF) [7] and receiver-based auto rate (RBAR) [8].

ARF observes the number of consecutive ACK frame receptions and misses the current transmit rate. Whenever it recognizes two consecutive losses, it reduces the rate by one increment and starts a timer. If this timer expires before the next loss or if ten consecutive ACKs are received, ARF sends one frame at the next higher rate and decides to continue at a higher or at the current rate, depending on whether this frame is acknowledged or not.

RBAR relies on an RTS/CTS frame exchange before a data transmission to give feedback about the channel condition. In the RTS frame, the sender suggests a tentative transmit rate using some heuristic based on previously used rates. The receiver then either accepts or modifies this suggestion based on its channel measurements and replies with a CTS frame containing the final rate to be used. Changing the suggested rate is costly, however, because in this case all nodes hidden from the receiver need to be informed of the final rate and frame lengths as well, so that they may update their network allocation vectors (NAVs). RBAR resolves this issue by sending the MAC header plus an additional reservation header at the tentative rate and only the rest of the frame at the final rate. Thus, it requires changes to the PLCP header to support two consecutive rate switches within a single frame.

More advanced and efficient rate control algorithms have been suggested for both approaches [9,10]. However, for the backward compatibility reasons given above, the implicit approach has a much higher relevance in practice.

One problem that needs to be overcome with the implicit approach is that a missing ACK may be caused both by frame corruptions due to a low SINR as well as by frame collisions due to high traffic in the network. In the latter case, however, a reduction of the transmit rate is not compensated by a lower frame error rate, but may on the contrary cause an additional increase of congestion in the network, as frames now take longer to be transmitted. It has been experimentally shown in Ref. [11] that rate adaptation that reacts to congestion may actually perform worse in high-traffic scenarios than if no rate adaptation is used at all. In particular, it performs worse than algorithms that distinguish between collisions and corruptions.

Another effect to be considered is that physical layer capture is more pronounced at lower transmit rates than at higher rates [4], which may be explained by the fact that at lower rates the

SINR requirements for successfully decoding a transmission are also lower. Thus, assuming the same relative signal strength between two concurrent transmissions, the probability of the stronger transmission being captured increases when its transmit rate is decreased.

4.2.4   Station Assignment

The assignment of stations to access points determines the share of the wireless and backhaul link capacity that each station can expect to receive from its access point. Often, the objective of station assignment is to maximize the fair share of each station, according to some fairness criterion, for example, max–min fairness. A station assignment strategy also has to consider the side effects of a particular assignment, as it may cause constellations in which interference and contention between nodes effectively reduce the available capacity.

Although the IEEE 802.11 standard does not define how stations should be assigned to access points, the most common strategy in practice is to let stations scan for access points on all channels and then associate to the one with the strongest signal, which should provide the best link quality. However, assuming all access points use the same transmit power for beacons and that path loss dominates all other factors, this translates into associating to the nearest access point. Thus, if users are not uniformly distributed, some access points experience a very high load while others are almost not utilized at all. Indeed, empirical studies of large wireless LANs (e.g., [12]) have confirmed this behavior.

Apart from the fact that the standard association mechanism is agnostic of the load of candidate access points, it is also suboptimal from a radio interference point of view. Assume a candidate access point is far away, but uses a higher transmit power than another much closer one. In this case, a station would associate to the farther access point and would thus require a higher transmit power itself, which unnecessarily increases both interference and energy consumption.

Previous studies (e.g., [13]) have shown that load-balancing can effectively reduce congestion at access points and increase the QoS experienced by stations. Load-balancing requires access points to provide stations with an indication of their current load level, which is commonly broadcasted in beacon frames. Potential load indicators include the number of currently associated stations, the mean signal strength of currently associated stations, and the bandwidth obtained by the next station that associates. Measurement results in Ref. [12] suggest that it is preferable to choose user generated traffic as load metric, rather than the number of associated users, as the two metrics have been found not to correlate very well. One of the more advanced load-balancing algorithms that is able to achieve a constant-factor approximation of the max–min fair rate (or alternatively: time) allocation is described in Ref. [14]. It considers the wireless link as well as the backhaul link and also copes with the online case of stations joining and leaving the network, which possibly requires a reassociation of some stations to other access points.

However, Ref. [2] provides evidence that load balancing does not always lead to optimal bandwidth allocations. The reason is that load-balancing proposals explicitly or implicitly assume that a station assignment to one access point does not affect the performance of neighboring access points. This assumption only holds if BSSs are sufficiently separated in space or frequency or both, though. Otherwise, a newly associated station may cause contention that reduces another BSSs performance. In such a case it may sometimes be better to actually switch off selected access points and balance stations over the remaining ones. This case is quite common especially in dense deployment scenarios.

There are various other aspects that should be considered by a station assignment algorithm. For example, Ref. [15] first described the effect that due to the CSMA/CA's tendency to provide per-frame-fair medium access (assuming the lack of effects like channel capture), a station transmitting at low data rates occupies the channel for a much longer time on average and thus effectively throttles other stations to the same rate as well. Thus, station assignment interacts with transmit power and rate assignments in a complex way.

A problematic aspect of station reassignment is that the association decision is currently performed by the stations rather than the access point, which has a better view of the load situation of its wireless and backhaul links and may coordinate with neighboring access points. One solution is to use the concept of "cell breathing" [16,17], in which access points reduce the transmit power of beacon frames if their load increases. This way, some stations are disassociated and prevented from reassociating. Along the same lines, an access point could choose to disassociate a station and refuse to let it reassociate.

Both methods are quite crude and do not allow an access point to control the next access point with which a station associates, especially because different stations may implement different association strategies. This obvious deficit will be remedied with the wide-spread adoption of the upcoming IEEE 802.11v standard amendment, which adds new mechanisms to direct stations to transfer to a specified access point (see Section 4.4.3).

4.2.5 Channel Selection

Concurrent transmissions may interfere with each other both on the same channel (co-channel interference) as well as on neighboring channels (adjacent-channel interference). A typical objective of channel selection is thus to assign operating channels to access points so that interference between nodes of neighboring cells is minimized.

At first sight, this problem is equivalent to that in FDMA-based mobile cellular networks, where it is popular to model cells as a hexagonal lattice and color them according to frequency reuse patterns to ensure a minimum spatial separation between cells using the same channel (Figure 4.4).

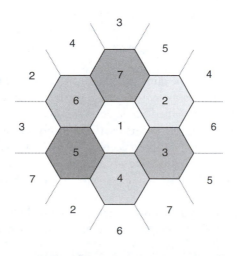

Figure 4.4 Typical frequency reuse pattern for a hexagonal-lattice cell model.

In fact, early works on channel selection in wireless LANs carried over the hexagonal cell model proposed very similar static channel assignment schemes.

Unlike mobile cellular networks, wireless LANs only have a small number of channels to choose from. For IEEE 802.11b/g-based wireless LANs (2.4 GHz ISM band), the standard specifies channels with a bandwidth of 22 MHz and a 5 MHz spacing between their center frequencies, starting at 2.412 GHz. The number of possible operating channels depends on regional radio regulations, which in the United States allows using 11 channels, in Europe* 13, and in Japan additionally a 14th channel at 2.484 GHz. Owing to their width and the narrow spacing, channels overlap with several neighboring channels causing mutual interference. To ensure interference-free operation, the standard therefore recommends choosing channels that are at least 25 MHz apart. This has led to the common practice of using only a subset of channels, usually 1/6/11 in the United States and 1/7/13 in Europe. These channels are often called orthogonal channels.

Similarly, the 802.11a standard (5 GHz ISM band) specifies a channel bandwidth of 20 MHz with 5 MHz spacing, but again recommends to use only every fourth channel for interference-free operation, starting with channel 36 at 5.180 GHz. Both in the United States and in Europe, regulations permit the use of channels 36–64 and 100–140, and in the United States additionally channels 149–161. Japan currently only allows the use of channels 36–48.

Obviously, with three orthogonal channels in the 2.4 GHz band, a conflict-free assignment of channels to access points is rarely possible in practice, especially considering that access point deployment is not restricted to a single plane. In a dense deployment with multiple contending wireless LANs, even the higher number of channels in the 5 GHz band is quickly exhausted. Channel bonding techniques, which combine two channels into one for increased bandwidth and are used in some proprietary 802.11a/g products as well as in the upcoming 802.11n standard, further add to the problem.

Yet, even if enough orthogonal channels were available to assign a unique channel to each access point in a network, it would be inappropriate to directly apply channel assignment approaches from cellular networks, due to several differences:

- Given the use of unlicensed frequencies, interference from other wireless networks may occur and disappear dynamically and on any or all the channels. Currently, these networks do not even have to follow a particular spectrum etiquette to reduce disturbances of their neighbors.
- If interference from other senders is present, it not only increases the BER, but if above a certain signal strength may also trigger the CCA and may thus completely interrupt a transmission.
- On the other hand, assigning the same channel to two neighboring access points may sometimes even be desirable, for example, to increase the temporal reuse of the channel if traffic load is low or not equally distributed across access points.
- Similarly, if the cells of two access points overlap, but no station is present in the region of overlap, using the same channel for both access points may be feasible without penalty.
- Finally, if stations are mobile, it might be advantageous to reassign channels to reflect change in load on access points and in interference due to adjustments of transmit power.

Making things worse, orthogonal channels are not always orthogonal in practice. Every transmitter also emits energy outside its current channel. These sideband emissions can only be

---

* Now also including France and Spain, which previously had much tighter restrictions.

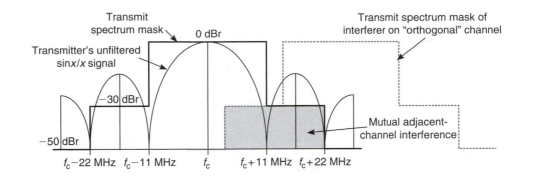

Figure 4.5 **Transmit spectrum mask and adjacent-channel interference. (Adapted from IEEE, Wireless LAN medium access control (MAC) and physical layer (PHY) specifications, ANSI/IEEE Std 802.11, August 1999.)**

attenuated to a certain extent, but not completely filtered out. The IEEE 802.11b standard specifies a transmit spectrum mask, according to which energy outside of ±11 MHz around a channel's center frequency f_c must be attenuated by at least 30 dB relative to the peak energy, outside of $f_c \pm 22$ MHz by at least 50 dB. However, this may still be enough to cause significant levels of interference on adjacent channels, as shown by the shaded area in Figure 4.5. A similar spectrum mask is specified for the OFDM modes of 802.11a/g. Table 4.1 reports the adjacent-channel interference on orthogonal channels for an 802.11a transmitter broadcasting on channel 52 [18].

A numerical example further illustrates the problem. Assume two 802.11b radios, one on channel 1 and the other on channel 6. Both have a receive sensitivity of −80 dBm, 2 dBi antennas, and are 10 m apart, which roughly corresponds to a free-space loss of 60 dB. They transmit at 2 Mb/s, for which the standard mandates an adjacent channel rejection (ACR) of 35 dB or better. The ACR describes the maximum difference between the power of an interfering signal on an adjacent channel (i.e., >25 MHz apart) and a receiver's sensitivity, for which the frame error rate is still below 8 percent. This may already be exceeded, if one of the radios transmits at 13 dBm:

$$(13\,\text{dBm} + 2\,\text{dBi} - 60\,\text{dB} + 2\,\text{dBi}) - (-80\,\text{dBm}) > 35\,\text{dB}$$

Note that the ACR may be much lower for higher transmit rates, for example, for 24 Mb/s it may be as low as 8 dB. At close distance, adjacent-channel interference can significantly degrade the performance of other radios, even if they operate on orthogonal channels. This is particularly

Table 4.1   Adjacent-Channel Interference for Transmitter Broadcasting on Channel 52

Ch.	36	40	44	48	52	56	60	64
dBm	−59	−59	−52	−27	0	−26	−53	−57

Source: From Cheng, C.-M., Hsiao, P.-H., Kung, H.T., and Vlah, D., IEEE Global Telecommunications Conference (GLOBECOM'06), San Francisco, CA, November 2006.

problematic if multiple radios are built into the same encasing, which is quite frequently the case in mesh networking.

However, with larger distances between nodes, lower transmit powers, and receivers with a better ACR, adjacent-channel interference may become negligibly small. This may, on the contrary, even allow using channels that are partially overlapping. Indeed it has been shown analytically and experimentally in Ref. [20] that in some scenarios spatial reuse and throughput may be increased by assigning partially overlapping channels.

The choice of the optimal channel thus strongly depends on the actual level of interference at each node. A fully distributed channel assignment algorithm that can consider partially overlapped channels is proposed in Ref. [21]. The algorithm is based on an annealed Gibbs sampler and is shown to converge to a state that minimizes global interference solely based on local interference measurements.

In Ref. [22], the authors contend that in densely deployed wireless LANs, even optimal static channel assignment cannot guarantee a fair allocation of radio resources over access points. The reason is that due to the low number of channels, some access points will always have to share a single channel while others do not. The authors, therefore, propose to use a channel-hopping approach, in which access points and their stations periodically change their operating channels, in a way that all access points receive the same average capacity share over a long term by equally distributing the time that access points have to share a common channel.

4.2.6 Access Point Placement

The placement of access points determines the constraints on coverage as well as the signal quality and network capacity at a given covered location. Classical access point placement imitates the placement of base stations in mobile cellular networks. Typical placement problems involve optimizations such as minimizing the number of required access points (i.e., the deployment, operation, and maintenance cost) given constraints on the available installation sites (availability of power outlets, wired LAN access), coverage, capacity, required amount of cell overlapping, limits on interference, etc. Thus, after an initial period in which access points were placed manually according to best practices, a plethora of tools was created to aid in performing site surveys and wireless LAN deployment planning.

However, the current trend in professionally deployed wireless LANs is to abandon the complex and inflexible placement of few access points by pico-cellular structures with large numbers of access points [23], which need to be carefully radio resource managed though. The benefit of this approach is increased capacity per square meter, QoS, and flexibility, as radio resource management is able to dynamically adapt to a changing radio environment, such as a changing user load and variations in signal propagation or external interference. Given the high cost-effectiveness of UMA, the trend to pico-cellular structures is likely to go much further than in mobile cellular networks.

Why can access point placement be considered a tuning parameter of radio resource management, as it is normally a part of the planning process? The reason is that if one assumes a dense deployment with large numbers of access points per area, it is possible to dynamically change the placement of access points by switching some of them on standby while activating others. Using this dynamic access point placement (DAPP), one could better react to changes in the user load and also improve the reliability of the network through failover access points.* Finally, by temporarily

---

* A similar strategy is employed in computing centers today: Use a large number of cheap, moderately reliable servers instead of a small number of expensive, but highly reliable servers.

borrowing access points from foreign networks, it would also be possible to dynamically adjust the coverage of a large-scale wireless LAN (see Section 4.3.2.2).

4.3   Radio Resource Management Architectures

A radio resource management system may be organized in various ways. This section provides a short taxonomy of radio resource management architectures and gives an overview of two example systems: one centralized, single domain system based on the Control and Provisioning of Wireless Access Points (CAPWAP) Protocol and a distributed, interdomain system based on the concept of virtualized wireless access networks.

4.3.1   Taxonomy

4.3.1.1   Network-Centric versus Client-Centric

In a standard wireless LAN, decision and control processes are distributed over all nodes in the network. For example, the access point determines the operating channel for its BSS, but each client station individually decides when and to which access points it associates, which transmit power and rate it uses for each frame, when to switch into power save mode, and so on. The advantage of this client-centric control approach is that the station which makes the decisions and executes them has all relevant information, such as the current signal strength and battery state, readily available, so that no extra communication overhead is required.

With a network-centric control approach, all decisions are made exclusively by the network that pushes the results onto the clients for execution. This approach is standard in mobile cellular networks and is becoming increasingly popular in professional wireless LANs as well. Therefore, the IEEE is working on extending the 802.11 standards with mechanisms for network control (see Section 4.4.3). The advantages of network-centric control are that the network is able to better predict the side effects of single decisions on the whole network and can therefore optimize the global network performance. Furthermore, network-centric control reduces the risk of faulty, misbehaving, or malicious clients impairing the performance of the network.*

4.3.1.2   Centralized versus Decentralized

Another architectural decision is whether radio resources are managed in a centralized or decentralized way. In decentralized architectures, each wireless access point is responsible for all resource management decisions in its own BSS, but coordinates its decisions in a peer-to-peer fashion with neighboring access points. An example for a decentralized approach is AutoCell by Propagate Networks [25]. A typical centralized architecture employs a single component, called wireless controller or intelligent switch, which manages a set of associated access points. Finally, hybrid approaches exist, such as federated architectures in which multiple controllers are each responsible for a domain of multiple access points and coordinate themselves to reduce interference between their respective domains.

---

* See Ref. [24] for a discussion of further issues related to network versus client-centric control.

4.3.1.3 Coordinated versus Uncoordinated

Current radio resource management regards interference from other wireless networks as external variable and therefore tries to optimize its own network's performance independently of (and without considering side effects to) other networks. Although such an uncoordinated approach is sensible in sparse deployment scenarios, it is of limited value in dense deployment scenarios, in which multiple wireless networks of different operators seriously degrade each other's performance and stability, as they contend for scarce, unlicensed frequency resources.

Coordinated radio resource management, in contrast, tries to jointly find configurations in which interference between neighboring wireless networks is actively reduced and the performance for all networks is increased. Such coordination may be explicit or implicit, depending on whether the radio resource managing entities of different networks are actively communicating with each other to exchange knowledge about their environment and to coordinate decisions or whether they try to infer a globally optimal configuration from what they can observe themselves about their environment. Intuitively, an explicit exchange of knowledge should provide more accurate results than what can be inferred by observation. Consider, for example, the case of hidden stations belonging to a neighboring network, for which it is unknown whether they may be migrated to other, more distant access points or not.

Furthermore, it is possible to distinguish between loose and tight coordinations. Loose coordination is a little invasive, as resource managing entities of different networks merely coordinate their use of operating channels and transmit powers to reduce interference. Tight coordination goes further in that it additionally allows to temporarily migrate stations of one network to an access point of another, if this is beneficial for the performance of both networks. Obviously, such coordination requires strong mechanisms against misuse, but is able to exploit the full potential of coordination.*

Figure 4.6 shows the amount of mutual contention in a scenario of 25 wireless networks with one access point and two stations each. The density is varied by translating all networks toward the center of the scenario. Densities are normalized, so that a density of zero percent corresponds to completely noninterfering networks. At 33 percent, network cells are just tangent to each other, but are nevertheless within each other's interference range. The density factors reported for San Francisco in Ref. [1] roughly correspond to a factor of 78 percent. The results show that tight coordination is able to more than halve the contention between the studied networks compared to optimal uncoordinated resource management [2].

4.3.2 Examples of Systems

4.3.2.1 Centralized Architectures Using the CAPWAP Protocol

In centralized network management architectures, some of the common management functionalities of access points are consolidated in a single component called access controller. Apart from radio resource management, this consolidation may also pertain to other aspects like distribution, handovers, QoS, and security. Depending on the degree by which functionality is taken over by the controller, one commonly distinguishes between three types of access points:

---

* See Section 4.3.2.2 for an example of how tight coordination might be achieved in practice.

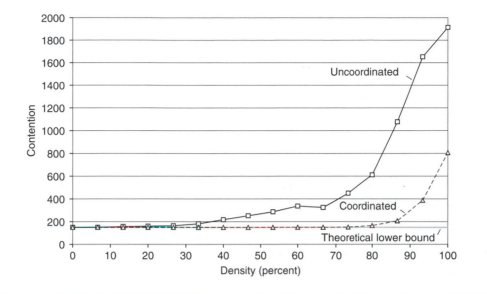

Figure 4.6 Comparison between coordinated and uncoordinated radio resource management at different deployment densities. (From Zdarsky, F.A., Martinovic, I., and Schmitt, J.B., On self-coordination in wireless community networks, International Conference on Personal Wireless Communications (PWC'06), Albacete, Spain, September 2006.)

1. Thick access points are full-featured, independently operable components, which may be configured by an access controller via their Simple Network Management Protocol (SNMP) interface.
2. Thin access points are reduced to performing real-time functions of the MAC layer, like per-frame transmit power and rate selection, control frame processing, and frame encryption. In contrast, all non-real-time tasks like dynamic channel selection, per-cell transmit power selection, load-balancing, and authentication are consolidated in an access controller.
3. Ultrathin access points are merely enhanced "signal converters," which transmit all frames passed to them by the access controller. All MAC layer functionality is performed centrally by the controller, even timing-critical ones.

To provide interoperability between centralized wireless network management solutions of different vendors, the IETF is working on the CAPWAP Protocol [26] for communication between the access controller and a group of access points. This protocol is supposed to be general enough to be able to incorporate other wireless technologies like IEEE 802.16 as well, but restricting itself to the management of thick and thin access points only.

Ultrathin access points are especially interesting in the context of radio resource management for sparse network deployments, because the completely centralized management allows controllers to schedule all frame transmissions temporarily and spatially in such a way as to completely avoid contention. Thus, all access points in the network may operate on a single channel and other channels are only needed to increase the capacity. In marketing slang, this approach is therefore sometimes called "channel blanket" to distinguish it from the more traditional cell-like approach. Furthermore, centralized management allows zero-delay handovers by simply switching between access points without the necessity of the standard authentication and association procedure. Thus, station "reassociation" may be performed on a per frame level. In dense wireless LAN deployments, however, this fully centralized management suffers from the same problems as all other uncoordinated approaches. It could even be more sensitive to nondeterministic contention from other networks on every channel as this makes the scheduling even more complex and error prone.

4.3.2.2 Virtualized Wireless Access Networks

Even the best centralized radio resource management cannot provide reliably high performance in the presence of interference from other, independently managed wireless networks. Therefore, the question is, whether interference may be mitigated through some form of coordination between networks of different domains, ideally a tight coordination.

One feasible approach is the concept of virtualized wireless access networks [27]. The basic idea is to enable densely deployed wireless LANs in a certain area to form a federation that coordinates itself across domain boundaries to minimize interference between wireless nodes. On top of this optimally managed substrate, each network operator may then instantiate virtual wireless LAN instances up to his share of resources contributed to the federation.

For a complete virtualization of the access network, it is necessary to provide mechanisms that allow users of one network to be temporarily connected via another network. This may be achieved through a careful combination of access point virtualization, tunneling, and traffic shaping. Using such an approach makes it possible to exploit the full potential of a tight coordination radio resource management approach. Furthermore, it could be used to implement DAPP as described in Section 4.2.6 and thus increase the resilience of the network as well as its flexibility with respect to network coverage.

4.4 Radio Resource Management Support in Standards

To be effective, radio resource management requires mechanisms to gather information about the current network state and to enforce decisions in the network. As the original IEEE 802.11 standard is little supportive in this respect, many workaround solutions have been proposed in the scientific literature (such as the cell breathing described above) and implemented as proprietary extensions by the industry, for instance within the Cisco Compatible Extensions (CCX) program. This section gives a brief overview about current and upcoming standardization efforts of the IEEE 802.11 task groups that are useful in the context of radio resource management.

4.4.1 IEEE 802.11h: Spectrum Management

The IEEE 802.11h [28] standard amendment specifies two spectrum management related functions: the TPC and the DFS. Both functions are mandatory in Europe and some other regions for all radios transmitting in the 5 GHz ISM bands. This is to avoid potential interference with aeronautic radar and satellite communications in the same frequency bands.

TPC enables stations to dynamically adjust their transmit power on a per-frame basis to the lowest acceptable transmit power. How the transmit power has to be adjusted is not specified by the standard, but regulations requires an average reduction of at least 3 dB. An informed change of transmit power requires a measurement of the radio link between two stations. Thus, TPC allows wireless nodes to request a TPC report from a neighboring node. This report contains two important informations: the transmit power with which the report is sent (in dBm) and the link margin (in dB). The former allows the requester to estimate the path loss from the reporting node, while the latter allows the reporting node to give feedback on the signal strength with which it receives frames from the requester, relative to its link margin requirements. Additionally, TPC introduces a new power capability information element to reassociation frames, using which stations trying to associate to an access point specify their minimum and maximum transmit power capabilities. The access point may use this information to reject stations with too high transmit power in order not to violate any radio regulatory constraints.

First, the primary task of DFS is to periodically check the channel for the presence of radar, in which case the channel has to be cleared immediately. To this end, an access point can silence all of its associated stations during its radar scan by including a Quiet information element in its beacon and probe response frames, which forces all stations to update their NAV to the specified period. If radar is detected, an access point may move its BSS to a different channel by including a channel switch announcement in one of its management frames. This announcement includes the channel to switch to and whether the change will be effective immediately or after a specified number of beacon periods.

Second, DFS enables access points to request a channel measurement from its stations to find an operating channel with low traffic and interference. The request specifies the channel to measure on the measurement period and the report type. Three different measurement reports may be requested: A basic report contains flags that inform which type of signals were detected during the measurement, for example, another BSS, a radar signal, or another, unknown signal. In a CCA report, the station specifies the fraction of time its CCA function has reported a busy channel during the measurement period. Finally, a receive power indication (RPI) histogram report returns a signal strength histogram of frames received during the measurement. Again, as with TPC, the standard does not specify if and how the reported results should be used in practice.

4.4.2   IEEE 802.11k: Radio Resource Measurements

As the only objective of 802.11h has been to fulfill regulatory requirements, its radio measurement support has been kept to a necessary minimum. From the perspective of radio resource management, the availability of a richer set of functions for network and radio resource measurements is desirable. The upcoming IEEE 802.11k [29] standard amendment adds many new measurement and report functions and also specifies the interfaces with which these may be accessed by higher layers. It builds on the request and response frame formats of 802.11h and augments them with new measurement types. Thus, a requester may still specify when and how long each measurement may be performed. It is now also possible to process several measurement requests concurrently.

One major category of measurements in 802.11k deals with characterizing a channel, a link, or an access point in terms of a long-term physical layer or link layer statistic. To the RPI histogram of 802.11h, it adds a noise histogram report, for which a channel is only sampled when the CCA signals an idle channel. On a higher level, a frame report lists the number of frames received on a specified channel during the measurement period, their average signal strength, etc. The frames may be listed separately for each transmitter and possibly also filtered by frame type and subtype. To gain better information on its stations, an access point may request a station statistics report, which contains a large number of values such as counts of successful and failed transmissions, retries, duplicates, fragments, checksum errors, etc. Furthermore, the access point may request a QoS metrics report from its stations, which contains similar success and failure counts for each traffic class, but also an average and histogram of the transmit queue delays. In return, access points announce the current channel load, access delays, and available admission capacity in their beacons.

The second category of measurements pertains to discovering certain types of nodes and retrieving information about them. One important application is the discovery of hidden stations. In a hidden station report, the reporting station lists all nodes that it knows have to exist, as it can overhear frames transmitted to them, but whose ACK replies it cannot receive. Furthermore, there is support for discovering neighboring access points, for example, to enable faster BSS transitions. By requesting a beacon report, a station may ask another station to return all beacon information it currently has, possibly after performing an active or passive scan. Alternatively, the station may request a neighbor report from the access point it is associated to, which contains a list of other access points of the same extended service set (ESS) that the access point is aware of. Finally, 802.11k supports location-based services, as a node may request a location configuration indication (LCI) report, which contains the geographical location (latitude, longitude, and altitude) of the reporting node.

4.4.3 IEEE 802.11v: Network Management

Although the measurement capabilities of the 802.11k standard are useful both in a network-centric and client-centric architectures, the upcoming IEEE 802.11v [30] standard amendment specifically targets the configuration and control requirements of a network-centric approach.

Most importantly, 802.11v specifies a new action management frame using which access points may direct single associated stations to reassociate to a specific other access point or a set of access points. This is extremely useful for explicit load-balancing and interference reduction. It also adds deferral management functions using which an access point may control the way that stations in its BSS perform their CCA, for example, by changing their EDT. Finally, 802.11v includes a feedback mechanism using which a station may recommend a transmit rate to its access point and vice versa. Thus, 802.11v completes the list of control mechanisms for the most important tuning parameters in a network-centric radio resource management architecture.

Another useful feature is multi-BSSID support, which reduces protocol overhead in the case of virtualized access points (logical access point instances operating concurrently on a single physical device). Each virtual access point instance has its own basic service set identifier (BSSID), which it normally has to announce in a separate Beacon frame. 802.11v now defines a new information element with which one virtual access point instance may announce multiple BSSIDs in a single beacon. In a similar direction, another new information element allows stations to request information from an access point on multiple service set IDs using just a single Probe Request frame. Both enhancements are especially useful, although not necessary, in the context of the previously described virtualized wireless access networks concept.

This list of new features planned for the upcoming IEEE 802.11v standard is by no means complete. Further interesting, but at the time of writing not yet very well-defined extensions include a scheme that allows neighboring access points to coordinate their time reference and resource management decisions, a reliable multicast scheme, remote firmware upgrades, and many others.

4.5 Conclusions and Open Research Issues

Radio resource management is an important prerequisite for reliable and high-quality services, but is especially challenging in UMA networks, which need to share radio resources with other networks. This chapter discussed the tuning parameters of IEEE 802.11-based UMA networks and their idiosyncrasies. Furthermore, it has described the current trend toward network-centric radio resource management approaches, which manifests itself in the standardization efforts for the CAPWAP Protocol and the amendments h, k, and v to the IEEE 802.11 family of standards.

A more worrying trend, however, is the ever-increasing density of UMA networks, which is both a result of their success as well as maybe a great inhibitor to their success over the long term, if they fail to deliver the expected QoS requirements due to excessive interference between networks. We therefore contend a paradigm shift toward radio resource management approaches, which coordinate themselves across administrative boundaries, to mitigate interference rather than tolerate it. However, it is not clear yet, whether such a coordination should be implicit or explicit to obtain optimal results. On the other hand, it has been previously shown that a tight coordination may result in much higher improvements than loose coordination alone. A first step toward such a coordination model has been made with the concept of virtualized wireless access networks.

Some interesting open research issues include

- Algorithms for DAPP that optimize the reliability and performance of a dense wireless LAN deployment.
- Unified radio resource management strategy that uses all of the tuning parameters described. Although many joint optimization approaches exist, it is not obvious whether they still work in an efficient and stable way if combined.
- Characterization and quantification of interference of dense network deployments on QoS, especially in architectures using centralized scheduling to implement a channel blanket.

As a final remark, it is interesting to note that in a UMA context, network planning seems to loose in significance relative to radio resource management. One reason is that it is not possible to choose optimal access point installation sites given that the locations and configurations of foreign access points cannot be controlled. Another is that radio resource management is able to compensate for suboptimal planning, as it may dynamically adapt the network to a changing radio environment. In dense network deployments, in which a DAPP is feasible, it should be possible to coordinate networks so that they perform at least as well as an optimally planned sparse network deployment.

REFERENCES

1. Aditya Akella, Glenn Judd, Srinivasan Seshan, and Peter Steenkiste. Self-management in chaotic wireless deployments. In International Conference on Mobile Computing and Networking (MobiCom'05), Cologne, Germany, September 2005.
2. Frank A. Zdarsky, Ivan Martinovic, and Jens B. Schmitt. On self-coordination in wireless community networks. In International Conference on Personal Wireless Communications (PWC'06), Albacete, Spain, September 2006.
3. Andrzej Kochut, Arunchandar Vasan, A. Udaya Shankar, and Ashok Agrawala. Sniffing out the correct physical layer capture model in 802.11b. In International Conference on Network Protocols (ICNP'04), Berlin, Germany, October 2004.
4. Kyle Jamieson, Bret Hull, Allen Miu, and Hari Balakrishnan. Understanding the real-world performance of carrier sense. In ACM SIGCOMM 2005, Philadelphia, PA, August 2005.
5. Jason A. Fuemmeler, Nitin H. Vaidya, and Venugopal V. Veeravalli. Selecting transmit powers and carrier sense thresholds in CSMA protocols for wireless ad hoc networks. In International Wireless Internet Conference (WICON'06), Boston, MA, August 2006.

6. Vivek P. Mhatre, Konstantina Papagiannaki, and Francois Baccelli. Interference mitigation through power control in high density 802.11 WLANs. In International Conference on Computer Communications (Infocom'07), Anchorage, AK, January 2007.

7. Ad Kamerman and Leo Monteban. WaveLAN-II: A high-performance wireless LAN for the unlicensed band. *Bell Labs Technical Journal*, 2(3):118–133, August 1997.

8. Gavin Holland, Nitin Vaidya, and Paramvir Bahl. A rate-adaptive MAC protocol for multi-hop wireless networks. In International Conference on Mobile Computing and Networking (MobiCom'01), Rome, Italy, July 2001.

9. Ceilidh Hoffmann, Mohammad Hossein Manshaei, and Thierry Turletti. CLARA: Closed-loop adaptive rate allocation for IEEE 802.11 wireless LANs. In International Conference on Wireless Networks, Communications and Mobile Computing (WirelessCom'05), Hawaii, June 2005.

10. John C. Bicket. Bit-rate selection in wireless networks. Master's thesis, Massachusetts Institute of Technology, Cambridge, MA, February 2005.

11. Sunwoong Choi, Kihong Park, and Chong kwon Kim. On the performance characteristics of WLANs: Revisited. In International Conference on Measurements and Modeling of Computer Systems (SIGMET-RICS 2005), Banff, Canada, June 2005.

12. Anand Balachandran, Geoffrey M. Voelker, Paramvir Bahl, and P. Venkat Rangan. Characterizing user behavior and network performance in a public wireless LAN. In International Conference on Measurements and Modeling of Computer Systems (SIGMETRICS'02), Marina Del Rey, CA, June 2002.

13. Anand Balachandran, Paramvir Bahl, and Geoffrey M. Voelker. Hot-spot congestion relief in public-area wireless networks. In IEEE Workshop on Mobile Computing Systems and Applications (WMCSA'02), Callicoon, NY, June 2002.

14. Yigal Bejerano, Seung-Jae Han, and Li (Erran) Li. Fairness and load balancing in wireless LANs using association control. In International Conference on Mobile Computing and Networking (MobiCom'04), Philadelphia, PA, October 2004.

15. Martin Heusse, Franck Rousseau, Gilles Berger-Sabbatel, and Andrzej Duda. Performance anomaly of 802.11b. In International Conference on Computer Communications (INFOCOM'03), San Francisco, CA, March 2003.

16. Paramvir Bahl, Mohammad T. Hajiaghayi, Kamal Jain, Sayyed Vahab Mirrokni, Lili Qiu, and Amin Saberi. Cell breathing in wireless LANs: Algorithms and evaluation. *IEEE Transactions on Mobile Computing*, 6(2):164–178, February 2007.

17. Yigal Bejerano and Seung-Jae Han. Cell breathing techniques for load balancing in wireless LANs. In International Conference on Computer Communications (Infocom'06), Barcelona, Spain, April 2006.

18. Chen-Mou Cheng, Pai-Hsiang Hsiao, H. T. Kung, and Dario Vlah. Adjacent channel interference in dual-radio 802.11a nodes and its impact on multi-hop networking. In IEEE Global Telecommunications Conference (GLOBECOM'06), San Francisco, CA, November 2006.

19. IEEE. Wireless LAN medium access control (MAC) and physical layer (PHY) specifications. ANSI/IEEE Std 802.11, August 1999.

20. Arunesh Mishra, Vivek Shrivastava, Suman Banerjee, and William Arbaugh. Partially overlapped channels not considered harmful. In International Conference on Measurements and Modeling of Computer Systems (SIGMETRICS'06), Saint Malo, France, June 2006.

21. Bruno Kauffmann, François Baccelli, Augustin Chaintreau, Konstantina Papagiannaki, and Christophe Diot. Self organization of interfering 802.11 wireless access networks. Technical Report 5649, INRIA, August 2005.

22. Arunesh Mishra, Vivek Shrivastava, Dheeraj Agarwal, Suman Banerjee, and Samrat Ganguly. Distributed channel management in uncoordinated wireless environments. In International Conference on Mobile Computing and Networking (MobiCom'06), Los Angeles, CA, January 2006.

23. Gregory Davi. Using picocells to build high-throughput 802.11 networks. *RF Design*, 27:16–23, July 2004.

24. Frank A. Zdarsky and Jens B. Schmitt. Handover in mobile communication networks: Who is in control anyway? In IEEE EUROMICRO Conference (Multimedia and Telecommunications Track), Rennes, France, September 2004.

25. Propagate Networks. AutoCell—the self-organizing WLAN. White Paper, 2003. http://www. propagatenet.com/news/docs/wpaper_autocell_soWLAN.pdf.

26. Pat Calhoun, Michael Montemurro, and Dorothy Stanley. Capwap protocol specification. IETF Draft, June 2007.

27. Frank A. Zdarsky, Ivan Martinovic, and Jens B. Schmitt. The case for virtualized wireless access networks. In International Workshop on Self-Organizing Systems (IWSOS'06), Passau, Germany, September 2006.

28. IEEE. Wireless LAN medium access control (MAC) and physical layer (PHY) specifications: Amendment 5: Spectrum and transmit power management extensions in the 5 GHz band in Europe. IEEE Standard, October 2003.

29. IEEE. Wireless medium access control (MAC) and physical layer (PHY) specifications: Amendment 9: Radio resource measurement. IEEE P802.11k/D4.0, March 2006.

30. IEEE. Wireless medium access control (MAC) and physical layer (PHY) specifications: Amendment v: Wireless network management. IEEE P802.11v/D0.02, March 2006.

Chapter 5

Security in IEEE 802.11-Based UMA Networks

Ivan Martinovic, Frank A. Zdarsky, Adam Bachorek, and Jens B. Schmitt

CONTENTS

Taking into consideration the tremendous deployment of wireless communications, one can easily see that IEEE 802.11 [1] networks have played a major role in supporting universal mobile access (UMA) during recent years. Contrary to the popularity of wireless LAN technology, the topic of its security gained a rather negative publicity. The tragic end of Wired Equivalent Privacy (WEP) and the simplicity of accomplishing various denial-of-service (DoS) attacks resulted in abandoning the security at the logical link layer. To regain trust in the IEEE 802.11 technology and to provide the major security goals such as authentication, integrity, and confidentiality of the user data, the solution was finally presented in 2004, when the IEEE successfully finished the 802.11i ratification process. A stronger user authentication, a new underlaying cipher, and a more reliable integrity verification are the significant changes provided by this standard. Nevertheless, it seems that the utilization of IEEE 802.11i security did not follow the same growth as the deployment of the IEEE 802.11 technology resulting in a number of proprietary security solutions. Such solutions can easily be applied within existing infrastructures and business models, but at the price of allowing various link-layer attacks and often misleading the user's sense of security.

This chapter introduces the new IEEE 802.11i security standard including not only its enhancements regarding user authentication, key management, and confidentiality but also its costs in terms of its message complexity, network delays, and still unsolved security vulnerabilities.

5.1   Introduction

In contrast to wired networks which allow for the physical separation of traffic, the broadcast nature of wireless communication has always been a major concern from the security perspective. To avoid passive and active attacks in WLANs, the IEEE 802.11 [2] standard initially defined security features collectively known as WEP. Its objective was to ensure the fundamental security goals such as authentication, confidentiality, and integrity of data similar to the security level offered within the wired networks. Shortly after its standardization in 1999, this security standard emerged as highly vulnerable and uncapable of achieving any of the intended security goals.

5.1.1   Tragic End of WEP

The WEP protocol relies on RC4 a hitherto popular symmetric cipher widely used in protocols due to its simplicity and good performance characteristics. The RC4 is a stream cipher and its security strongly depends on the choice of the initialization vector (IV), which together with a secret key generates a pseudorandom stream called keystream. To encrypt data, a keystream is usually XORed with the corresponding plaintext. Since an IV is not secret and it is commonly transmitted as a plaintext, selecting the same IV should be avoided. Using the same IV results in the same keystream, which consequently enables decrypting a ciphertext without knowing the secret key.

Even though the WEP standard recommends the usage of a different IV for each packet, its poor design along with faulty implementations resulted in the tragic end of its security. In their seminal paper, Fluhrer et al. [3] showed several weaknesses in the RC4's key scheduling algorithm (i.e., in the generation of a keystream) and theoretically described an attack based on abusing IVs. As a result of their analysis, the authors were able to create ciphertext-only attacks by collecting enough transmitted ciphertexts encrypted using the keystreams generated from weak IVs, which statistically revealed parts of the secret key. Although such attacks might be avoided by a cautious design of keystream derivation methods, i.e., by applying cryptographic methods to randomize the IVs,* the underspecified design of WEP itself implied an even faster way of disclosing of the encrypted data. Not only the IV's length of 24 bit emerged as too short which lead to frequent reoccurrences of the same values, but also faulty implementations of the design caused an even higher repetition rate of equal IVs. For example, the authors of Ref. [5] report that many PCMCIA WLAN cards reset IV to 0 every time they are reinitialized, as a result of which a keystream based on a low-valued IV occurs more often during the lifetime of the corresponding secret key. The first practical implementation of the attack that abuses the RC4 key scheduling weakness was presented by Stubblefield et al. [6]. Shortly after, various tools for breaking WEP were also available on the Internet (e.g., WEPCrack, Airsnort, and WEPOff). In 2005, it was reported that at one of the security conferences, the FBI demonstrated the breaking of WEP in three minutes. Interestingly, although it can be assumed that now a days WEP can be broken by your neighbor, the majority of WLAN networks still relieson WEP protection. In the same work, the authors present a more efficient version of an attack called fragmentation attack, which is also effective against a frequent rekeying, the method used as a simple countermeasure against the original attack. Finally, in 2007 another tool [8] was available for attacking WEP using an enhanced version of the weak IV attack, and showed that time required for breaking WEP could even be achieved within a minute.

The integrity protection offered by WEP shared the same destiny. From the beginning there was a problem of how the integrity check is created and verified. The concept was based on CRC-32 checksums, which are commonly used to detect accidental changes of data during transmission, but it has no cryptographic properties to protect data against intentional manipulation [5].

To summarize, the security provided by WEP can be described as nonexistent [6], i.e., using WEP one should assume no link-layer security and thus, consider all systems connected via IEEE 802.11 as external, place an access point (AP) outside the firewall, and for the security sensitive data use higher layer security protocols such as IPSec, or Virtual Private Networks (VPNs).

The end of WEP left wireless clients without a standardized protection giving rise to a dispersion of proprietary mechanisms. To provide security solutions for their wireless products, many IEEE 802.11 technology vendors implement protection mechanisms based on various unrealistic assumptions. An example is the access control based on a station's hardware address (Media Access Control (MAC) address or also called Ethernet address) under the assumption that such an address is a static and an unchangeable identifier. On the contrary, most of the WLAN drivers allow manipulation of hardware addresses, therefore misleading a user in its sense of given security. All that an attacker requires to successfully overcome such control is to monitor the wireless channel and impersonate the address of an already associated station. Another similar protection frequently found in various security recommendations is to hide the Service Set ID (SSID) which is used as a network name for network discovery and usually broadcasted by APs within the Beacon frame. The idea is to use SSID as a shared secret, so that only stations knowing the network's name would be able to join it. In reality, this approach offers no protection at all and can be described as security by obscurity. In Ref. [9], a number of such solutions are analyzed and discussed in more detail. The security they offer fully complies with the title of the paper "Your 802.11 Wireless Network Has No Clothes."

---

* In Ref. [4], Ron Rivest, the author of RC4 responded to the aforementioned IV weaknesses and suggested the solution.

5.1.2 Intermediate Solution: Wi-Fi Protected Access

Facing the security problems left by WEP and the increasing dispersion of proprietary solutions, an industry consortium of WLAN equipment vendors established the Wireless Fidelity (Wi-Fi) Alliance and launched an interoperability certification program for the IEEE 802.11 conformant products. This Wi-Fi certification program was merely intended as an interim solution by introducing the so-called Wi-Fi Protected Access (WPA) industry standard. To avoid the demand for new hardware, WPA inherited RC4 as underlying cipher from his predecessor, yet with certain improvements. The key length increased to 128 bit, the length of IV doubled to 48 bit, but most importantly WPA defined a new key mixing mechanism using a cryptographic hash function for the computation of IVs. This mechanism together with the new integrity check and a frequent rekeying (dynamic changing of a secret key to avoid statistical analysis) is incorporated within the Temporal Key Integrity Protocol (TKIP). Another feature of WPA is the support of the Extensible Authentication Protocol (EAP), which is an authentication framework embedding many currently approved authentication mechanisms (e.g., Transport Layer Security [EAP-TLS]). To simplify its usage, WPA offers a pre-shared key (PSK) mode of authentication which, similar to WEP, only requires a passphrase while circumventing an extensive mutual authentication process.

Actually, most of the features provided by WPA are part of the IEEE 802.11i standard with except for the underlaying cipher. The goal behind the introduction of WPA was to provide a fast solution to replace WEP without requiring more than a driver update. Consequently, WPA still suffers from certain weaknesses especially with respect to integrity protection, as well as its susceptibility to dictionary attacks against the weak passphrases in the PSK mode of operation.

5.2 WLAN Has New Clothes: IEEE 802.11i

The expected solution to the hitherto existing security inadequacies was finally presented in 2004 when the IEEE successfully ratified the promising IEEE 802.11i security standard. The cornerstones of this standard are the separation of user authentication and message protection, as well as the concept of a robust secure network (RSN). This concept is based upon a security framework composed of several known and approved protocols and techniques to ensure a robust protection of wireless communication within so-called RSN associations (RSNAs). As link-layer connections between RSN-enabled network entities, RSNAs offer port-based access control through IEEE 802.1X [10], which on its part defines the fundamental model for the support of authentication services such as enhanced mutual authentication and key management via EAP [11].

The major enhancement regarding data confidentiality and integrity is the introduction of a new cryptographic algorithm named Counter Mode with Cipher Block Chaining MAC Protocol (CCMP). On the basis of the Advanced Encryption Standard (AES), CCMP provides strong data encryption and reliable data origin authenticity. Taking the full set of security requirements of the ratified IEEE 802.11i into account, the Wi-Fi Alliance released the WPA2 certification program at the end of 2004 to support WLAN product interoperability.

In the following sections we focus on RSN entities and their interaction as defined in the IEEE 802.11i standard.

5.2.1 Port-Based Network Access Control

The IEEE 802.1X standard [10] specifies a general framework for the provision of authentication services by means of a port-based access control mechanism. This model includes three fundamental components. Usually as part of the distribution system, the AS is the network element that provides authentication service to an entity called authenticator through the back end of the network system. A supplicant, in this context, is a network component that shares a wired or wireless point-to-point link with the authenticator and seeks to get authenticated by the means of a particular authentication service.

Now in the context of IEEE 802.11i, the roles of an authenticator and a supplicant are adopted by an AP and a wireless station (STA), respectively, and the port-based access mechanism may be briefly described as follows (see Figure 5.1 for the illustration of the IEEE 802.1X port-based network access control mechanism used in a RSN). The AP holds two communication ports, an uncontrolled authentication port and a controlled service port. As long as the authentication process including the key management between the STA and the AS has not yet finished successfully, the AP only permits authentication-related traffic via its uncontrolled port and blocks any other traffic on the controlled port. Once the phase of authentication and key management is successfully completed, the authenticated STA is granted access to the network via the controlled service port of the AP. Commonly used and recommended protocols for the authentication process are EAP and the Authentication Dial-In User Service (RADIUS) protocol [12].

Whereas EAP encapsulates the messages of any suitable higher layer authentication method between the STA and the AP, the RADIUS protocol on its part attends to encapsulate the EAP method's message exchange between the AP and the AS. The combination of both protocols provides an authentication between the STA and the AS while the AP rather acts as forwarder of messages via its uncontrolled authentication port from STA to AS and vice versa.

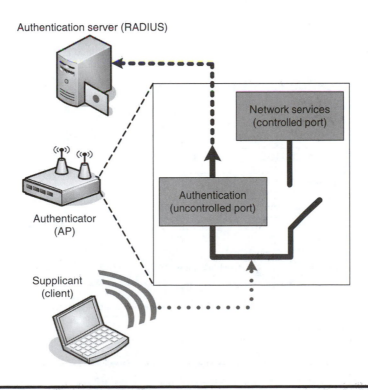

Figure 5.1   IEEE 802.1X port-based network access.

5.2.2 Mutual Authentication

As already stated in the preface of Section 5.2, an RSNA relies on the IEEE 802.1X port-based access control mechanism to provide authentication services, whereas the EAP represents a flexible authentication framework that can be adapted to a wide variety of authentication methods. While EAP defines the general packet formats and message types utilized to encapsulate the concrete authentication procedure, EAP methods on their part perform the authentication transaction and are responsible for the procurement of the necessary key material used to cryptographically protect subsequent data exchange and thus grant a confidential and authentic communication.

In general, an authentication can be based on passwords, smart cards, certificates, or other credentials verifying the proper identity of the communicating entities. EAP, in this context, abstracts away from the encapsulated authentication method and enables the AP to forward authentication messages between the STA and the authentication infrastructure in the back end of the network, typically consisting of a unique RADIUS server (RS).

Figure 5.2 depicts the three-part message flow of 802.1X/EAP in conjunction with RADIUS/EAP as the carrier protocols for specific authentication methods. In the first part, the user identity of the STA is requested by the AP and the response is forwarded to the AS, which is supposed to validate this identity against an existent user account defining available methods and credentials needed to proceed with the authentication process. This authentication phase may be initiated either by the STA sending an EAP over LAN (EAPOL) start frame before a Request identity frame or alternatively by the AP sending the Request identity message immediately. In the second part, the actual authentication method (as defined for the user account) is initiated by the AS. Henceforth, any message of the chosen authentication method is encapsulated

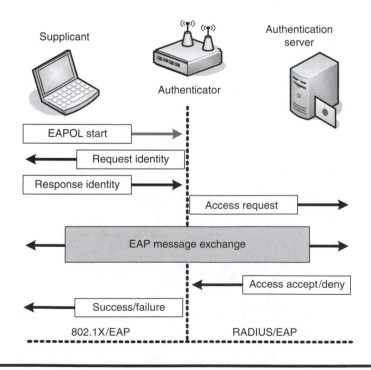

Figure 5.2   EAP authentication message flow.

within EAP-Request/-Response frames between the STA and the AP as well as within RADIUS Request/-Challenge frames between the AP and the AS. When the authentication method finishes its message exchange, the AS transmits an EAP-Success or EAP-Failure message to the STA via the AP depending on whether the authentication was successful or not.

As the 802.11i standard does not specify which particular authentication method to employ when implementing an RSN, it is up to the organization or users to decide which one fits best into their existing or target network environment. The most commonly used authentication method is TLS which is considered one of the most secure EAP methods due to its strong cryptographic background using public key certificates. Also, TLS is one of the only five EAP methods currently comprised in the WPA2 certification program of the Wi-Fi Alliance. As it is being widely deployed by numerous WLAN vendors, EAP-TLS has finally emerged as the dominant authentication protocol for trusted IEEE 802.11i RSN support. In the first instance, it benefits from its resistance against Man-In-The-Middle (MITM) and dictionary attacks as well as its support for protected cipher suite negotiation, mutual authentication, and session key derivation.

5.2.3 Key Hierarchy

Generally speaking, security provisioning is based upon secret keys. In RSN environments, all keys have a limited lifetime and are organized in a key hierarchy (Figure 5.3).

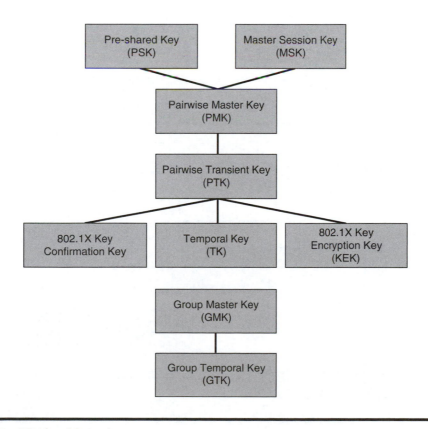

Figure 5.3 RSN key hierarchy.

Central to this hierarchy is a 256-bit cryptographic key called Pairwise Master Key (PMK) which is obtainable in two ways. Either it is derived from a static PSK, which has to be manually installed on each device prior to communication, or from the PMK, which may be derived from the result of any method applied in the authentication phase, e.g., the shared session key (called the Master Session Key [MSK]) as the output of the EAP-TLS authentication process. By means of a pseudorandom function, the PMK is then used to generate the Pairwise Transient Key (PTK), a temporal key (TK) for unicast traffic protection from which further encryption and integrity keys like the EAPOL-Key Confirmation Key (KCK), EAPOL-Key Encryption Key as well as the TK are extracted.

In addition to these unicast keys, in an RSN there may also exist two group keys, the Group Master Key (GMK) and the temporal Group Transient Key (GTK) as a derivation of the GMK using another pseudorandom function. The GTK is used for broadcast and multicast traffic protection.

5.2.4 Key Management

Following a successful EAP-authentication, the cryptographic keys still need to be derived and installed on the supplicant and authenticator. The key management phase includes two types of handshakes, a 4-way-handshake and optionally a group key handshake. Both handshakes are depicted in Figure 5.4.

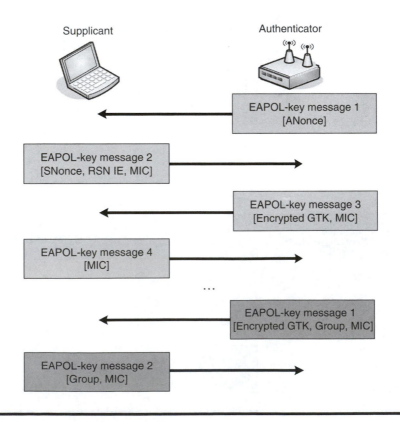

Figure 5.4   4-way handshake message flow.

As the very first step after the mutual authentication process, the 4-way-handshake is initialized by the authenticator to confirm that both authentic entities possess a current PMK, to confirm the cipher suite selection, to derive a fresh PTK from the PMK, and to install the encryption and integrity keys as well as the GTK. To carry out the corresponding message exchange, EAPOL RSN key message frames are used.

During a 4-way-handshake, four of those frames are exchanged between the STA and the AP. It is initiated by a first completely unprotected message including a random number ANonce sent by the AP. After generating its own SNonce and receiving the ANonce, the STA is able to use both random numbers along with additional parameters to derive the PTK and all TKs from the PMK. This allows for protecting the subsequent messages with a message integrity code (MIC) computed using the KCK. When the AP receives this integrity protected message containing the SNonce along with the STA's RSN information element (RSN IE), which confirms the cipher suite selection, it can not only derive the PTK and all TKs on its part but is also able to verify that the STA is in possession of the current PMK and has derived the TKs properly. Within the subsequent message, the AP includes a KCK-computed MIC and a GTK encrypted with the KEK. The receipt of this frame again lets the STA verify that the AP holds the PMK. The transmission of the fourth and last frame allows the STA to announce that the derived TK will be installed. At this point, both entities have proved their knowledge of the previously negotiated PMK to each other and derived the TK material needed to protect subsequent data exchange. Thus, after the successful completion of the 4-way-handshake both entities are mutually authenticated and the STA is qualified to be granted access to the network resources via the controlled service port of the AP.

In contrast, the rarely used group key handshake generally plays a secondary role and supports multicast or broadcast application traffic. By means of a two-way exchange of integrity protected EAPOL-key messages, the AP and the concerned STAs may negotiate a new GTK in security jeopardizing conditions to preserve their ability to receive protected broadcast or multicast messages. More precisely, the AP simply derives a new GTK, encrypts it with the temporal KEK, and passes it to any affected STA which in turn acknowledges the receipt by a subsequent EAPOL-key message.

5.2.5 Confidentiality and Integrity

Now that the STA has successfully completed the mutual authentication as well as key management phase, it is supposed to make use of the derived TK as a shared secret to secure the wireless communication with the corresponding AP.

To establish secure connections, the 802.11i security amendment includes two corresponding cipher suites TKIP based upon RC4 (confer WPA) and CCMP. Because the support of the former is merely classified as optional by the standard, we concentrate on CCMP as the mandatory cipher suite for RSN-conformant network environments. It is also less vulnerable to attacks than TKIP and is therefore the recommended choice in terms of robust security establishment in business and advanced private wireless networks.

CCMP is based on a generic encryption block cipher mode of AES and uses a single 128-bit cryptographic key (which is the TK) for both encryption and integrity protection. As per specification, the TK is a per-session key which means it is valid for the entire duration of a station-to-AP association. These features along with the precomputation of certain cryptographic parameters allow for a reduced complexity. Besides, the cryptographic scheme of CCMP includes integrity protection of the packet payload as well as of a portion of the packet header by means of a cipher block chaining message authentication code (CBC-MAC). Replay attack resistance is guaranteed by a

48-bit packet number, which is used as an IV for the cryptographic algorithm and included in the CCMP part of the frame header. Incrementing the packet counter by one with every encrypted frame lets the TK outlive any conceivable association period and thus limits the need to renew the TK to adverse conditions like the compromise of any credentials. In short, CCMP represents a high-level and surpassing security scheme, which is considered as central to the RSN provisioning of confidentiality and data authenticity.

5.2.6   Pre-Authentication and PMK Caching

In addition to the previously mentioned security features, the IEEE 802.11i security amendment introduces two mechanisms to better cope with station mobility and to increase network performance. The mechanisms are pre-authentication and PMK caching, also known as PMK security association (PMKSA) caching.

PMKSA caching conduces to the ability of nearly seamless resumption of previously established secure communication sessions. Therefore, a supplicant and the corresponding authenticator have to store the shared secret, which is negotiated by EAP or derived directly from PMK. A reason for session resuming might be the connection loss of a station to its associated AP due to, for instance, weak radio signal. When reconnecting to the network, a supplicant may prove its eligibility to rejoin the security association by supplying the appropriate ID out of a list of available PMKIDs to the authenticator. The caching-enabled authenticator may then verify the ongoing security association depending on whether he finds a match in his PMKID list or not. In the former case, the fact of having cached the negotiated shared secret prevents a station from repeating the entire authentication process and allows for fast reassociation with the corresponding AP by merely renewing the PTK out of the PMK via another 4-way-handshake. However, if the handshake fails or the authenticator fails to verify the PMKID, the full 802.1X and EAP authentication process must be repeated.

Harnessing the feature of pre-authentication, a wireless station is enabled to roam more seamlessly between adjacent APs of an extended service set provided that PMKSA caching is supported. If so, a station may initiate an authentication process with an authenticator in advance using an existing security association with another AP. Through caching of the established SAs along with the corresponding PTKs, this station may then roam between the authenticated APs whenever it needs to, again without having to repeat the entire authentication process. Besides the default network discovery operation, the obligatory final step in authentication for the station to pass through remains the 4-way handshake.

5.2.7   Summary of the RSN Connection Process

Summarizing at a glance all the mechanisms introduced in the previous subsections, this subsection provides an overview of the complete RSN connection process.

Basically, the act of connecting to any kind of wireless networks consists of a sequence of successive phases, each with a different purpose. Dividing the entire process into its individual steps and discarding the actual secured data transfer phase, four main connection-related phases can be identified as shown in Figure 5.5.

The first phase corresponds to the active or passive network discovery procedure (henceforth scanning phase) of a mobile station seeking to find appropriate network APs. Here, a STA either passively scans the wireless medium for Beacon frames periodically transmitted by APs, or it actively sends Probe Request frames requesting nearby APs to answer with Probe Responses. Both

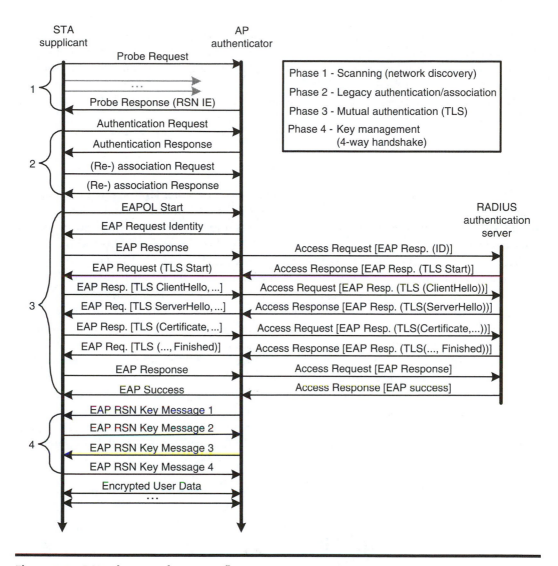

Figure 5.5   RSN phases and message flow.

AP messages are expected to comprise the so-called RSN information element (RSN IE), which announces the AP's RSN capabilities regarding the cipher and key management suite as well as pre-authentication support. As active scanning is the preferred scanning method, the scanning phase in this context is delimited by the first captured probe request addressed to the dedicated AP and the authentication request initiating the subsequent phase.

The second phase includes the 802.11 legacy authentication and association part that merely serves for backward compatibility and, allows an STA to connect to the uncontrolled port of the AP (only authenticated and associated stations are able to send data frames). This phase begins with the Authentication Request frame sent by the STA and ends with the Acknowledgment frame confirming the (Re-) Association Response of the AP. Again, the RSN IE embedded in the STA's (Re-) Association Request frame provides information about its capabilities and helps the AP to decide whether to accept or to reject the STA's admission request. As already mentioned, the EAP-TLS combination is one of the most frequently used authentication methods, and therefore we take it as an example of a third phase, which completes EAP-TLS message exchange between the STA and the AS processed via the AP.

At last, the key management phase with the execution of the 4-way handshake concludes the RSNSA establishment process.

5.2.8   Cost of IEEE 802.11i

All things considered, IEEE 802.11i offers a complete security suite. However, it should also be evident that such a substantial enhancement involves additional processing and communication complexity. This section discusses some results of real-world measurements that have been conducted using three off-the-shelf devices as supplicants (two Laptops and one PDA) with currently available state-of-the-art implementations of the IEEE 802.11i standard [13]. Furthermore, two prevalent APs as authenticators were selected, Proxim AP-4000 (≈300 USD) and Linksys WRT54g (≈60 USD). These APs provide a rough estimate on the contingent discrepancy in respect of the cost/performance ratio between expensive professional equipment and rather affordable devices dedicated for residential use. The IEEE 802.11i RSN infrastructure was implemented with an AS running Ubuntu Linux 6.06, Kernel v.2.6.15, and FreeRADIUS v.1.1.0.

Figures 5.6 and 5.7 show the latencies resulting from the EAP-TLS mutual authentication phase and the 4-way handshake (key management phase), respectively. They also correspond to phases 3 and 4 in Figure 5.5. As can be seen, the EAP-TLS protocol on two devices takes ≈300 ms to finish. Only STA1 encounters an extremely long delay of ≈3–4 s. The reason for such high delays is due to an implementation anomaly where the device does not seem to respond to the EAPOL-Identity Request frame sent by the APs. Instead, it only proceeds after sending an EAPOL Start frame itself

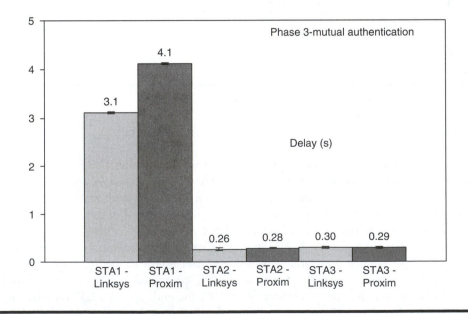

Figure 5.6   EAP-TLS authentication delays.

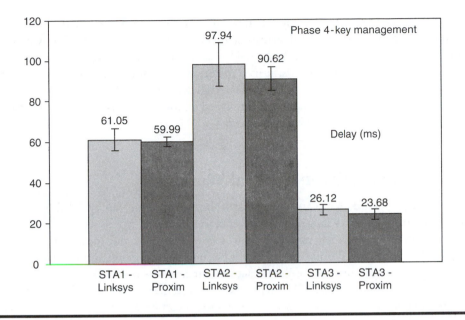

Figure 5.7   4-way handshake delays.

(as mentioned in Section 5.2.2 and depicted in Figure 5.2, EAP authentication can be initiated either by the supplicant or the authenticator). This implementation inflexibility has a high price in terms of authentication delay. Similarly, the delays originated from the 4-way handshake also let us assume that different implementations of IEEE 802.11i standard are the reasons for the high variation.

With respect to delay-reducing mechanisms like pre-authentication and PMK caching, which are most important for a handover scenario within the same distribution set, two out of three supplicants have implemented both features. Only STA3 did not support them and therefore had additional cost of ≈300 ms for EAP-TLS authentication every time it changed an AP within the same distribution set.

To further minimize authentication delays, Ref. [14] introduces the concept of proactive key distribution (PKD), which tries to circumvent the mutual authentication phase. In their approach, the session key derived by both the STA and the AP is distributed among the adjacent APs and used whenever the STA chooses to switch an AP. The resulting analysis shows that the delay can be reduced to 50 ms. Two additional methods are presented in Ref. [15], which exclude the key management phase from the overall connection process. The first technique enables an STA to accomplish the 4-way-handshake with all adjacent APs immediately after connecting to the network so as to preclude another key exchange in handover scenarios. In contrast, the second technique makes arrangements to make up for the 4-way-handshake ex post, i.e., after the handover.

Considering overall IEEE 802.11 network latencies, empirical studies in Refs. [16–18] substantiate that more than 90 percent of involved IEEE 802.11 delays are to be attributed to the network discovery procedure and detection of network loss. Particularly in Refs. [16,17], various WLAN products were analyzed and a significant implementation diversity interms of highly variable delays has been identified. Having a major effect on the overall delay, both, the scanning and the connection loss detection delay, are subject to optimization within the IEEE 802.11r standard. On the other hand, the 802.11i security standard has already been ratified, leaving less room for improvement. This could change the overall landscape of latencies within 802.11i secured networks, making the latency caused by security a major challenge for competitive implementations. This in turn could have a further impact on the overall deployment of secure wireless networks. It would not be the first time to see security being turned off for better performance, even if problems with the latter is only a matter of implementation.

5.3 WLAN Security in the Real World

The new security standard exemplary solves most of the security issues left by WEP. Its cost in regard to network performance also shows justifiable price and there are not much reasons not to apply it as soon as possible. However, its deployment is rather stagnant and most of the WLANs still rely on WEP or at most WPA protection. This is because WPA2 often requires new hardware and the extension of the already existing infrastructure. Furthermore many devices have not yet been certified according to the standard. As a result, wireless Internet service providers (WISPs), which are also the major protagonists of UMA technologies, incorporate proprietary security solutions that can easily be implemented within their infrastructure and business models, providing higher usability and lower complexity for customers, but on the other hand expecting from the customers to take care of security themselves. In this section we describe security problems prevalently found in present WLANs, especially in a popular scenario of wireless hotspots. Before going into more detail, we tackle the problem of availability, which especially in wireless networks is the most vulnerable among all security goals.

5.3.1 Attacks on IEEE 802.11 Availability

In contrast to the feasibility of attacks as described in Section 5.1.1, which is the result of unspecific design guidelines and an incorrect use of cryptographic primitives, many attacks on availability are due to the inherent broadcast nature of wireless communication. The IEEE 802.11 technology operates in industrial, scientific, and medical (ISM) radio frequency bands. These are the frequencies used for unlicensed operation of personal short-range devices such as cordless phones, Bluetooth gadgets, baby monitors, and wireless cameras. Hence, the impaired availability of communication can result not only from malicious disruption of radio signals, so-called jamming, but also from densely crowded radio bands, indicated as interference. Therefore, the issue of communication availability can be considered a general problem of the physical layer of any wireless technology. A modulation schemes for increasing jamming resistance of radio communication exist, jamming feasibility mostly depends on the hardware capabilities of an attacker.

Nevertheless, frequency jamming has never been considered a grave security problem. The reason is that jamming of the physical layer prevents any communication on the wireless channel, but for being effective it must be continuously present which increases the risk of the attacker's exposure. In a security context, more interesting attacks are those executed silently and assisting in attacking other security goals. Therefore, the MAC layer has attracted more attention. The IEEE 802.11 management and media access protocols are found as highly vulnerable to simple, yet very effective DoS attacks. Most of them have their roots in unauthenticated management and control frames, which are used for connection management and channel control. To better understand the

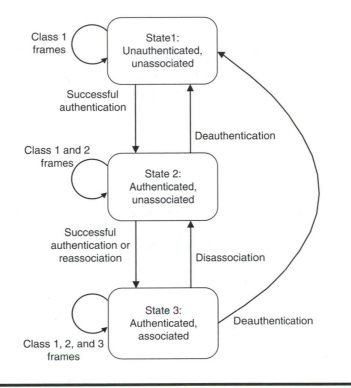

Figure 5.8 IEEE 802.11 state machine.

nature of those attacks, we briefly describe the corresponding frames and the state machine of IEEE 802.11 [19], which is presented in Figure 5.8.

Beside data frames used to transport higher layer protocol data, the IEEE 802.11 standard defines control and management frames. They are divided into three classes:

■ Class 1: Control frames within this class provide operations for channel acquisition, positive acknowledgment of received frames and carrier-sensing (e.g., Request-to-Send and Clear-to-Send [RTS/CTS] and Acknowledgment [ACK]). Management frames are used for supervisory functions like network discovery and network joining or leaving procedures (e.g., Beacon frame, Authentication Req./Resp., and Deauthentication frame).
■ Class 2: Management frames are responsible for starting and ending associations and for supporting mobile stations in moving within the distribution system (e.g., Association Req./Resp. and Reassociation Req./Resp.).
■ Class 3: A Class 3 management frame signals the end of an authenticated relationship (Deauthentication). Also, all data frames (frames carrying user data) belong to this class.

Which frame is allowed to be sent or received depends on the class and state of the connection. The IEEE 802.11 state machine allows Class 1 frames to be sent and received in every state. This is due to their function to provide basic services for network discovery and connection management as opposed to data frames, which are allowed to be transmitted only in the state 3. If an AP detects that a frame from a certain class is sent from a sender being in a wrong state, it resets its current state and forces the sender (by sending either deauthentication or disassociation frame) to reauthenticate or reassociate. The fact that no authentication of control and management frames is provided serves as the major source of vulnerabilities exploited by DoS attacks. An attacker has a wide spectrum of different, yet very simple and efficient attacks based on sending different control and management frames misusing the hardware addresses of authenticated client. With IEEE 802.11i the situation has not changed for the better. The IEEE 802.11i authentication and key exchange are executed within state 3 (i.e., utilizing data frames), providing the adversary with further possibilities for mounting attacks even on IEEE 802.11i secured networks (examples on attacking IEEE 802.11i are given in Section 5.3.4).

In Ref. [20], a variety of such attacks is discussed and their feasibility is experimentally shown. Their simple, cheap, and rather silent execution, yet heavy impact on the network availability made them much more prominent than frequency jamming attacks. The most effective attack is to send a forged deauthentication frame to an AP with the address of an associated station. As a result, the station's connection state is set to the initial state where no data traffic is allowed and it is forced to repeat the complete association process. Today, this attack is still one of the most frequently used DoS attacks on all IEEE 802.11-based networks. In the case of using IEEE 802.11i, this attack produces even higher (re-)association delays.

There are also various attacks abusing unauthenticated control frames. Usually, wireless stations switch into the low power mode, also called sleeping mode, in order to extent their battery life. Periodically, the station awakes and sends a PS-poll frame to so as retrieve any of its frames buffered at the AP. To attack the wireless station, the attacker impersonates PS-poll frames by using the spoofed hardware address of its victim and then sends these frames to the AP. Consequently, this results in a frame loss for the authentic station since the AP clears its buffer after buffered frames are polled.

To cope with such attacks, in 2005 the IEEE 802.11w standard for the authentication of management frames has been proposed. However, until no standardized solution is implemented, these vulnerabilities will always enable attacks based on frame impersonation. In the next subsections, we discuss implications of such attacks.

5.3.2 Wireless ARP Spoofing Attack

Today, hotspot offer Internet access in various public environments, e.g., coffee shops, shopping malls, airports, restaurants, etc. The registration is usually done by Web-based payment, after which the hardware address of the used wireless device is allowed to connect to the Internet. Although it is known that such a business model is highly subsceptible to link layer attacks, much more interesting is the simplicity by which such attacks can be prepared and executed. In such environments customers are usually in transit and can hardly distinguish between legitimate and fake network access points. Therefore, their wireless devices used for browsing the Internet are subsceptible to the injection of fake web pages commonly used for stealing user credentials, e.g., phishing. The nature of hot spots where the joining clients are new and initially do not possess much information about the network infrastructure, makes this environment very attractive to attack a legal AP and to inject a fake Web page asking the user for his credentials.

Attacks harnessing such kind of circumstances are of ten times based on Address Resolution Protocol (ARP) spoofing. The ARP is used to resolve IP addresses to hardware addresses. By sending ARP replies containing a fake target hardware address, an attacker can simply redirect a client's data traffic to itself. This is why the ARP has served as the basis for many different MITM and DoS

Protocol ver.	Type	Subtype	To DS	From DS	More frag	Retry	Pwr mgt	More data	WEP	Order

Figure 5.9 802.11 frame control field.

attacks mostly targeting wired networks. In the following, we discuss how simple it is to adapt this kind of attack to peculiarities of wireless environments [21].

By manipulating certain IEEE 802.11 frame characteristics, an attacker can successfully send fake ARP packets with randomly chosen MAC addresses, can redirect the traffic, and can even be hard to detect due to the attack's low overhead. Figure 5.9 shows a generic frame control field, which is a part of every 802.11 frame. The two one-bit flags ToDS and FromDS are used to indicate whether the frame is sent to or received from the distribution system. In infrastructure mode, any frame sent from a wireless client will have the ToDS bit set and FromDS bit cleared. Those frames are checked by an AP to assure that the sender is an authenticated and associated station. If the frame has the FromDS bit set, the AP believes that the frame was sent by a different AP (since a distribution system can contain several APs). The AP does not know all the stations within the distribution system and cannot check if the sender's MAC belongs to a known and associated entity. As a result, by setting FromDS bit, an attacker can send arbitrary frames without being intercepted by the AP. These fake frames will, therefore, not be forwarded to the distribution system, which renders protection mechanisms inside the network difficult to apply.

Most common protections against ARP spoofing is a static ARP where the MAC-to-IP mapping can only be changed manually. Although very efficient within small infrastructures, this solution is not suitable for more dynamic environments. Especially in wireless networks where joining clients are new and initially do not know the network configuration, this solution cannot be implemented without introducing additional complexity. Furthermore, different monitoring and traffic analyzing tools used inside wired networks to check if ARP replies provide valid hardware addresses are also not effective. These mechanisms focus on networks in which traffic can be physically controlled. In contrast, a wireless environment with its broadcast nature makes neither of these solutions practically applicable.

5.3.3 Evil Twin Attack

While the first attack targets wireless clients and can eventually be solved by the IEEE 802.11w standard, the attack described in this subsection is based on abusing operational anomalies of certain APs and it hijacks wireless clients from the beginning of their association.

When executing a DoS attack, the aim of an attacker is to exhaust a server's resources, which may then result in the server's inability to provide any service at all. This can remain the main goal of an attack or it can also serve as a starting point for other attacks. In a wireless environment, this kind of attack is well known and traditionally executed by installing a rogue AP transmitting with a higher signal strength, and by using the same network name as the legitimate AP. In the association phase, wireless clients would per default prefer to associate with an AP emitting a stronger signal and therefore choose the rogue AP. Today, many APs therefore implement a rogue AP detection mechanism, which monitors wireless channels for transmissions of APs with unknown/untrusted hardware addresses. Therefore, one should assume that this kind of attacks are not feasible anymore on modern APs. With the aim of investigating this matter, in Ref. [21], six different APs dating from 2003 to 2006 were chosen based on their popularity and price. The APs were then subject to resource depletion attacks by flooding them with fake association requests. Many of low-priced and middle-priced APs exhibited operational anomalies resulting in high response delays and frame losses enabling an attacker to install a rogue AP on the same channel with the same hardware address. Due to better performance properties, the rogue AP was first to respond to clients' requests. The consequence was the possibility to launch a simple yet effective attack where the rogue AP is fully disguised with the same hardware address of the legitimate AP and it operates on the same channel; any rogue AP detection mechanism is not able to counter act the attack.

The following steps show the simplicity of starting such an attack:

1. An attacker is endued with a laptop running a Web server and two wireless interfaces. One of the interfaces is set to operate in master mode to enable the access functionality (henceforth fake AP) while another one is used to start the flooding attack.
2. Hardware address (MAC address) and the network name of the fake AP is set to correspond to the configuration of the legitimate AP.
3. Attacker starts flooding the legitimate AP. After the latter has increased its response delay, the fake AP starts answering every request sent by wireless clients.
4. Attacker captures HTTP requests and responds with a fake Web page asking for user registration (it can also choose to respond to any other control packet like ARP, DNS, Dynamic Host Configuration Protocol [DHCP], etc.).

This attack was successfully accomplished on many low-priced and middle-priced APs. Only the most expensive APs were immune to this kind of attack due to their better hardware capabilities. This is an important fact because often the low price of IEEE 802.11 technology is considered to be one of its most frequently mentioned advantages.

5.3.4 Attacks on IEEE 802.11i

Even if a WLAN is protected by IEEE 802.11i, the threat emerging from most of the attacks discussed in Section 5.3.1 still persist. This has an impact on the protocol execution as defined by IEEE 802.11i because the key exchange data is still transferred within unauthenticated frames. In Ref. [22], the authors have analyzed the 4-way handshake protocol and described a new DoS attack called 4-way handshake blocking. It is based on forging the first message of the 4-way handshake and providing different security parameters to the wireless station. Thereupon, the wireless station calculates PTK from the wrong data and the 4-way handshake ends up with inconsistent values. To mitigate this attack and to assure that the handshake is nonblocking, the authors discuss solutions based on keeping all the received parameters until the new PTK has been verified. This again can lead to memory exhaustion attacks. For more information and additional interesting attacks on IEEE 802.11i we refer the reader to Ref. [23].

5.3.5 Conclusion and Open Issues

The various confidentiality and integrity vulnerabilities of WEP and the simplicity of launching impersonation attacks by manipulating a sender's MAC address accounted for bad reputation of

IEEE 802.11 security. To regain trust in this widespread technology, in 2004 the IEEE 802.11i ratification process was successfully finalized. The new standard provides a security framework composed of several known and approved protocols to ensure a robust protection of wireless communication. Nevertheless, the IEEE 802.11i security standard only focuses on securing user data, i.e., it provides security for the data frames used to transport higher layer protocol data. The control frames used for channel control and management frames used for connection administration like network discovery and association procedures have not been included in the IEEE 802.11i protection framework. The reason seems to be twofold. First, the tragic end of WEP left wireless clients without standardized protection giving rise to dispersion of proprietary solutions, hence the interoperability certification program (e.g., WPA, WPA2) and the standard has been impatiently awaited by both the industry and the public. Second, the control and management frames impact the availability of IEEE 802.11 networks which is, especially in wireless networks, the weakest chain among all security goals. Frequency jamming attacks at the physical layer are an indigenous property of wireless communication and therefore the importance of providing availability protection at the wireless link layer is inherently downgraded. Nevertheless, there is an important difference between physical layer attacks, which focus on attacking the channel capacity, hence denying any communication, and link-layer attacks, which affect the services provided by an AP and connection states of wireless stations.

The new, still to be ratified IEEE 802.11w standard may protect clients against various aforementioned attacks. Nevertheless, the problem of unauthenticated control frames and the assumption of the fair-play nature of the Carrier Sense Multiple Access (CSMA) mechanism within a wireless environment still remains. However, it is worth considering whether the traditional security, which is based on a binary relationship between entities, provides sufficient means for coping with such problems. The research on a cooperative behavior between wireless stations, enforcing fairness, as well as on detecting and limiting misbehavior could in this case provide more appropriate solutions.

REFERENCES

1. IEEE 802.11i/D10.0. Security Enhancements, Amendment 6 to IEEE Standard for Information Technology. IEEE Standard, April 2004.
2. IEEE 802.11. Telecommunications and information exchange between systems-Local and metropolitan area networks-Specific requirements-Part 11: Wireless LAN Medium Access Control (MAC) and Physical Layer (PHY) Specifications. IEEE Standard, July 1999.
3. S. Fluhrer, I. Mantin, and A. Shamir. Weaknesses in the key scheduling algorithm of RC4. In SAC'01: Revised Papers from the 8th Annual International Workshop on Selected Areas in Cryptography, August 2001.
4. RSA Security Response to Weaknesses in Key Scheduling Algorithm of RC4. http://www.rsa.com/rsalabs/node.asp?id=2009 (last access: 07-17-2007).
5. N. Borisov, I. Goldberg, and D. Wagner. Intercepting mobile communications: The insecurity of 802.11. In MobiCom'01: Proceedings of the 7th Annual International Conference on Mobile Computing and Networking, July 2001.
6. A. Stubblefield, J. Ioannidis, and A. D. Rubin. Using the Fluhrer, Mantin, and Shamir Attack to break WEP. In Proceedings of the Network and Distributed System Security Symposium (NDSS 2002), February 2002.
7. A. Bittau, M. Handley, and J. Lackey. The final nail in WEP's coffin. In SP '06: Proceedings of the 2006 IEEE Symposium on Security and Privacy (S&P'06), August 2006.

8. E. Tews, A. Pychkine, and R.-P. Weinmann : aircrack-pwt. http://www.cdc.informatik.tu-darmstadt.de/aircrack-ptw/ (last access: 07-15-2007).

9. W. A. Arbaugh, S. Shankar, J. Wang, and K. Zhang. Your 802.11 network has no clothes. In Proceedings of the 1st IEEE International Conference on Wireless LANs and Home Networks, December 2001.

10. IEEE 802.1X. IEEE Standard for Local and Metropolitan Area Networks-Port-Based Network Access Control. IEEE Standard, June 2001.

11. Extensible Authentication Protocol (EAP). Request for Comments: 3748, June 2004.

12. B. Aboda and P. Calhoun. RADIUS Support For Extensible Authentication Protocol (EAP). RFC 3579, 2003.

13. I. Martinovic, F. A. Zdarsky, A. Bachorek, and J. B. Schmitt. Introduction of IEEE 802.11i and measuring its security vs. performance tradeoff. In Proceedings of the 13th European Wireless Conference, Paris, France, April 2007.

14. H. Duong, A. Dadej, and S. Gordon. Proactive context transfer and forced handover in IEEE 802.11 wireless LAN based access networks. *SIGMOBILE Mob. Comput. Commun. Rev.*, 9(3):32–44, 2005.

15. M. Kassab, A. Belghith, J.-M. Bonnin, and S. Sassi. Fast Pre-authentication based on Proactive Key Distribution for 802.11 Infrastructure Networks. In WMuNeP '05: Proceedings of the 1st ACM workshop on Wireless Multimedia Networking and Performance Modeling, pp. 46–53, June 2005.

16. H. Velayos and G. Karlsson. Techniques to reduce IEEE 802.11b MAC layer handover time. Technical Report TRITA-IMIT-LCN R 03:02, KTH, Royal Institue of Technology, Stockholm, Sweden, April 2003.

17. A. Mishra, M. Shin, and W. Arbaugh. An Empirical Analysis of the IEEE 802.11 MAC Layer Handoff Process. *SIGCOMM Comput. Commun. Rev.*, 33(2):93–102, 2003.

18. I. Ramani and S. Savage. SyncScan: Practical fast handoff for 802.11 infrastructure networks. In Proceedings of the IEEE INFOCOM Conference, March 2005.

19. M. Gast. *802.11 Wireless Networks: The Definitive Guide*. O'Reilly, 2005.

20. J. Bellardo and S. Savage. 802.11 Denial-of-service attacks: Real vulnerabilities and practical solutions. In Proceedings of the USENIX Security Symposium, August 2003.

21. I. Martinovic, F. A. Zdarsky, A. Bachorek, C. Jung, and J. B. Schmitt. Phishing in the wireless: Implementation and analysis. In Proceedings of the 22nd IFIP International Information Security Conference (SEC 2007), May 2007.

22. C. He and J. C. Mitchell. Analysis of the 802.11i 4-way handshake. In Proceedings of the 2004 ACM Workshop on Wireless Security, pp. 43–50, October 2004.

23. C. He and J. C. Mitchell. Security analysis and improvements for IEEE 802.11i. In Proceedings of the 12th Annual Network and Distributed System Security Symposium (NDSS'05), pp. 90–110, February 2005.

Chapter 6

Mobility Management between UMA Networks and Cellular Networks

Daqing Xu and Yan Zhang

CONTENTS

Mobility is everywhere, in the train, in the office, in the meeting room, in the car, in the airport, and so on. Most users want to communicate anytime, anywhere, and without any configuration hurdles. These users want session continuity between various access networks (cellular networks [GSM or GPRS or 3G] and UMA [unlicensed mobile access] networks [WLAN or Bluetooth or WiMAX]), giving them a continuous and seamless connection to the best available network. This chapter describes the solutions that enable seamless mobility between cellular networks and UMAN (UMA networks) for voice and data connections.

6.1 Introduction

In the last five to ten years, UMA network (UMAN) and cellular network have developed very rapidly with the Internet. UMA networks can provide relatively high data rates at low prices in the short range, and cellular networks can provide relatively low data rates at high prices in the long range. The UMA networks and cellular networks will coexist and complement each other in the future. It is no doubt that making phone calls by WLAN (Voice-over-IP [VoIP]) is much cheaper than over cellular or fixed phone networks. UMA technology provides a way to access the core global system for mobile communications (GSM) network through Wi-Fi. This sounds attractive from the point of view of cellular network operators, who could thus extend their network coverage through Wi-Fi hot spots with minimal additional investment; Wi-Fi hotspots are much cheaper to deploy than GSM base stations. UMA could thus decrease the load of legacy GSM and general packet radio service (GPRS) access networks. At the same time, the operator still holds control over the connection, and therefore can charge upon that. Most players in the FMC (fixed and mobile convergence) space see UMA as the long-term future technology.

The main element in UMA is a UMA network controller (UNC), which provides a similar basic functionality as a conventional base station controller (BSC) to handle authentication and encryption and data integrity. The UNC enables mobile devices to access circuit switched (CS) services via A interface with mobile switching centers (MSCs) and to access GPRS service via Gb interfaces with serving GPRS support nodes (SGSNs). Also, the UNC maintains session control during handoff (Figure 6.1). The UNC includes a security gateway (SGW) that terminates secure remote access tunnels from the handset. The SGW interfaces with the authentication, authorization, and accounting (AAA) proxy-server via the Wm interface and the AAA proxy-server retrieves user information from the home location register (HLR). The UMA functional architecture is shown in Figure 6.1.

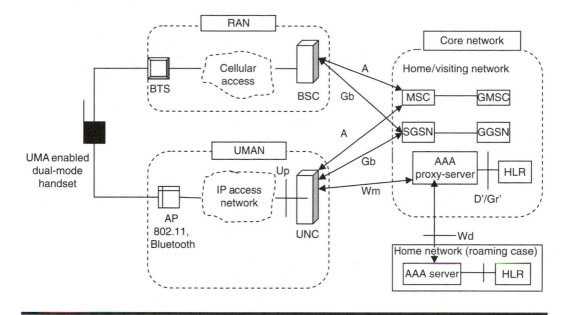

Figure 6.1 UMAN interworking with cellular networks.

A dual-mode (cellular and UMA) handset (Figure 6.1) can be set up in one of four preferences:

1. GSM only (i.e., never uses the UMAN mode)
2. GSM preferred (i.e., uses GERAN/UTRAN [GSM EDGE radio access network/universal terrestrial radio access network] mode where possible, switching to UMAN mode only when the GSM network is not available)
3. UMAN preferred (i.e., uses UMAN mode where possible)
4. UMAN only (i.e., switches to UMAN mode immediately after the phone starts up and registers on the GERAN/UTRAN mode, never switches to GERAN/UTRAN mode)

The descriptions of techniques and specifications presented in this chapter are based on Refs. [1,2].

6.2   UMAN Protocol

6.2.1   *Standard GERAN Protocols*

The UMA architecture uses the following standard GERAN protocols without any modifications (Figures 6.1 through 6.5):

- GSM mobility management, connection management (CM), and higher layer protocols are used without any changes in the mobile station (MS) or the mobile switching center (MSC).
- GSM voice encoding carried over IP between the MS and UNC.
- GPRS logical link control (LLC) and higher layer protocols are used without any changes on the MS and SGSN.
- A-interface protocols are used without any changes between the MSC and UNC.
- Gb-interface protocols are used without any changes between the SGSN and UNC.
- Wm interface protocols are used without any changes between the UNC and the AAA server.

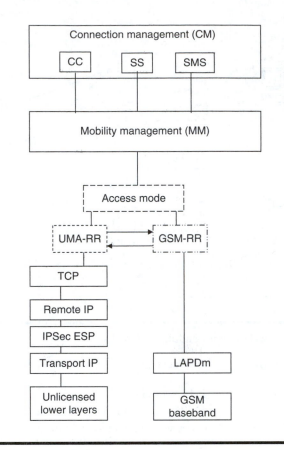

Figure 6.2 Dual-mode MS signaling architecture for CS domain.

Further, all GSM and GPRS protocols on the MS are unaffected when they are operating over the GERAN base station subsystem (BSS).

6.2.2 Standard Unlicensed Radio Access Protocols

The UMA architecture uses the following WLAN and Bluetooth protocols without any modifications:

- 802.11 protocols for PHY and Medium Access Control, including functions for connection, authentication, encryption, data transfer, and traffic prioritization.
- Bluetooth protocols for PHY, Baseband, SDP, etc., including functions for discovery, paging, authentication, encryption, ACL and data and voice traffic transfer. Additionally, BNEP is used to provide Ethernet emulation over Bluetooth ACL links. IEEE 802.11 and Bluetooth have minimum radio performance requirements.

6.2.3 Standard IP-Based Protocols

The UMA architecture uses the following standard IP-based protocols without any modifications (Figures 6.4 and 6.5):

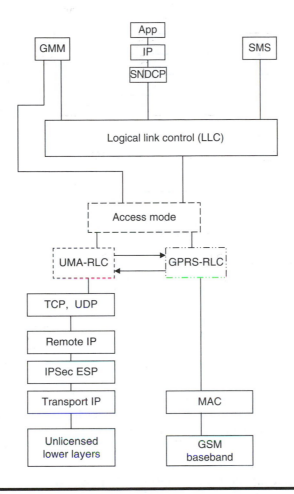

Figure 6.3 Dual-mode MS signaling architecture for packet switched (PS) domain.

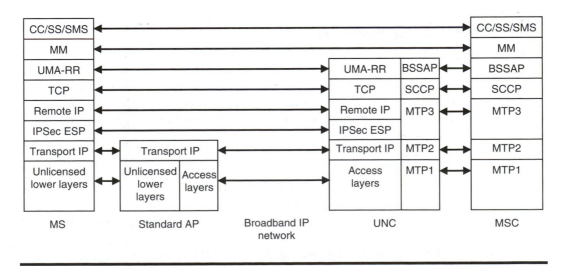

Figure 6.4 UMAN signaling protocol architecture for CS.

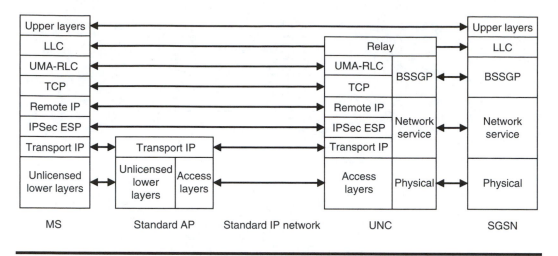

Figure 6.5 UMAN signaling protocol architecture for PS.

- IP over standard lower layers
- TCP to provide a tunnel for GSM/GPRS signaling and SMS
- IPSec encapsulating security payload to provide a secure tunnel for GERAN user and control plane traffic
- IKEv2, Extensible Authentication Protocol-Subscriber Identity Module (EAP-SIM), and Extensible Authentication Protocol-Authentication and Key Agreement for authentication and for establishing and maintaining a security association (SA) between MS and UNC
- UDP for IPSec NAT traversal
- UDP for GPRS data transfer
- RTP/UDP for transfer of GSM voice frames over IP transport

6.2.4   UMA-Specific Protocols

6.2.4.1   UMA-RR

UMA Radio Resource (UMA-RR) protocol provides a radio resource management layer, which is the peer of GSM-RR, in the MS. It is designed to take advantage of the characteristics of the unlicensed radio link as they are quite different from that of the GERAN radio link. Specifically, it provides the following functions (Figure 6.4):

- Registration with UNC
- Set up of bearer path for CS traffic between the MS and UNC
- Handover support between GERAN and UMAN
- Perform functions such as GPRS suspension, paging, ciphering configuration, application level keep-alive, etc.
- Support for identification of access point (AP) used for UMA access

6.2.4.2 UMA-RLC

UMA Radio Link Control (UMA-RLC) protocol provides the following services (Figures 6.3 and 6.5).

- Delivery of GPRS signaling and SMS messages over a secure tunnel
- Paging, flow control and GPRS transport channel management
- Transfer of GPRS user plane data

6.3 Mobility Management Standard

Mobility management consists of location management and handover procedure; location management includes location registration, update, and call delivery subprocedure.

6.3.1 Discovery and Registration Procedure

When an MS supporting UMA first attempts to connect to a UNC based on a UMA subscription, it needs to identify the serving UNC. To do this, it first connects to a provisioning UNC and then discovers a default UNC, which in turn can redirect the MS to a serving UNC.

The registration procedure is performed between the MS and the UNC (default or serving UNC). It serves the following purposes:

- It informs the UNC that an MS is now connected through a particular AP and is available at a particular IP address. The UNC keeps track of this information for the purpose of paging, for example, from the core network.
- It provides the MS with the operating parameters associated with the UMA service.

The discovery and registration procedure is performed through the following steps (Figure 6.6). If MS has a provisioned or derived fully qualified domain name (FQDN) of an SGW/UNC, it first performs a DNS query to resolve FQDN to an IP address.

1. If MS has a provisioned IP address for the provisioning SGW, it establishes a secure tunnel to the provisioning SGW. All signaling and user plane data over the Up interface between MS and UNC shall be sent through the SA and shall be protected by the IPSec tunnel. Mutual authentication is accomplished using EAP-SIM or EAP-AKA and IKEv2.
2. If the MS has a provisioned IP address for the provisioning UNC, MS sets up a TCP connection to a well-defined port on the provisioning UNC. It then queries the provisioning UNC for default UNC using a URR Discovery Request. The message contains GERAN/UTRAN Cell Info, AP Identity, MS Identity, international mobile subscriber identity (IMSI), etc. Because the UMAN coverage area overlaps the GERAN coverage, the UNC assigns a UMAN cell ID based on the above message.
3. The provisioning UNC returns a URR Discovery Accept message using the location information provided by MS to provide the FQDN or IP address of the default UNC and its associated default SGW.
4. Alternately, the provisioning UNC may return a URR Discovery Reject indicating the reject cause.

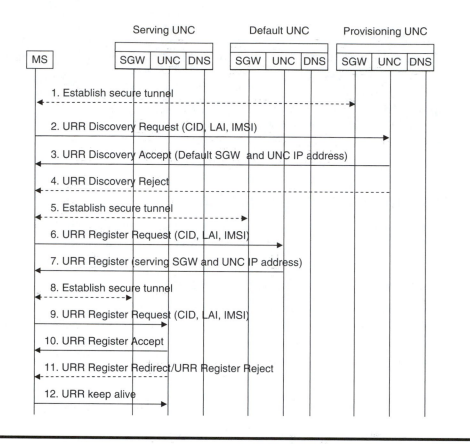

Figure 6.6 Discovery and registration.

5. The MS shall then set up a secure tunnel to the default SGW, and sets up a TCP session to a well-defined port on the default UNC.
6. The MS shall attempt to register on the default UNC by transmitting a URR Register Request. The message contains GERAN/UTRAN Cell Info, AP Identity, MS Identity, IMSI, UMA Services Required, etc.
7. If the default UNC wishes to redirect the MS to another serving UNC, it shall respond with a URR Register Redirect providing the FQDN or IP address of the target serving UNC and associated SGW. Alternatively, the default UNC may reject the registration and in this case the default UNC shall respond with a URR Register Reject indicating the reject cause. Alternately, the default UNC may return a URR Register Accept to accept the registration, and then go to step 10.
8. If the MS was redirected, then it shall set up a secure tunnel to the serving SGW. And then it sets up a TCP connection to a well-defined port on the serving UNC.
9. The MS shall attempt to register on the serving UNC by transmitting a URR Register Request. The message contains GERAN/UTRAN Cell Info, AP Identity, MS Identity, IMSI, etc. The UMAN cell ID value is stored in the MS registration context. The core network must be configured with the new UMAN cell ID value; however, the GERAN cell ID and the corresponding UMAN cell ID can share the same location information.

10. If the serving UNC accepts the registration attempt, it shall respond with a URR Register Accept. The message contains
    a. Cell description comprising BCCH ARFCN, PLMN color code, and base-station color code
    b. Location-area identification comprising the mobile country code, mobile network code, and location-area code corresponding to the UNC cell
    c. Cell identity identifying the cell within the location area
    d. Application level keep-alive timer value
11. Alternately, serving the UNC may reject the request or redirect the MS to another serving UNC.
12. The keep-alive process is a mechanism to indicate that the MS is still registered to the UNC.

6.3.2   Registration Update Procedure

After the MS has successfully registered to a UNC, the MS shall update the UNC if the AP or the overlapping GERAN/UTRAN coverage has changed. This update is performed using the registration update procedure, which is performed through the following steps:

1. Whenever an MS changes the AP, it sends a URR Register Update Uplink to the UNC with the updated AP ID information. If the MS requires updating the UNC with a new list of UMA services required, then it sends a URR Register Update Uplink message to the UNC including the new UMA services required list.
2. The UNC may optionally send a URR Register Redirect when it wants to redirect the MS based on updated information.
3. The UNC may also optionally deregister the MS on receiving an update by sending a URR Deregister to the MS.

The registration update downlink procedure is used by the UNC to update information in the MS such as system information or status of location services.

6.3.3   Mobile-Originated Speech Call

A mobile-originated speech call procedure (successful case) is performed through the following steps (Figure 6.7):

1. Upon request from the user to originate a call, the MS sets up a new URR connection, if needed, and sends the CM Service Request to serving UNC in URR Uplink Direct Transfer.
2. The serving UNC establishes an signaling connection control part (SCCP) connection to core network (CN) and forwards a CM Service Request to the CN.
3. The CN may optionally authenticate the MS using standard GERAN authentication procedures.
4. The CN may optionally update the ciphering parameters in the serving UNC using the Cipher Mode Command. The ciphering mode is negotiated during connection establishment.
5. The serving UNC signals the permitted ciphering algorithms to MS using the URR Ciphering Mode Command. The MS stores this information for possible future use after a handover to GERAN.

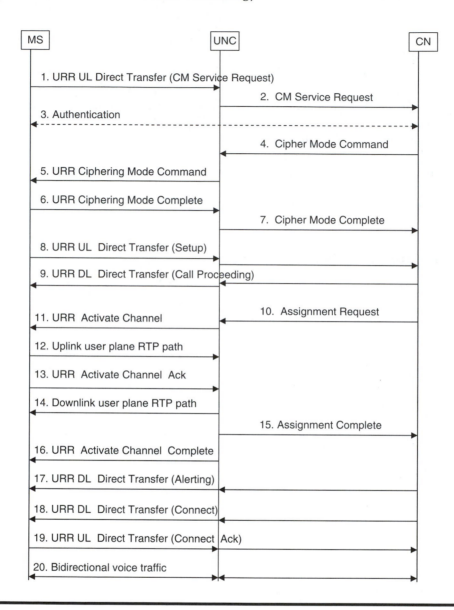

Figure 6.7 Mobile-originated speech call.

6. The MS signals the selected ciphering algorithm in a URR Ciphering Mode Complete message back to the serving UNC.

7. The serving UNC signals the selected ciphering algorithm to the CN using Cipher Mode Complete.

8. The MS sends the Setup message providing details on the call to the CN and its bearer capability and supported codec. This message is contained within the URR Uplink Direct Transfer between the MS and the UNC. The UNC forwards the Setup message to the CN.

9. The CN indicates that it has received the call setup and it will accept no additional call-establishment information, via the Call Proceeding message. The message is transferred to the UNC, which forwards the message to the MS in a URR Downlink Direct Transfer.

10. The CN requests the serving UNC to assign call resources using Assignment Request.
11. The serving UNC sends a URR Activate Channel to the MS including bearer path setup information such as
    a. Channel coding
    b. UDP port and the IP address for the uplink stream
    c. Voice sample size
    d. Cipher mode (for use in case of subsequent handover to GERAN)
12. The MS now establishes the RTP path to the serving UNC. The MS has not connected the calling party to the audio path.
13. The MS sends a URR Activate Channel Ack (Acknowledge) to the serving UNC indicating the UDP port and IP address for the downlink stream.
14. The serving UNC establishes the downlink RTP path between itself and the MS. The serving UNC may start sending idle RTP/UDP packets to MS.
15. The serving UNC signals to the CN that call resources have been allocated by sending an Assignment Complete message.
16. The serving UNC signals the completion of the bearer path to the MS with a URR Activate Channel Complete message. An end-to-end audio path now exists between the MS and the CN. The MS can now connect the calling party to the audio path.
17. The CN signals that the called party's phone is ringing, via the Alerting message. The message is transferred to UNC which forwards the message to MS in URR Downlink Direct Transfer.
18. The CN signals that the called party has answered, via the Connect message. The message is transferred to UNC which forwards the message to MS in the URR Downlink Direct Transfer. It connects the calling party to the audio path.
19. The MS sends Connect Ack in response, and the two parties are connected for the voice call. This message is contained within URR Uplink Direct Transfer between MS and UNC. UNC forwards Connect Ack message to CN.
20. Bidirectional voice traffic flows between MS and CN through the serving UNC is available.

6.3.4   Mobile-Terminated Speech Call

A mobile-terminated speech call procedure (successful case) is performed through the following steps (Figure 6.8):

1. A mobile-terminated call arrives at the CN. The CN sends a Paging Request to the UNC identified through the last Location Update. The IMSI of the mobile that is paged is always included in the request.
2. The serving UNC pages the MS a using URR Paging Request message. The message includes the TMSI, if available, in the request from the CN; else it includes only the IMSI of the mobile.
3. The MS responds with a URR Paging Response including the MS class mark and ciphering key sequence number.
4. The serving UNC establishes an SCCP connection to the CN and forwards the Paging Response to the CN.
5. The CN may optionally authenticate the MS using standard GERAN authentication procedures.

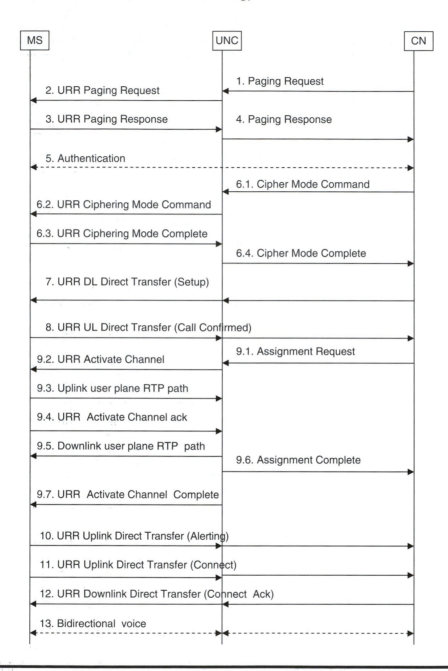

Figure 6.8   Mobile-terminated speech call.

6.1. The CN may optionally update the ciphering configuration in the MS, via the serving UNC, using a Cipher Mode Command. The ciphering mode is negotiated during connection establishment.

6.2. The serving UNC signals the permitted ciphering algorithms to the MS using a URR Ciphering Mode Command. The MS stores this information for possible future use after a handover to GERAN.

6.3. The MS signals the selected ciphering algorithm in the URR Ciphering Mode Complete message back to the serving UNC.

6.4. The serving UNC signals the selected ciphering algorithm to the CN using the Cipher Mode Complete message.

7. The CN initiates a call setup using the Setup message sent to the MS using a URR Downlink Direct Transfer.

8. The MS responds with Call Confirmed using a URR Uplink Direct Transfer after checking its compatibility with the bearer service requested in the Setup and modifying the bearer service as needed.

9.1. The CN requests the serving UNC to assign call resources using Assignment Request.

9.2. The Serving UNC sends a URR Activate Channel to the MS including bearer path setup information such as
   a. Channel coding
   b. UDP port and IP address for the uplink stream
   c. Voice sample size
   d. Cipher mode (for use in case of subsequent handover to GERAN)

9.3. The MS now establishes the RTP path to the serving UNC. The MS has not connected the calling party to the audio path.

9.4. The MS sends the URR Activate Channel Ack to the serving UNC indicating the UDP port and IP address for the downlink stream.

9.5. The serving UNC establishes the downlink RTP path between itself and the MS. The serving UNC may start sending idle RTP/UDP packets to the MS.

9.6. The serving UNC signals to the CN that the call resources have been allocated by sending an Assignment Complete message.

9.7. The serving UNC signals the completion of the bearer path to the MS with the URR Activate Channel Complete message. An end-to-end audio path now exists between the MS and the CN. The MS can now connect the calling party to the audio path.

10. The MS signals that it is alerting the user, via the Alerting message. The CN sends a corresponding Alerting message to the calling party.

11. The MS signals that the called party has answered, via the Connect message. The CN sends a corresponding Connect message to the calling party. The MS connects the user to the audio path.

12. The CN acknowledges via the Connect Ack message. The two parties on the call are connected on the audio path.

13. Bidirectional voice traffic flows between the MS and CN through the serving UNC are available now.

6.3.5 Handover from GERAN to UMAN

This procedure assumes that the MS is on an active call on GERAN or UTRAN, and the mode selection preference is UMAN preferred, and the MS has detected UMA coverage and successfully registered on the UMAN, allowing the MS to obtain system information relating to the UMAN cell. The handover from GERAN to UMAN is always triggered by the MS, based on its local measurements of UMA coverage signal quality as well as any uplink quality indications received from the UMAN. When the MS detects UMA coverage, it sets up a connection with the UMAN, thereby establishing an IPSec secure association with the UNC-SGW. The handover procedure from cellular network to UMAN is performed through the following steps (Figure 6.9):

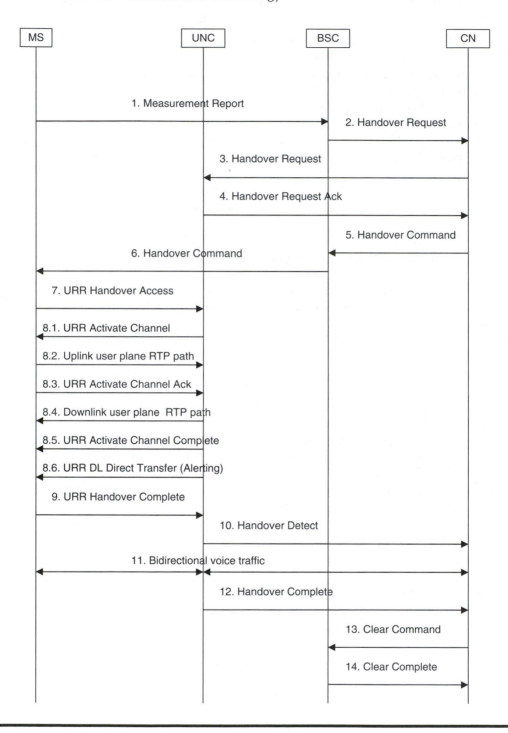

Figure 6.9   Handover from GERAN to UMAN.

1. The MS now begins to include the UMA cell information in the Measurement Report to the GERAN. The MS reports the highest signal level for the UMA cell {ARFCN, BSIC}.

2. On the basis of the MS measurement reports and other vendor-specific BSS algorithms, the GERAN decides to handover to the UMA cell, using an internal mapping of {ARFCN, BSIC} to CGI. It is only used in GERAN/UTRAN and CN for identifying a target cell for handover to UMAN. The GERAN starts the handover preparation by sending a Handover Required message to the CN, identifying the target (UMAN) cell.

3. The CN requests the target UNC to allocate resources for the handover, using the Handover Request.

4. The target UNC acknowledges the Handover Request, using Handover Request Acknowledge, indicating it can support the requested handover, and provides a Handover Command that indicates the radio channel to which the mobile station should be directed.

5. The CN forwards the Handover Command to the GERAN, completing the handover preparation.

6. The GERAN sends a Handover Command to MS to initiate handover to UMAN. The Handover Command includes other parameters, information about the target UMAN such as BCCH ARFC, PLMN color code, and BSIC. The MS does not switch its audio path from GERAN to UMAN until handover completion, i.e., until it sends the URR Handover Complete, to keep the audio interruption short.

7. The MS accesses the serving UNC using the URR Handover Access message, and provides the entire Handover Command received from the GERAN. The MS enters URR Dedicated state. The handover reference in the handover command allows the serving UNC to correlate the handover to the Handover Request Acknowledge message earlier sent to the CN and identify the successful completion of the handover.

8.1. A UMA traffic channel assignment procedure is started. The serving UNC sets up the bearer path with the MS. The serving UNC sends a URR Activate Channel to the MS including bearer path setup information such as
   a. Channel coding
   b. UDP port and the IP address for the uplink stream
   c. Voice sample size
   d. Cipher mode (for use in case of subsequent handover to GERAN)

8.2. The MS now establishes the RTP path to the serving UNC. The MS has not connected the calling party to the audio path.

8.3. The MS sends a URR Activate Channel Ack to the serving UNC indicating the UDP port and IP address for the downlink stream.

8.4. The serving UNC establishes the downlink RTP path between itself and MS. Serving UNC may start sending idle RTP/UDP packets to MS.

8.5. The serving UNC signals the completion of the bearer path to the MS with the URR Activate Channel Complete message. An end-to-end audio path now exists between the MS and the CN. The MS can now connect the calling party to the audio path.

8.6. The CN signals that the called party's phone is ringing, via the Alerting message. The message is transferred to the UNC, which forwards the message to the MS in the URR Downlink Direct Transfer.

9. The MS transmits the URR Handover Complete to indicate the completion of the handover procedure at its end. It switches the user from the GERAN user plane to the UMAN user plane.

10. The serving UNC indicates to the CN that it has detected the MS, using a Handover Detect message. The CN may now switch the user plane from source GERAN to target UMAN.
11. Bidirectional voice traffic is now flowing between the MS and CN, via the serving UNC.
12. The target UNC indicates that the handover is complete, using the Handover Complete message, if this is not already done in step 10.
13. Finally, the CN tears down the connection to the source GERAN, using Clear Command.
14. The source GERAN confirms the release of GERAN resources allocated for this call, using Clear Complete.

6.3.6 Handover from UMAN to GERAN

This sequence assumes that the MS is on an active call on the UMAN and the conditions and mode selection preferences are listed as the following:

- UMAN-preferred, and the MS begins to leave UMA coverage, based on its local measurements of the UMA coverage signal quality as well as any uplink quality indications received from the UMAN
- GERAN-preferred, and a GSM PLMN becomes available

The handover procedure from the UMAN to a cellular network is performed through the following steps (Figure 6.10):

1. The UNC may send a URR Uplink Quality Indication based on certain criterion.
2. The MS sends a URR Handover Required message to the serving UNC indicating the Channel Mode and a list of target GERAN cells identified by CGI, in order of preference for handover, and includes the received signal strength for each identified GERAN cell. This list is the most recent information obtained from the GSM RR sublayer and could have been stored before GSM RR enters the hibernate mode.
3. If the serving UNC selects a target GERAN cell, the handover to the GERAN procedure is performed. Serving UNC starts the handover preparation by signaling to the CN the need for handover, using Handover Required, and including the GERAN cell list provided by the MS. The UNC may include only a subset of the cell list provided by the MS.
4. The CN selects a target GERAN cell and requests it to allocate the necessary resources, using Handover Request.
5. The target GERAN builds a Handover Command message providing information on the channel allocated and sends it to the CN through a Handover Request Acknowledge message.
6. The CN signals the serving UNC to handover the MS to the GERAN, using the Handover Command message, ending the handover preparation phase.
7. The serving UNC transmits a URR Handover Command to the MS including the details sent by GERAN on the target resource allocation.
8. The MS transmits Handover Access containing the handover reference element to the BSC.
9. The target GERAN confirms the detection of the handover to the CN, using the Handover Detect message.
10. The CN may at this point switch the user plane to target the BSS.
11. The GERAN provides physical information to the MS, i.e., Timing Advance, to allow the MS to synchronize with GERAN.

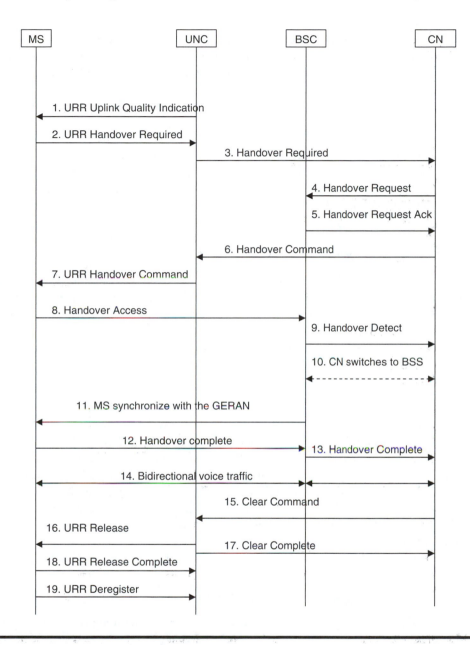

Figure 6.10 Handover from UMAN to GERAN.

12. The MS signals to the GERAN that the handover is completed, using Handover Complete.
13. The GERAN confirms to the CN about the completion of the handover, via the Handover Complete message. If the user plane has not been switched in step 10, the CN switches the user plane to target the BSS.
14. Bidirectional voice traffic is now flowing between the MS and CN, via GERAN.
15. On receiving confirmation regarding the completion of the handover, the CN indicates to the serving UNC to release any resources allocated to the MS, via the Clear Command.

16. The serving UNC commands the MS to release resources, using a URR Release message.
17. The serving UNC confirms resource release to the CN using a Clear Complete message.
18. The MS confirms resource release to the serving UNC using a URR Release Complete message.
19. The MS may finally deregister from the serving UNC, using a URR Deregister message.

The inter-radio access technology handover procedure from the UMAN (i.e., MS capable of operating in all of the UMA, GSM/GERAN, and UTRAN modes) is always triggered by the MS, based on its local measurements of UMA coverage signal quality as well as any uplink quality indications received from the UMAN and the network selection algorithm. The procedure is performed similarly to the handover procedure from UMAN to GERAN.

6.3.7 Paging Procedure

The UNC initiates this procedure when it receives a Paging Request over the A-interface or over the Gb-interface. The MS to be paged is identified by the identity received in the request from the CN. If the request includes the TMSI then the UNC should include the TMSI as the mobile identity.

6.3.7.1 Packet Paging for GPRS Data Service

The procedure of packet paging for a GPRS data service is performed through the following steps:

1. The core network sends a packet switched (PS) page for a mobile station that is currently GPRS attached via the UMAN. The paging message will include either PTMSI (Packet Temporary Mobile Subscriber Identity) or IMSI as a mobile identifier.
2. The UNC verifies that the MS is registered for UMAN service and forwards the corresponding URLC-PS-Page message to the MS using a TCP signaling connection. The message includes Mobile Identity IE as the standard GPRS. It will be either PTMSI, if available, or IMSI.
3. The MS sends any LLC PDU (Protocol Data Unit) to respond to the page, activating a channel as needed. The uplink LLC PDU is forwarded to UNC using the standard mechanism for GPRS data.
4. UNC forwards LLC PDU to core network via Gb interface as standard GPRS.

6.3.7.2 Packet Paging for Circuit Mode Service

The procedure of packet paging for a circuit mode service is performed through the following steps:

1. The core network sends a CS page for a UMA-registered mobile station via the Gb interface. The mobile station is currently GPRS attached via the UMAN. The paging message will include either TMSI or IMSI as a mobile identifier.
2. After verifying that the MS is registered for UMAN services, the UNC forwards the corresponding URR Paging Request message to the MS using a signaling TCP connection. The message includes Channel Needed and Mobile Identity IEs as the standard GSM. The Mobile Identity will be either TMSI or IMSI depending on the original base station system GPRS Protocol CS page.
3. The MS initiates the standard CS page response procedure via the UMAN. The MS enters a URR-dedicated state.

4. On receiving a URR Paging Response, the UNC should establish the signaling connection to the MSC and forward the contents of the URR Paging Response to the MSC.

6.3.8 UMA–UMA Handover (Non-Normative)

The UMA to UMA handover shall be controlled by the MS. The MS shall support double connections to the AP. Both PAN (Bluetooth network) sessions shall never be connected to the IP layer simultaneously. The MS should be able to perform inquiry procedures if the link quality on the current connection drops. The MS should be able to connect to a second AP and establish a second PAN session. The MS should be able to switch between the new PAN session and the old one, without affecting the IP layers. The procedure is performed through the following steps (Figure 6.11):

1. A PAN session is ongoing.
2. The link quality drops below a threshold and MS shall start inquiry.
3. The inquiry response is received, but the link quality is poor.

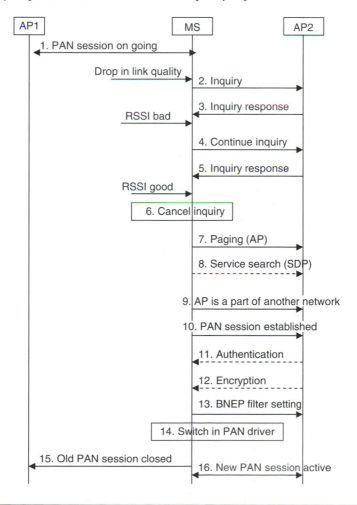

Figure 6.11 UMA–UMA handover.

4. The inquiry is continued.
5. The inquiry response is received and the link quality is good.
6. The inquiry is canceled.
7. The MS pages the AP.
8. The MS shall be able to perform an SDP search on AP. MS could cache results from previously performed SDP searches.
9. A SDP search revealed that AP was a part of another network. The Service name in the SDP record is applied.
10. A PAN session is established.
11. The AP or MS could start authentication.
12. The AP or MS could start encryption.
13. The MS shall send filter settings to AP.
14. The MS shall perform a switch in the PAN driver.
15. The old PAN session is closed.
16. A new PAN session is active.

6.4 Implementation Solution of Seamless Mobility Management

6.4.1 Interworking Model for UMAN and Cellular Networks

On the basis of architectural principles, mobility management can be divided into tightly and loosely coupled systems. The unlicensed access networks appear to the cellular core network as another cellular access network, thus the multi-access system is built using the existing cellular core network; UMA belongs to a tightly coupled integration (Figure 6.12). In a tightly coupled architecture, it is possible to reserve capacity for WLAN-cellular handoff, thus improving performance that reduces the dropped call rate. 3GPP has specified how 3GPP systems and WLANs can interwork. The aim of a loosely coupled system is to extend 3GPP packet-based services to the WLAN access

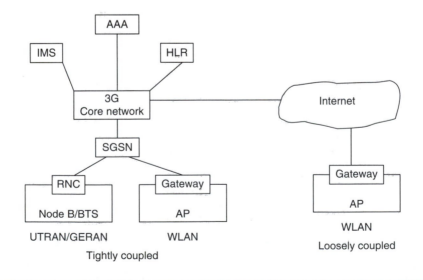

Figure 6.12 3G and WLAN integration architecture.

environment. Unlike the UMA architecture, which is mainly designed for CS services, the loosely coupled system architecture is designed for packet-based services by connecting the WLAN to the same external IP network used by the GPRS access network.

6.4.2 Voice and Data over UMAN and Cellular Networks

In an enterprise, the wireless extension to the LAN offers a convergent access infrastructure for both voice and data. In most cases, these VoWLAN phones will not be pure phones but will combine Wi-Fi, cellular connectivity, and VoIP support; they will be progressively replaced by dual-mode phones (Wi-Fi/Cellular). At home, for customers subscribing to a VoIP service using a SIP-based call server, the VoWLAN SIP phones are available on the market [9], allowing them to make or receive phone calls as they are in a WLAN coverage at their homes. To make the seamless use of these dual-mode phones a reality, UMA solutions can be deployed.

Although the situation is evolving rapidly, mobile phones with Bluetooth connectivity are currently far more common and cheaper than Wi-Fi/cellular handsets. These phones can also be used to carry voice over Bluetooth if they support the cordless telephony profile (CTP) standard.

6.4.3 Real-Time Handover with UMA Technology

UMA technology allows real-time handover between access networks, and allows mobile subscribers to roam seamlessly between various access networks. It is suitable primarily for mobile operators and can be used for residential customers, at home or in public hot spots, or in combination with an enterprise (IP) solution.

UMA allows a user to be reachable on one device using one phone number, regardless of where he or she happens to be. UMA solutions accomplish this by the following points:

- Automatic selection of the most convenient access network (usually offering the cheapest tariff), for example, either the cellular access network or the Wi-Fi/Bluetooth access network, which generally includes a broadband network using VoIP
- Rerouting incoming calls to the appropriate access network

With UMA, when the handset is within the range of a access point, the voice traffic is tunneled through WLAN (or possibly Bluetooth) to UNC. Once out of the range of an AP, a handover between the UNC and the BSC/radio network controller is performed. This UNC equipment is part of the mobile network and is therefore controlled by the mobile operator.

6.4.3.1 UNC

The UNC manages all subscriber access to mobile voice, data, and multimedia services from various WLAN locations and facilitates automatic seamless roaming and handover between mobile radio access networks and WLAN. The UNC provides the following functionalities:

- Interworking between the IP access network and the mobile core network through existing A and Gb interfaces
- User access security and registration; UNC functions equivalent to GSM RR and GPRS RLC such as for paging and handovers
- Tunneling of GSM information (voice, data, and signaling) to/from the end-user terminal (e.g., bearer establishment, management, etc.)

6.4.3.2 Intelligent Client Software

To facilitate user experience and to best meet seamless mobility requirements, the end-user terminal is equipped with intelligent client software. This software client automatically selects the required access technology and performs seamless handover between the available access technologies according to the operator's and user's preferences and policies.

6.4.4 Test Bed and Measurement Result

The mobility management performance of the UMA system between GSM and UMA was tested by Arjona and Verkasalo [3] in an environment as shown in Figure 6.1. The testing methodology consisted of establishing a voice call and afterward moving in and out of the UMA coverage. All the measurements were carried out in an indoor environment. To measure handover times, signaling was captured from interfaces in the network. A Nethawk GSM Analyzer [4] was used to capture packets in the A interface between the core network and the UNC, and the A interface between the core network and the BSC. Additionally, Wireshark Protocol Analyzer software [5] was used to capture packets in the interface between the UNC and the UMA client. Packet data performance tests consisted of multiple iterations of file downloads from a file server in the Internet. The server is known to have high-speed bandwidth, where the tests took place.

6.4.4.1 Measurement Results

6.4.4.1.1 Handover Performance Results

The result for the handover measurements from GSM to UMA is 271 ms; and the result for the handover measurements from UMA to GSM is 759 ms [3], in accordance with the signaling phases described in Sections 6.3.5 and 6.3.6.

The measurements reveal that the possible voice break in UMA–GSM handovers are in line with the typical breaks in GSM inter-BSC handovers (120–220 ms) [6]. Likewise, the UMA software in the MS is likely to implement packet loss concealment algorithms. Packet loss concealment algorithms hide transmission losses in an audio system.

6.4.4.1.2 Packet Data Performance Results

The results from the packet data performance tests are summarized in the following equation [3]:

Average TCP throughput = 268 kbps

The average throughput is not as high as expected. However, the average throughput of 268 kbps is considerably higher than what is available via GPRS or EDGE in GSM networks. Current GSM and EDGE networks provide throughputs around 30 and 120 kbps, respectively. The UMA data throughput is very similar to the data rates available in WCDMA (3G) networks. From the operator's point of view, UMA can provide an additional value to its current services with a minimal investment. Some of the current bottlenecks [3] that can limit data performance are

- Mobile terminal processing power.
- Gb interface (between UNC and SGSN), which is normally based on T1 lines with a maximum bandwidth of 2 Mbps to be shared among all subscribers served by the UNC.

■ Subscriber's data throughput limit defined in the HLR by the operator. This limit is usually assigned based on the service subscription type and can vary from very low rates such as 64 kbps to several megabits (e.g., 8 Mbps for HSDPA subscribers).
■ The GPRS theoretical data limit is around 600 kbps.

6.5 Open Issues

UMA development work was transferred to 3GPP in April 2005 and renamed GAN [7]. In addition to improving and extending the UMA standard, the GAN specification allows any generic IP-based access network to provide connectivity between the handset and the GAN controller (GANC or UANC), through the Up interface.

The IP multimedia subsystem (IMS) is a key element of the Third Generation (3G) Architecture [8]. 3G is a unification of the cellular world and the Internet, and IMS is the mechanism that enables IP-level services. IMS is access independent. It can be used on top of fixed, cellular, or unlicensed access networks. UMA can be used to support the move toward a converged IMS-compliant infrastructure. Once the UMA service concept is established, an evolution to a broadband service becomes a logical next step. This allows the provider to take intermediate steps toward a more homogeneous SIP-based architecture, for which IMS is the reference. IMS opens up new opportunities for a wide range of additional services, which are video telephony, video on demand, instant messaging, and others. In addition, IMS allows operators to offer the same services through different access networks; it also allows services to be directly provided by the enterprise.

UMA solution is now available with current wireless access technologies (mostly Wi-Fi) and will evolve to take into account new radio technologies such as WiMAX. The introduction of these new radio accesses will be facilitated by the IMS architecture, which offers a consistent set of services that are independent of the access technology. Unbounded mobility lets user-centric services to be important today.

6.6 Conclusions

UMA is a cellular network-based solution. UMA delivers GSM and GPRS mobile services over unlicensed spectrum technologies like WLAN essentially by encapsulating GSM signaling in IP, and routing it to a gateway (UNC) in the cellular network. UMA enables subscribers to roam, and supports seamless handover between cellular networks and public/private UMAN networks using dual-mode handsets. The implementation solution of seamless mobility management and open issues are also presented in this chapter. The measurements showed that the throughput is twice or more than the data rates available in GSM networks, and the handover times between UMA and GSM are similar to typical inter-BSC handovers in the GSM system. One of the advantage of UMA is that subscribers receive the same user experience as they transfer between cellular and WLAN networks. In UMA, it is possible to reserve capacity for WLAN-cellular handoff, thus improving performance (reducing the dropped call rate). Another advantage for UMA is its standardization, which could help to promote widespread adoption and interoperability. The limitations are that many enterprise customers prefer enhanced features such as corporate dialing plans, coverage plans, voice mail, and other features. Additionally, UMA-based solutions may not integrate easily with other IP services. The authors think that UMA can be viewed as a long-term technology for cellular operators to bring unlicensed access solutions to the market.

Abbreviations

AAA	authentication, authorization, and accounting
ACL	asynchronous connectionless link
AKA	Authentication and Key Agreement
AP	access point (802.11 or Bluetooth)
ARFCN	an RF channel number
BCCH	broadcast control channel
BNEP	bluetooth network encapsulation protocol
BSC	base station controller
BSIC	base station identity code
BSS	base station subsystem
BSSAP	base station system application part
BSSGP	Base Station System GPRS Protocol
BTS	base transceiver station
CC	call control
CGI	cell global identification
CM	connection management
CN	core network
CS	circuit switched
DL	down link
DNS	domain name system
EAP	Extensible Authentication Protocol
EAP-SIM	Extensible Authentication Protocol-Subscriber Identity Module
EDGE	enhanced data rates for global evolution
ESP	encapsulating security payload
FQDN	fully qualified domain name
GAN	generic access network
GANC	generic access network controller
GERAN	GSM EDGE radio access network
GMM	GPRS mobility management
GPRS	general packet radio service
GSM	global system for mobile communications
HSDPA	high speed downlink packet access
IKEv2	internet key exchange
IMS	IP multimedia subsystem
IMSI	international mobile subscriber identity
IPSec	IP Security
LAI	location area identity
LAPDm	link access procedure on dm channel
LLC	logical link control
MAC	Medium Access Control
MS	mobile station
MSC	mobile switching center
MTP	message transfer part
NAT	network address translation
PAN	personal area network

PDU	protocol data unit
PHY	physical layer
PLMN	public land mobile network
PS	packet switched
PTMSI	packet TMSI
RF	radio frequency
RLC	radio link control
RR	radio resource
RSSI	received signal strength indication
RTP	Real-Time Protocol
SCCP	signaling connection control part
SDP	Session Discovery Protocol
SGSN	serving GPRS support node
SGW	security gateway
SIP	session initiation protocol
SMS	short message service
SNDCP	Sub-Network Dependent Convergence Protocol
SS	security sublayer
TCP	Transmission Control Protocol
TMSI	temporary mobile subscriber identity
UDP	User Datagram Protocol
UL	Uplink
UMA	unlicensed mobile access
UMAN	unlicensed mobile access network
UMA-RLC	UMA radio link control
UMA-RR	UMA radio resource
UMTS	universal mobile telecommunication system
UNC	UMA network controller
URLC	UMA radio link control
URR	UMA radio resource
UTRAN	universal terrestrial radio access networks
VoIP	Voice-over-IP
VoWLAN	Voice over IP over WLAN
WCDMA	wideband code division multiple access
WiMAX	World Interoperability for Microwave Access
Wi-Fi	wireless fidelity
3GPP	third generation partnership project

REFERENCES

1. UMA Architecture Technical Specification, Stage 2, R1.0.4, May 2, 2005.
2. UMA Protocols, Stage 3, R1.0.4, May 2, 2005.
3. Andres Arjona and Hannu Verkasalo, Unlicensed mobile access (UMA) handover and packet data performance analysis, Second International Conference on Digital Telecommunications (ICDT'07).
4. Nethawk GSM Analyzer, www.nethawk.fi.
5. Wireshark Protocol Analyzer, www.wireshark.org.

6. 3rd Generation Partnership Project (3GPP), Radio subsystem synchronization, *3GPP Technical Specification*, TS 05.10 V8.12.0, August 2003.
7. 3GPP TS 43.318, Generic access to the A/Gb interface, Stage 2, January 2005.
8. 3GPP TS 23.228, IP Multimedia Subsystem (IMS), Stage 2, June 2007.
9. Philippe Laine and Eric Moisset, Unbounded mobility: Always connected, anywhere, Technology White Paper of ALCATEL.

PROTOCOLS AND SECURITY

Chapter 7

Protocols and Decision Processes for Vertical Handovers

Jie Zhang, Enrique Stevens-Navarro, Vincent W.S. Wong,
Henry C.B. Chan, and Victor C.M. Leung

CONTENTS

7.1   Introduction

Over the past decade, Internet services have become pervasive and are expected to be accessible any-time and anywhere. Although third generation (3G) cellular systems allow mobile users to access the Internet with considerable bandwidth and better quality of service (QoS) management, they are still not able to satisfy the great bandwidth requirements of many emerging multimedia applications. Moreover, the costs of Internet access via 3G systems may be prohibitively expensive in comparison with many users' expectation that "Internet is free," because these systems operate on expensive licensed spectra. As a result, several alternate wireless systems that operate in unlicensed frequency bands have been developed for high-speed Internet access according to different usage models. For example, the emerging Worldwide Inter-Operability for Microwave Access (WiMAX) technologies, based on IEEE 802.16e-2005 standard and scalable orthogonal frequency division multiple access (OFDMA), can offer various mobile services with QoS-provisioning for citywide areas [1]. Furthermore, WiFi technologies based on IEEE 802.11x standards are already widely deployed and providing high-speed wireless local area network (WLAN) connectivity in millions of offices, homes, and public locations at relatively low costs as they operates over unlicensed spectra. It is expected that WiMAX will be a competitive alternative to 3G systems for wide area networking (WAN) due to its high bandwidth and low cost, while WiFi is already widely accepted as an important complement to current 3G systems over local hot spots.

On the other hand, rapid advancements of digital signal processing technologies are enabling portable mobile terminals (MTs) to be equipped with multiple network interfaces so that they can access more than one wireless networks simultaneously. So, spurred by both technological advances and market demands, next generation (NG) networks are expected to integrate various access technologies and provide always best connected (ABC) services to users [2]. In other words, NG networks will be heterogeneous networks.

Owing to their abilities to exploit unlicensed spectra and the potential cost savings that result, it is of interest to integrate WiMAX and WiFi networks. Such an integrated network is simply referred as a WiMAX/WiFi network here. The architectures for 3G/WLAN integration have been extensively studied, including both the tightly coupled and loosely coupled alternatives [3]. These architectures can also be extended to enable WiMAX/WiFi integration. The WiMAX Network Reference Model (NRM) consists of three functioning entities: MT, access services network (ASN), and core services network (CSN) [1]. ASN serves as the point of entry for MTs to access various services via the WiMAX network. It connects multiple base stations (BSs) that provide radio access. CSN provides network functions such as IP address allocation and authentication, authorization, and accounting (AAA). ASN and CSN can interconnect in a meshed manner. In the tightly coupled WiMAX/WiFi network architecture, WiFi access points (APs) can be connected to WiMAX BSs through WiMAX radio links so that all data packets to/from the WiFi network are routed via the WiMAX network [4,5]. In this case, the WiFi network serves as the access network and its APs are connected to the Internet via the WiMAX network. Another alternative for a tightly coupled WiMAX/WiFi network is for the WiMAX BSs and WiFi APs to be connected to a common ASN and through a controlling ASN gateway (ASN GW) to a common IP network and IP multimedia subsystem (IMS) to share the same AAA functionality [6]. In this case, MTs may access the CSN via either WiMAX or WiFi, and services that are accessible via WiMAX are also accessible via WiFi using the same credentials, subscription profiles, etc. On the other hand, in the loosely coupled WiMAX/WiFi network architecture the WiMAX and WiFi networks are each connected to the Internet independently via their own ASN GWs and border routers, respectively, and they interwork by exchanging signaling messages over the Internet. Generally, the loosely coupled architecture is more flexible as it allows for independent deployment and traffic management of both networks, but it complicates interworking, for example, to authenticate a user accessing one network while her credentials reside in another network.

How to handle the mobility of users is an important issue in mobile communication systems, which involves providing the reachability of mobile users and maintaining their ongoing connections when they move between different network attachment points (i.e., BSs or APs) in the same network or between networks. This problem becomes more complicated in heterogeneous systems. In conventional networks, users only move between network attachment points with homogeneous underlying technologies, and mobility management only involves the link layer or can be easily solved in the network layer if the BSs or APs belong to different subnets. However, mobile users in WiMAX/WiFi networks can move between domains employing heterogeneous access technologies. Network or upper layer approaches are required to provide uniform mobility management regardless of the technologies used in the underlying networks. The process of switching connections between heterogeneous networks is referred as vertical handover. In contrast, the process of switching connections in homogeneous networks is referred as horizontal handover. Vertical handover introduces many new challenges for protocol design and decision making in mobility management.

First, the system must determine when to perform a vertical handover. Traditional handover decision schemes may not function effectively in WiMAX/WiFi networks. In general, these schemes are based on comparisons of the radio signal strength (RSS) or signal-to-interference ratio (SIR) measurements from different network attachment points [7]. These techniques are not applicable to vertical handover in WiMAX/WiFi networks because the direct comparisons of RSS/SIR from heterogeneous attachment points employing different physical layer standards are not useful, and the coverage areas of the heterogeneous networks may overlap such that handovers may not be mandated by the needs to maintain connectivity; that is, vertical handovers in WiMAX/WiFi networks may not necessarily be compulsory. For example, a vertical handover from a wide-coverage WiMAX cell to a local-coverage WiFi hot spot can be optional. So, the handover decision should not only decide when but also whether to hand over and optimize the user satisfaction by making an appropriate network selection [8].

Second, when an MT moves between different domains, it should execute several operations to complete the handover, such as re-authenticating with the new network, updating the addresses in the home network, and transferring its context information to the new domain. Overall, these operations will incur a higher signaling cost and a longer delay than horizontal handovers, which will result in QoS degradations such as connection interruption or even connection termination. Also, the MT needs to inform the corresponding node (CN) that is communicating with it about the handover to keep the connection going smoothly. Because of these operations during a vertical handover, some ongoing packets may be delayed or even lost. To minimize these negative impacts, the handover process must be designed very carefully.

Finally, besides the handover decision, it is also necessary to decide whether a new connection or handover connection should be admitted into a domain to optimize the system efficiency while meeting the QoS requirements of the users. Although there are many admission control schemes used in homogeneous networks (e.g., WiMAX), they need to be modified to adapt to the different

QoS metrics and provisioning methods of different access networks. Also, compulsory and optional handovers must be distinguished to facilitate the decision making.

Several novel solutions have been proposed to deal with these new challenges. For example, some network selection schemes have been proposed to evaluate the service offered by WiMAX/WiFi networks to help an MT choose the best service. Some handover triggering algorithms have also been specifically designed to help an MT decide when to initiate a handover. A number of vertical handover schemes have also been proposed. Their operation scope varies from network layer to application layer. Also, several extensions of admission control policies have also been proposed for WiMAX/WiFi networks. In this chapter, we discuss the design challenges of vertical handover support in WiMAX/WiFi networks, survey existing solutions, and describe in more detail our proposed solutions. In Section 7.2, we consider vertical handover decision methods and provide a location-based vertical handover decision algorithm. In Section 7.3, we consider the vertical handover procedures and present an SIP-based soft-handover (S-SIP) scheme. In Section 7.4, we study admission control schemes for vertical handover and propose an evaluation model. Section 7.5 concludes the chapter and describes some open research issues.

7.2 Vertical Handover Decision

As one of the goals of NG networks is to provide ABC service to users equipped with multimode MTs, the handover decision schemes in WiMAX/WiFi networks are required to not only maintain the connections but also optimize users' satisfaction of services. There are a lot of factors that may influence user's satisfaction, such as network throughput, packet delay, access charges, and power consumption. Moreover, the priority of these factors also depends on user applications and user preferences. A lot of network selection schemes have been proposed recently to help users select the best network. In this section, we first summarize the existing vertical handover decision schemes. A novel handover decision scheme is then presented, followed by a summary of the section.

7.2.1 Related Work

The user's satisfaction of a service depends on three aspects: (1) the service characteristics of the access network, such as the QoS provision, charging rate, and MT power consumption to access the network; (2) the application requirements and associated constraints, such as the QoS requirements and remaining battery power; and (3) the user preference, for example, some users may prefer a high connection quality while some users may prefer low charging rates. Generally, a vertical handover decision scheme seeks to set up an efficient model to represent each user's satisfaction based on these factors.

A number of network selection schemes have been proposed recently. Before introducing these schemes, we first discuss the factors that influence the satisfaction degree of a network. According to the survey of QoS in mobile communications [9], there are eleven factors that can reflect the QoS offered by a wireless network. As shown in Figure 7.1, these factors can also be classified into six categories: availability, throughput, timeliness, reliability, security, and cost. The availability of a network is reflected by the RSS/SIR and the coverage area of the network. With a higher RSS or SIR, packet errors and hence packet retransmission ratio can be reduced. Throughput is another very important factor. With a high throughput, data applications can be serviced more efficiently and multimedia data can be transmitted with a higher resolution/quality. The timeliness of the service is especially important for real-time applications such as Voice-over-IP (VoIP), where packet delay and delay jitter must be kept below some limits for acceptable quality at the receiver. A high

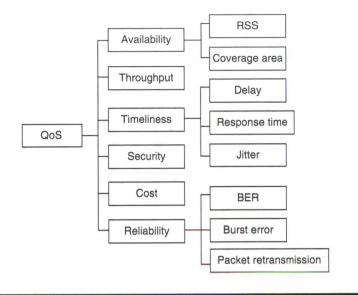

Figure 7.1 **Network selection factors.**

degree of security can ensure the confidentiality of user's information, which is extremely important in financial transaction applications. The last factor, cost, determines the affordability of the service to the users. Obviously, users want the cost to be as low as possible while meeting some minimal performance objectives.

The network selection scheme proposed in Refs. [10,11] represents the degree of a user's satisfaction on a network as a cost function and the network with the lowest cost is preferred. In Ref. [10], three factors are considered for network selection: bandwidth, charging rate, and power consumption. These factors are classified into positive factors and negative factors—the positive factors are ones for which a larger value is favored, such as bandwidth, whereas the negative factors are ones for which a lower value is preferred, such as charging rate and power consumption. This scheme is based on a simple assumption: given two networks A and B, where B has double the throughput and charging rate of network A, the user should consider them with the same satisfaction degree. On the basis of this assumption, this chapter proposes that the factors should be normalized logarithmically rather than linearly. As a cost function is used to evaluate networks, if the network with the lowest value is preferred, the positive factors are normalized by a negative natural log function and the negative factors are normalized by a natural log function. The cost of a network is the weighted sum of the normalized values. The work in Ref. [11] extends that in Ref. [10] with more factors and also considers the requirements of various applications for the network selection. It considers the case when an MT has multiple ongoing applications, each having a certain QoS requirement. This scheme assigns weights to all ongoing applications and each application has different weights for the factors. The cost function is then represented as the weighted sum of the costs for all the applications.

Another network selection scheme that helps users select the optimal networks among different networks according to their preferences is proposed in Ref. [12]. This scheme adopts an analytic hierarchy process (AHP) algorithm to help users derive the weights on all factors by means of paired comparison. It uses a number from 1 to 9 to rank the relatively importance of each pair of factors. In this way, the user can set up a $N \times N$ matrix M, where N is the number of factors. Here, the value $M_{i,j}$ is the relative importance of factor i over factor j. Thus, the value $M_{j,i}$ is the reciprocal of value $M_{i,j}$ and $M_{i,i} = 1$. When the matrix is set up, the weights of all factors can be derived by finding the eigenvector of the matrix M. With the weights of all factors, the satisfaction degree on these networks can be calculated based on grey relational analysis (GRA). GRA is a method that decides the best option by determining the similarity between each option and the ideal option. The more the similarity, the more preferable the option is. In this case, the ideal/worst network is a network with best/worst values of all factors among the candidate networks. For each factor of the candidate network, the normalized distance value is the distance of the factor between the candidate network and the worst network divided by the distance of factor between the ideal network and the worst network. The satisfaction degree of a candidate network can be represented as the weighted sum of the relative distances. The network with the highest satisfaction degree is always chosen as the target access network.

A linear function is used in Ref. [13] to normalize the factors. It assumes that each application has a minimum requirement and a maximum requirement on each factor. A network should not be considered if any of these factors cannot satisfy the minimum requirement of the application. The values between the minimum and maximum requirements are linearly normalized between 0 and 1. The overall satisfaction degree on the service is the weighted sum of the linearly normalized factors. Finally, the network with the maximum weighted sum is preferred.

Unlike the above schemes of calculating the satisfaction degree of a network, handover decisions based on fuzzy logic are proposed in Refs. [14–16]. For each network, they evaluate the fuzzy rate of factors and calculate the handover decision based on some proposed rules.

As vertical handover usually takes a longer time to complete and causes a larger QoS degradation than horizontal handover during the handover process. Therefore, a vertical handover is beneficial only if the short-term service degradation can be compensated by the long-term service enhancement, measured in terms of users' utilities. In Ref. [10], the minimum duration that an MT should stay in a WiFi hot spot to compensate for the downward handover degradation is calculated and used as the delay-trigger for a downward handover after the MT enters into the WiFi hot spot. In Ref. [14], a downward handover is triggered when the RSS of a WLAN exceeds a threshold, which is determined by fuzzy logic based on the MT's speed and the WLAN traffic load. The comparison of the above evaluation functions can be found in Ref. [17].

7.2.2 Location-Based Vertical Handover Decision Algorithm

In the papers reviewed above, Refs. [11–13] focus on network service comparison but do not consider service degradation experienced during handovers while Refs. [10,14,15] take service degradation into account by considering the RSS and MT velocity. However, these papers fail to consider all the factors that determine the sojourn period of an MT in the WiFi hot spot, such as the WiFi coverage area and the moving pattern of the MT. For example, the sojourn period of an MT moving along a straight line at a constant speed of 1 m/s through a WiFi hot spot with a 100 m diameter varies between (0,100) s depending on its direction. These variations are further increased if the MTs change directions or pause within the hot spot.

With advances in low-cost implementations of global positioning system (GPS) receivers, it is foreseeable that future mobile devices will be equipped with GPS receivers with positioning accuracy within 1 m [18]. Taking advantage of this positioning capability, we propose a novel vertical handover decision scheme [19]. The key to the proposed scheme is to predict the mobility pattern of an MT based on its movement and location topology, and utilize this information for handover decisions. We formulate the optimal handover decision problem as a Markov decision process (MDP) and solve the problem by dynamic programming.

Consider an integrated WiMAX/WiFi network architecture in which a location service server (LSS) is used to assist MTs to access different networks [15]. Periodically, each GPS-equipped MT reports its position to the LSS, and in turn the LSS replies with information on available WiFi hot spots near the MTs location. These messages include the essential information such as the identity of WiFi hot spots and their service parameters (i.e., bandwidth, security, coverage area). Once the MT learns about the existence of a nearby WiFi hot spot that can offer a better service, it will set its GPS receiver to track its own position at regular time intervals (say every second) and request the handover-helping service by notifying the LSS about its movement information and satisfaction utilities on both networks. Based on this information, the LSS will return the handover decision rules to the MT and the MT will decide whether and when to execute a downward handover from the WiMAX to the WiFi network based on its current state. Similarly, an upward handover is executed when the MT moves out of the WiFi hot spot coverage before call completion.

To provide a rational decision rule, the LSS should be aware of both the coverage area of the WiFi hot spot and the mobility pattern of the MT. In combination, these two factors allow the LSS to estimate the MT's sojourn period in the WiFi hot spot. Owing to the existence of obstacles such as buildings and trees, the coverage area of the domain may not be circular. Therefore, we propose that the MTs should report their positions to the network whenever the RSS falls below a predefined threshold. Using these reports, the network can learn the true coverage areas of the WiFi hot spots after collecting a sufficient number of positioning reports. To predict the MT's movement, we need an appropriate mobility model. Many mobility models have been proposed previously to model the movements of mobile users, such as Fluid-Flow model [20] and Gauss–Markov model [21]. Here, we adopt the Markovian model due to its generality. This model assumes that the current state of an MT is highly correlated with its previous location and its velocity [21]. According to Gauss–Markov model, the current movement of an MT can be represented as $S = (L, V)$, where L and V denote the MT's location and velocity, respectively, and the location of the MT at the next time unit is $L + V\Delta T$, where ΔT is the length of the time unit. In each time unit, the velocity V changes with a certain probability. In this way, the state S forms a Markov chain and the transitional probability between V can be calculated based on the MT's movement records.

In each time unit, the MT receives a certain reward, which is equal to the satisfaction degree on the current connection. Generally, the MT will receive a higher reward when connecting to a WiFi than to a WiMAX network. However, due to the short-term QoS degradation and handover cost caused by the handover procedure, the MT will receive the lowest reward during the handover period. Assuming that the duration of the call follows an exponential distribution with an average duration of $1/\lambda$, the probability that the connection stays active after ΔT is $e^{-\lambda \Delta T}$. If the connection terminates, it will no longer receive any reward. Therefore, the reward received in the next time unit should be discounted with $e^{-\lambda \Delta T}$. Given the above, the handover procedure can be formulated as a discounted MDP, where the state is the combination of the movement state of the MT and the network that it is attached to. The optimal handover decision can be solved by some dynamic programming methods such as the value iteration algorithm or the policy iteration algorithm [22].

7.2.3 Summary

In this section, we have presented some current schemes to facilitate vertical handover decisions by mobile users. It is generally agreed that the services provided by alternate wireless networks can be evaluated by some factors and these schemes focus on modeling the user's satisfaction degree according to these factors. Because each vertical handover has a cost due to momentary QoS degradation and signaling load increase during the handover process, the handover decision should also consider the user's sojourn time in the WiFi hot spot to ensure that the gain due to the handover more than offsets the cost. Therefore, given the heterogeneous network integration framework, we have extended the previous work by proposing to consider user's location and mobility in vertical handover decisions to optimize the user's satisfaction.

7.3 Vertical Handover Management

So far, numerous mobility management solutions have been proposed to handle handovers in NG heterogeneous wireless networks. In this section, we briefly discuss how these schemes operate as well as their characteristics. We also introduce the new seamless SIP-based handover (S-SIP) scheme proposed in Ref. [23] to enable seamless vertical handover between heterogeneous wireless networks.

7.3.1 Related Work

Existing solutions for mobility management or handover can be classified according to the layer that they operate on [24,25].

7.3.1.1 Network Layer Solution

As a network layer solution, mobile IP (MIP) [26] is regarded as a good candidate for mobility management in NG networks. Although MIPv6 can support more functions and achieve better performance, it requires the widespread deployment of IPv6, which is not expected to be realized in the near future. Here, we consider only MIPv4. In MIPv4, each network deploys two mobility supporting agents: the home agent (HA) and the foreign agent (FA). A roaming MT must communicate with the HA of its home network and the FA of its currently visiting network to keep its location and address known by all networks. When the MT moves to a new network, it will keep its IP address and register it with the FA of the new network, Also, it will obtain a care of address (CoA) from the FA in the new network (the CoA is usually the IP address of the FA) and then report it to the HA of its home network. So, the HA of the home network can encapsulate all packets destined for the MT and tunnel them to the FA of the visiting network using the CoA as the destination address. The FA will decapsulate the tunneled packets and send them to the MT. This mechanism causes the well-known triangular routing problem, where packets for the MT have to be forwarded through the HA of the home network while packets that the MT sends to the CN can be sent directly. The non-optimal delivery of the packets for the MT may cause a high data transmission cost as well as longer packet delivery time. Some solutions have been proposed to eliminate this problem. The triangular routing problem is solved by the route optimization method in Ref. [27], whereby the MTs CoA is cached at the CN so that the CN can send packets directly to the MT by tunneling them to the CoA. The cache is updated when update messages are received from the HA of the MTs home network. Note that the CN cannot accept CoA updates from the MT directly due to security considerations (the CN cannot trust an MT with an unauthenticated address; it can only trust the HA with a fixed authenticated IP address). Another problem is that when an MT roams between networks, it cannot receive packets until the address update is completed. The packets sent to the MT within this period are lost or deferred. This kind of in-flight packet losses can be reduced by setting up a temporary connection between the FAs in the old and new visited networks [27]. In this manner, the number of in-flight packet losses can be greatly reduced. To further reduce the delay caused by handover detections, handovers can be predicted according to MT's moving pattern [28].

In MIPv4, when an MT moves among networks, it will frequently update its CoA to the HA in its home network, which will incur a high signaling cost as well as long handover delay. When the MT roams in a foreign domain, this kind of cost and delay can become substantial because the MT may be far from the home network. Some schemes have been proposed to localize the signaling procedures for micro-mobility within the foreign domain. Generally, these solutions can be classified into two groups: tunnel-based schemes [29,30], which define hierarchical structures to handle different levels of mobility, and routing-based schemes [31,32], which keep the MTs CoA unchanged and update the routing table in the visited domain to follow the MTs movements.

Compared with link layer solutions, network layer solutions can support global mobility over IP networks with no dependency on lower layers. However, they require that HA and FA be established in each network in the system. Although WiMAX networks have HAs and FAs installed in CSNs and ASNs by default [1], there is no such requirement in WiFi networks. Furthermore, CNs should also be modified to implement route optimization. The reservation of IP address for each MT is also fairly costly in the IPv4 network due to shortage of IP addresses. The potentially long handover latency due to the need to update CoA with the HA is a major problem of MIPv4.

7.3.1.2 Transport Layer Solution

Some solutions have also been proposed to handle mobility and handover in the transport layer. A TCP-Migrate Protocol is proposed in Ref. [33]. In this protocol, a unique token number is negotiated between the MT and the CN once a TCP connection is established. The token numbers are usually calculated based on the addresses/ports of the connection. When the MT moves into a new network, it will notify the CN of its new IP address/port with the token number attached. Authenticated by the token number, the CN will update the information of the connection and packets will be sent to the new address/port afterward. In Refs. [34,35], a mapping layer between the IP layer and TCP layer is proposed. Whenever the MT moves to a new network, it will notify the CN of the new address. These packets will be intercepted by the mapping layer, which will translate the destination/source addresses of the outgoing/incoming packets to/from the MT into the current address of the MT. However, if the MT and the CN roam at the same time, the connection will be terminated. Note in this case that both parts acquire new addresses but they each inform the other with an old destination address so the process cannot succeed. For this kind of simultaneous roaming, it is proposed in Refs. [34,35] to set up a third party subscription/notification server to deal with the problem. When the MT roams into a new network, it will send the new address to both the CN and the notification server. So, when simultaneous roaming happens, the MT and CN can track the new addresses of their counterparts from the notification server so that the connection can be maintained. Mobility can also be supported in the transport layer using the Mobile Stream Control Transmission Protocol (M-SCTP) [36]. M-SCTP supports dynamic reconfiguration of multihomed SCTP connections, that is, an SCTP connection can be associated with multiple addresses, one of which is the primary one used by the connection. When the MT detects a new network, for example, when it moves into a WiFi hot spot while in a WiMAX cell, it will acquire an address from the new domain, set it as the secondary address, and notify the CN. When the handover actually happens, the MT sets the secondary address as the primary address, gives up the old primary address, and also notifies the CN. So, when the CN cannot get the ACK from the primary address of the MT, it will retry with the secondary address.

Compared with network layer solutions, transport layer solutions provide end-to-end mobility management without imposing special requirements on the network operations. These solutions are protocol-specific, which means that each solution only works for a specific transport layer protocol. In other words, it requires the modification of the protocol stack in all hosts, no matter mobile or fixed. It also takes a longer time to transmit handover-related messages between the MT and CN at the transport layer; therefore, the handover delay can be relatively long.

7.3.1.3 Application Layer Solution

Recently, both IETF and Third Generation Partnership Project (3GPP) adopt the Session Initiation Protocol (SIP) [37] as the application-layer signaling protocol to set up QoS-provisioned connections in packet-switched networks [38]. It allows two or more participants to establish a session consisting of multiple media streams. Besides, SIP also enables a connection to be set up in the middle of a session and provides various mobility supports. In SIP, all MTs are identified with logical addresses in the form of e-mail addresses. This information is stored in the registration server and the SIP agent. If an MT moves to another network, it updates its location information in the registration server and the SIP proxy in its home network. Figure 7.2 shows the procedure of setting up a new session. To set up a new session, CN A needs to send an Invite message to the home

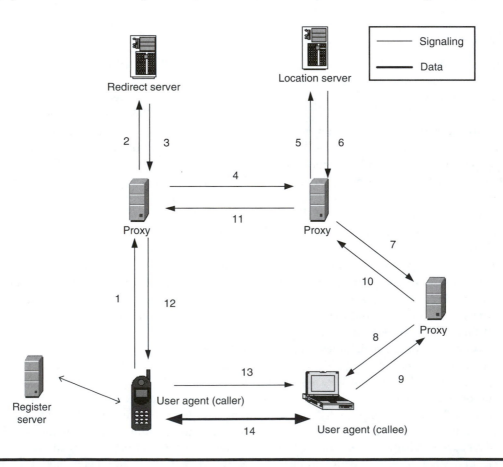

Figure 7.2 SIP architecture.

network of MT B. The address of the home network can be found from the public SIP address of the MT B. This request is captured by the SIP proxy in the home network of MT B. After checking the registration server, the proxy in the home network of MT B knows that currently MT B is not in the home network and acquires its current contact address in network C. Then, the SIP proxy will return a Moved Temporarily message to CN A with the current contact address of MT B attached. So, CN A will send another Invite message to MT B directly. After reading the session setup information in the Invite message, MT B returns an OK message to negotiate the session setup with CN A. CN A will send an Ack message to confirm the acceptance of the modification. When the negotiation is done, the session with several connections will be directly set up between CN A and MT B. If the MT moves when it is in a session, it also notifies the CN of its new contact address and the updated session description by a Re-invite message. Once getting this message, the CN will reestablish the connection to the new location.

Compared with other solutions, the SIP-based solution can achieve true end-to-end mobility management without modifying the network architecture or end-user protocol stacks. Moreover, SIP offers excellent extensibility and scalability due to its operation at the highest layer and use of text-based control messages. Besides, SIP is also the signaling protocol used for session control in the IMS [39] for mobile networks. Therefore, SIP is considered as an attractive candidate to support mobility in NG networks.

Although SIP is free from drawbacks of the lower-layer solutions, it incurs considerable handover delays due to the exchange of application layer messages. It has been proven that such delay is unacceptable for real-time multimedia services [40]. Therefore, many enhanced SIP-based schemes focus on reducing the handover delays. In Ref. [41], a hierarchical mobile SIP (HMSIP) is proposed to localize signaling messages for intradomain roaming. In this scheme, a specific global proxy is set up to handle the mobility within one domain besides those set up for every network. When an MT moves between different networks of one domain, it will register its address to the global proxy instead of the CN. The CN only knows the address of the global agent and sends packets there. Then, the global proxy will redirect the incoming packets from the CN to the current contact address of the MT. HMSIP focuses on intradomain mobility management, where all networks have similar characteristics so that they can cooperate easily. Generally, it is more difficult to reduce interdomain handover delay for the following reasons. First, the underlying domains may be heterogeneous, thus an end-to-end session needs to be reestablished during handover. Second, interdomain handover involves more procedures such as re-authentication so the handover delay is much longer. To address the long handoff delay in interdomain roaming, an MT needs to establish a security associations (SA) with all neighboring domains of its current domain [41]. In this way, it can make authentication locally when it roams into another domain so that the handoff procedure can be shortened. A handover scheme has also been proposed in Ref. [42] to make an SIP-based handover more seamless. When an MT enters into a new domain, it will report the new contact address to both the CN and the associated proxy in the old domain. The proxy in the old domain will then set up a temporary session between the old proxy and the MT to forward the receiving packets. This session will be maintained until the MT can stably receive the packets from the CN. SIP-based vertical handover management is particularly useful for multimedia sessions.

7.3.2 Seamless SIP-Based Handover Scheme

We proposed a seamless SIP-based handover scheme (S-SIP) in Ref. [23] to support seamless interdomain roaming. Unlike the previous schemes, it makes use of a make-before-break handover mechanism and handover prediction based on location tracking technologies.

We explain the S-SIP scheme using the following example. Basically, an MT and a CN exchange some SIP messages to set up a session. Then, both of them know the SIP address and contact address of each other. As shown in Figure 7.3, the MT detects that the RSS of the current network has decreased dramatically and decides to hand over to another network to maintain the session. It first authenticates itself with the new network and acquires an IP or domain address through the required protocols. Here, we assume that the new domain is using the Dynamic Host Configuration Protocol (DHCP) to assign IP addresses. After obtaining the new contact IP address, the MT sends an Invite message with a special Join header [43] to the CN (via the new interface if needed). This Join header contains all the relevant information about the ongoing call; for example, it sets the parameters like the call ID and tags of the previous connection. In this way, the CN knows that the new connection wants to join the ongoing connection. After negotiation, another connection is established between the MT and the CN. Note that these two connections follow different paths. Normally, a conference session is created when a third party joins a two-party conversation [43]. But, in this case, the CN knows the new participant and the MT of the previous conversation is logically at the same location because their SIP URIs are the same. Therefore, the CN will concurrently send packets to both interfaces. After the transaction, the MT and the CN can communicate through the two connections independently and the MT will discard any duplicate packets. When the new connection is set up, the MT will send a Bye message to the CN to terminate the connection via the old domain. It also updates the contact address in the home registration server with the new contact address.

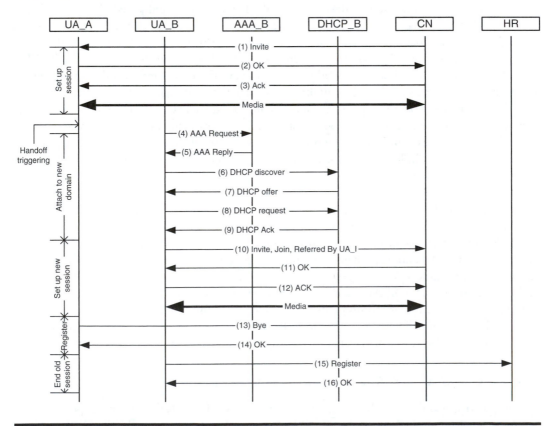

Figure 7.3 Handover procedure of S-SIP.

As handovers take place near the boundary of the working network, the MT should initiate the location tracking function only when the RSS of the network is lower than a certain threshold. Furthermore, it should also execute the handover whenever it predicts that it will leave the network soon. Specifically, as the positions are sensed at regular intervals, the MT should initiate the handover if it expects that it will leave the network within a duration that is less than the sum of the detecting period and the handover delay. Obviously, the prediction may not always be correct. If the MT leaves the network earlier than the estimated time, the handover will lead to packet losses. If it leaves later than the estimated time, it will utilize more resources than necessary. Hence we should maintain a balance, which also depends on the traffic type and user's preference. Consequently, we suggest to make the threshold adjustable.

The S-SIP scheme has several advantages over other SIP-based schemes. First, compared to the basic SIP-scheme and the other two improved interdomain schemes, it realizes real seamless handovers by applying the make-before-break mechanism to guarantee no packet loss while the other schemes only shorten the packet loss period. In the basic SIP, the packets sent out before the arrival of the signaling messages to the CN are lost. In HMSIP, the packets sent out before the location update signals reach the global proxy are lost. The fast SIP handoff scheme reduces the packet lost period to be the period of signaling from the MT to the proxy of the last working network. The mechanism in Ref. [41] reduces only the authentication period. Therefore, none of the above schemes can support real seamless handover.

Second, S-SIP requires no modification on the network entities as well as the underlying protocols. HMSIP requires the set up of a global proxy for each domain to manage the location update and data redirection for MTs in the entire domain. The fast SIP handover requires all APs in the networks to be equipped with the back-to-back user agents (B2BUA), which may not be preferred by some operators. The proposal in Ref. [41] requires modifications to the existing authentication systems of all domains, which may not be easily agreed by all parties. Moreover, it incurs a heavy signaling cost because a SA must be established with each neighboring network irrespective of the movement of the MT.

7.3.3 Summary

In this section, we have studied various handover management schemes implemented in different layers. On the basis of the study, the M-SCTP-based handover management solutions are considered as good candidates for multimode terminals in NG heterogeneous wireless systems, while the SIP-based application layer solutions are regarded as attractive candidates to support handover over IMS. We also have introduced a novel S-SIP scheme to support seamless interdomain handover. Using the Join signaling message, the proposed S-SIP scheme can realize seamless handovers by simultaneously maintaining two ongoing connections between the MT and the CN so as to avoid packet losses during the handover. We have also discussed how the S-SIP scheme can outperform other similar schemes with respect to both performance and implementation feasibility.

7.4 Admission Control on Vertical Handover

Many recent research studies have focussed on vertical handover and admission control between cellular networks and IEEE 802.11 WiFi networks. However, with the increasingly widespread

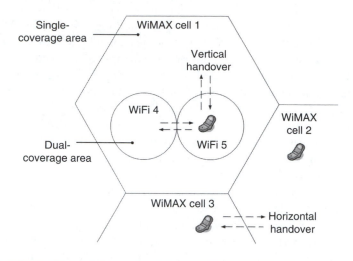

Figure 7.4   Integrated WiMAX/WiFi system.

adoption of unlicensed mobile broadband access technologies such as WiMAX for wireless widearea wireless networking, admission control for vertical handovers in integrated WiMAX/WiFi network is becoming an important area of investigation. For admission control, handovers are assumed to have a higher priority over new connections. If one access network does not have enough resources to support a handover connection request, the request can be transferred to the other network, if available. On the basis of the model in Ref. [44] for cellular/WLAN interworking, we evaluate an integrated WiMAX/WiFi system. Two admission control strategies from cellular networks [45], namely cutoff priority and a simplified fractional guard channel, are considered. The new connection blocking and handover dropping probabilities are calculated by using repeated substitutions.

We consider an integrated WiMAX/WiFi system where one or more WiFi hot spots may be deployed inside each cell of the WiMAX system. We assume that the WiMAX and WiFi operators have an agreement to share network resources (i.e., unused capacity) to guarantee the connection-level QoS, and to reduce the handover dropping probability. In this system, there are two specific coverage areas to consider: the single-coverage area, and the dual-coverage area. These areas are shown in Figure 7.4. In this context, coverage means service availability. Horizontal and vertical handovers can occur in different coverage areas.

7.4.1   Related Work

To achieve a seamlessly integrated WiMAX/WiFi network, there are several technical challenges that have to be addressed. Here, we focus on the vertical handover and admission control. In Ref. [46], a vertical handover scheme that considers the impact of different service fees in integrated WiMAX/WiFi networks is presented. In Ref. [47], bandwidth allocation and admission control are investigated in WiMAX/WiFi networks by using a novel game-theoretic approach. In Refs. [48,49], common radio resource management and joint admission control are considered, respectively. While in Ref. [48] network parameters such as service type, user mobility, location and load information, and service cost are taken into account, in Ref. [49] user's preferences are proposed for the admission decision. In Ref. [50], a mechanism for QoS support to enable ABC service in WiMAX/WiFi networks is presented. Admission control for multimedia services is considered in which the connections are admitted whenever there are resources to support at least the mean data rate of the connection. Finally, in Ref. [51], vertical handovers are used to balance the traffic load, and to manage radio resources efficiently among the different mobile access networks.

7.4.2 Admission Control and Resource Sharing in WiMAX/WiFi Networks

7.4.2.1 System Model

We consider a WiMAX system with M^{m} cells. Let A_i^{m} be the set of cells adjacent to cell i, W_i^{m} be the set of WiFi hot spots inside the coverage of cell i, A_k^{w} be the set of WiFi hot spots adjacent to hot spot k, and D_k^{w} be the set containing the overlaying cell of hot spot k (i.e., a dual-coverage area). As an example, from Figure 7.4, we have that $M^{\mathrm{m}} = 3, A_1^{\mathrm{m}} = \{2, 3\}, W_1^{\mathrm{m}} = \{4, 5\}, A_4^{\mathrm{w}} = \{5\}$, and $D_4^{\mathrm{w}} = \{1\}$. The new connection arrival processes to cell i and WiFi hot spot k are Poisson with rates λ_i^{m} and λ_k^{w}, respectively, which are independent of other arrival processes. The channel holding time of a connection in cell i (i.e., the time that a user is using resources in cell i) is an exponentially distributed random variable with mean $1/\mu_i^{\mathrm{m}}$. The channel holding time in hot spot k is exponentially distributed with mean $1/\mu_k^{\mathrm{w}}$. Both are independent of earlier arrival times and connection duration times. Each cell i of the WiMAX system has a capacity of C_i^{m} units of bandwidth, while each WiFi hot spot k has a capacity of C_k^{w} units of bandwidth. For simplicity, a connection may request only one unit of bandwidth (i.e., one channel).

At the end of a holding time, a connection in cell i of the WiMAX system may terminate and leave the system with probability q_{iT}^{m}, or move within the system and continue in an adjacent cell or WiFi hot spot with probability $1 - q_{iT}^{\mathrm{m}}$. The probability that a connection continues and moves to an adjacent cell of cell i or hot spot k inside cell i is given by

$$P_i^{\mathrm{m}}\text{mobility} = 1 - q_{iT}^{\mathrm{m}} = \sum_{j \in A_i^{\mathrm{m}}} q_{ij}^{\mathrm{m}} + \sum_{k \in W_i^{\mathrm{m}}} q_{ik}^{\mathrm{m}} \qquad (7.1)$$

where q_{ij}^{m} is the probability to attempt a horizontal handover to adjacent cell j and q_{ik}^{m} is the probability to attempt a vertical handover to WiFi hot spot k inside cell i.

On the other hand, at the end of a holding time of a connection in WiFi hot spot k, it may terminate and leave the system with probability q_{kT}^{w}. The probability that the connection continues and moves to the overlaying cell or to an adjacent hot spot of hot spot k is given by

$$P_k^{\mathrm{w}}\text{mobility} = 1 - q_{kT}^{\mathrm{w}} = \sum_{l \in A_k^{\mathrm{w}}} q_{kl}^{\mathrm{w}} + \sum_{i \in D_k^{\mathrm{w}}} q_{ki}^{\mathrm{w}} \qquad (7.2)$$

where q_{kl}^{w} is the probability to attempt a horizontal handover to adjacent WiFi hot spot l and q_{ki}^{w} is the probability of attempt a vertical handover to overlaying cell i.

7.4.2.2   Admission Control and Resource Sharing Schemes

We define T_i^m and T_k^w as the reservation parameters of cell i and WiFi hot spot k for handover connections, respectively. These parameters act as the admission policy to provide handover priority over new connections. Two admission control schemes are considered: cutoff priority and a simplified fractional guard channel. Let n_i be the number of connections in cell i. When cell i is in any of the states $n_i \leq C_i^m - T_i^m$, it accepts new and handover connections under both admission policies. When cell i is in any of the states $n_i > C_i^m - T_i^m$, only handover requests are accepted for the cutoff priority scheme. For the fractional guard channel scheme, all handover requests are accepted when free channels are available; each new connection request is accepted with probability w_i^m if a free channel is available. The same admission policies apply to the WiFi hot spots with probability w_k^w. The resource sharing capability provided by the agreement among different access networks operates as follows: if a handover request is not accepted in one network, then the request is transferred to the other network. In the WiMAX system, only when the user is inside the dual-coverage area can the connection be transferred to the WiFi network. Connections that are within a WiFi hot spot can always be transferred to the corresponding overlaying cell in the WiMAX system.

7.4.2.3   Traffic Equations

The occupancy of a cell evolves according to a birth–death process independent of other cells. The process for cell i evolves with birth rate ρ_i^m for the unreserved states (i.e., $0 < n_i \leq C_i^m - T_i^m$), and α_i^m for the reserved states (i.e., $C_i^m - T_i^m < n_i \leq C_i^m$). The death rate of cell i in state n_i is $n_i \mu_i^m$. The total traffic offered to cell i in state n_i is

$$\rho_i^m = \lambda_i^m + \psi_i^m, \quad n_i \leq C_i^m - T_i^m \tag{7.3}$$

$$\alpha_i^m = \lambda_i^m \omega_i^m + \psi_i^m, \quad n_i > C_i^m - T_i^m \tag{7.4}$$

$$\psi_i^m = \sum_{j \in A_i^m} v_{ji}^m + \sum_{k \in W_i^m} v_{ki}^w + \sum_{l \in W_i^m} \gamma_{li}^w \tag{7.5}$$

with $\omega_i^m = 0$ for the cutoff priority and $0 \leq \omega_i^m \leq 1$ for the fractional guard channel. ψ_i^m corresponds to all aggregated handover and shared traffic and is given by the terms v_{ji}^m is the horizontal handover rate of cell j offered to cell i, for adjacent cells i and j, v_{ki}^w is the vertical handover rate of WiFi hot spot k offered to overlay cell i, and γ_{zj}^w corresponds to all handover traffic that is not accepted in hot spot z and hence transferred to cell j:

$$v_{ji}^m = \lambda_j^m \left(1 - B_j^m\right) q_{ji}^m + \sum_{x \in A_j^m} v_{xj}^m \left(1 - B_{h_j}^m\right) q_{ji}^m$$

$$+ \sum_{y \in W_j^m} v_{yj}^w \left(1 - B_{h_j}^m\right) q_{ji}^m + \sum_{z \in W_j^m} \left[\gamma_{zj}^w \left(1 - B_{h_j}^m\right) q_{ji}^m\right] \tag{7.6}$$

$$v_{ki}^{\mathrm{w}} = \lambda_k^{\mathrm{w}} \left(1 - B_k^{\mathrm{w}} \right) q_{ki}^{\mathrm{w}} + \sum_{x \in A_k^{\mathrm{w}}} v_{xk}^{\mathrm{w}} \left(1 - B_{h_k}^{\mathrm{w}} \right) q_{ki}^{\mathrm{w}}$$

$$+ \sum_{y \in D_k^{\mathrm{w}}} v_{yk}^{\mathrm{w}} \left(1 - B_{h_k}^{\mathrm{w}} \right) q_{ki}^{\mathrm{w}} + \sum_{z \in D_k^{\mathrm{w}}} \left[\zeta_{zk}^{\mathrm{m}} \left(1 - B_{h_k}^{\mathrm{w}} \right) q_{ki}^{\mathrm{w}} \right] \tag{7.7}$$

$$\gamma_{zj}^{\mathrm{w}} = v_{jz}^{\mathrm{w}} B_{h_z}^{\mathrm{w}} + \sum_{l \in A_z^{\mathrm{w}}} v_{lz}^{\mathrm{w}} B_{h_z}^{\mathrm{w}} \tag{7.8}$$

where B_j^{m} and $B_{h_j}^{\mathrm{m}}$ are the new connection blocking and handover dropping probabilities in cell j, respectively, and B_k^{w} and $B_{h_k}^{\mathrm{w}}$ are the new connection blocking and handover dropping probabilities in WiFi hot spot k, respectively. Note that in Equation 7.7, ζ_{zk}^{m} is the proportion of all handover traffic that is not accepted in cell z and hence transferred to hot spot k, then

$$\zeta_{zk}^{\mathrm{m}} = \left(v_{kz}^{\mathrm{w}} B_{h_z}^{\mathrm{m}} + \sum_{l \in A_z^{\mathrm{m}}} v_{lz}^{\mathrm{m}} B_{h_z}^{\mathrm{m}} \right) R_{zk} \tag{7.9}$$

where R_{zk} is the coverage factor between WiFi hot spot k and overlay cell z. The coverage factor R_{zk} considers the coverage ratio between the radio coverage area of hot spot k and the radio coverage area of cell z with $0 < R_{zk} \leq 1$.

From the birth–death process in cell i and the detailed balance equations, $P_i^{\mathrm{m}}(n_i)$, the stationary distribution that cell i is in state n_i, is given by

$$P_i^{\mathrm{m}}(n_i) = \frac{\left(\rho_i^{\mathrm{m}} \right)^{n_i}}{G_i^{\mathrm{m}} n_i! \left(\mu_i^{\mathrm{m}} \right)^{n_i}}, \quad n_i \leq C_i^{\mathrm{m}} - T_i^{\mathrm{m}} \tag{7.10}$$

$$P_i^{\mathrm{m}}(n_i) = \frac{\left(\rho_i^{\mathrm{m}} \right)^{C_i^{\mathrm{m}} - T_i^{\mathrm{m}}} \left(\alpha_i^{\mathrm{m}} \right)^{n_i - C_i^{\mathrm{m}} + T_i^{\mathrm{m}}}}{G_i^{\mathrm{m}} n_i! \left(\mu_i^{\mathrm{m}} \right)^{n_i}}, \quad n_i > C_i^{\mathrm{m}} - T_i^{\mathrm{m}} \tag{7.11}$$

$$G_i^{\mathrm{m}} = \sum_{n_i=0}^{C_i^{\mathrm{m}} - T_i^{\mathrm{m}}} \frac{1}{n_i!} \left(\frac{\rho_i^{\mathrm{m}}}{\mu_i^{\mathrm{m}}} \right)^{n_i} + \sum_{n_i=C_i^{\mathrm{m}} - T_i^{\mathrm{m}} + 1}^{C_i^{\mathrm{m}}} \frac{\left(\rho_i^{\mathrm{m}} \right)^{C_i^{\mathrm{m}} - T_i^{\mathrm{m}}} \left(\alpha_i^{\mathrm{m}} \right)^{n_i - C_i^{\mathrm{m}} + T_i^{\mathrm{m}}}}{n_i! \left(\mu_i^{\mathrm{m}} \right)^{n_i}} \tag{7.12}$$

Similarly, the occupancy of WiFi hot spot k evolves with birth rates ρ_k^{w} and α_k^{w} based on the state n_k, and death rate $n_k \mu_k^{\mathrm{w}}$. The total traffic offered to hot spot k in state n_k is

$$\rho_k^{\mathrm{w}} = \lambda_k^{\mathrm{w}} + \psi_k^{\mathrm{w}}, \quad n_k \leq C_k^{\mathrm{w}} - T_k^{\mathrm{w}} \tag{7.13}$$

$$\alpha_k^{\mathrm{w}} = \lambda_k^{\mathrm{w}} \omega_k^{\mathrm{w}} + \psi_k^{\mathrm{w}}, \quad n_k > C_k^{\mathrm{w}} - T_k^{\mathrm{w}} \tag{7.14}$$

$$\psi_k^{\mathrm{w}} = \sum_{l \in A_k^{\mathrm{w}}} v_{lk}^{\mathrm{w}} + \sum_{j \in D_k^{\mathrm{w}}} v_{jk}^{\mathrm{w}} + \sum_{g \in D_k^{\mathrm{w}}} \zeta_{gk}^{\mathrm{w}} \tag{7.15}$$

with $\omega_k^{\mathrm{w}} = 0$ for the cutoff priority, and $0 \leq \omega_k^{\mathrm{w}} \leq 1$ for the fractional guard channel. ψ_k^{w} corresponds to all aggregated handover and shared traffic and is given by the terms v_{lk}^{w} is the horizontal handover rate of adjacent WiFi hot spot l offered to hot spot k, and v_{jk}^{w} the vertical handover rate of overlaying cell j offered to hot spot k:

$$v_{lk}^{\text{w}} = \lambda_l^{\text{w}} \left(1 - B_l^{\text{w}}\right) q_{lk}^{\text{w}} + \sum_{x \in A_l^{\text{w}}} v_{xl}^{\text{w}} \left(1 - B_{h_l}^{\text{w}}\right) q_{lk}^{\text{w}}$$

$$+ \sum_{y \in D_l^{\text{w}}} v_{yl}^{\text{w}} \left(1 - B_{h_l}^{\text{w}}\right) q_{lk}^{\text{w}} + \sum_{z \in D_l^{\text{w}}} \left[\zeta_{zl}^{\text{m}} \left(1 - B_{h_l}^{\text{w}}\right) q_{lk}^{\text{w}}\right] \tag{7.16}$$

$$v_{jk}^{\text{w}} = \lambda_j^{\text{m}} \left(1 - B_j^{\text{m}}\right) R_{jk} q_{jk}^{\text{m}} + \sum_{x \in A_j^{m}} v_{xj}^{\text{m}} \left(1 - B_{h_j}^{\text{m}}\right) R_{jk} q_{jk}^{\text{m}}$$

$$+ \sum_{y \in W_j^{\text{m}}} v_{yj}^{\text{w}} \left(1 - B_{h_j}^{\text{m}}\right) R_{jk} q_{jk}^{\text{m}} + \sum_{z \in W_j^{\text{m}}} \left[\gamma_{jz}^{\text{w}} \left(1 - B_{h_j}^{\text{m}}\right) q_{jk}^{\text{m}}\right] \tag{7.17}$$

We point out that the stationary distribution that WiFi hot spot k is in state n_k, $P_k^{\text{w}}(n_k)$ is similar to Equations 7.10 through 7.12, but with parameters $\rho_k^{\text{w}}, \alpha_k^{\text{w}}, \mu_k^{\text{w}}, C_k^{\text{w}}, T_k^{\text{w}}$, and G_k^{w} defined for the WiFi system. In the WiMAX system, using $P_i^{\text{m}}(n_i)$, the new connection blocking probability B_i^{m} in cell i is

$$B_i^{\text{m}} = \sum_{n_i = C_i^{\text{m}} - T_i^{\text{m}}}^{C_i^{\text{m}} - 1} P_i^{\text{m}}(n_i) \left(1 - \omega_i^{\text{m}}\right) + P_i^{\text{m}} \left(C_i^{\text{m}}\right) \tag{7.18}$$

and the handover dropping probability $B_{h_i}^{\text{m}} = P_i^{\text{m}} \left(C_i^{\text{m}}\right)$. Similarly, B_k^{w} in WiFi hot spot k is

$$B_k^{\text{w}} = \sum_{n_k = C_k^{\text{w}} - T_k^{\text{w}}}^{C_k^{\text{w}} - 1} P_k^{\text{w}}(n_k) \left(1 - \omega_k^{\text{w}}\right) + P_k^{\text{w}} \left(C_k^{\text{w}}\right) \tag{7.19}$$

and the handover dropping probability $B_{h_k}^{\text{w}} = P_k^{\text{w}} \left(C_k^{\text{w}}\right)$.

7.4.2.4 Numerical Results

We evaluate a wireless system consisting of a WiMAX system with $M^{\text{m}} = 3$, and $\left|W_i^{\text{m}}\right| = 2$ as shown in Figure 7.4. We compare the integrated WiMAX/WiFi system with and without resource sharing. All cells in the WiMAX system have capacity $C_i^{\text{m}} = 30$ units of bandwidth, and all WiFi hot spots have capacity $C_k^{\text{w}} = 60$. Channel holding times in the cells of the WiMAX system have means $1/\mu_i^{\text{m}} = 1$ minute, and channel holding times in the WiFi hot spots have means $1/\mu_k^{\text{w}} = 4$ minutes. The coverage factor $R_{jk} = 0.75$. The mobility of users given by Equation 7.1 in the WiMAX system is $q_{iT}^{\text{m}} = 0.4$, and the mobility in the WiFi hot spots, given by Equation 7.2 is $q_{kT}^{\text{w}} = 0.4$. These values define a mobility level of 60 percent (i.e., 60 percent of users perform handovers).

Figure 7.5 shows the new connection blocking probability at cell-level in the integrated WiMAX/WiFi system when the arrival rate λ_i^{m} is increased from 5 to 20 connections per minute

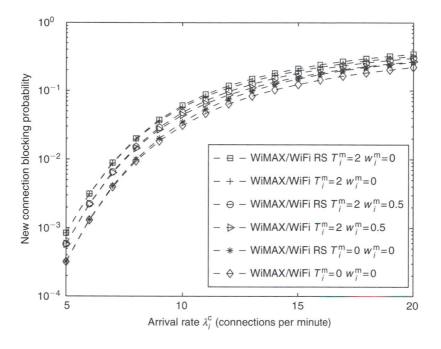

Figure 7.5 New connection blocking probability.

in all cells, and $\lambda_k^w = 5$. The results labeled as $T_i^m = 2$, $\omega_i^m = 0$ correspond to the admission policy cutoff priority; $T_i^m = 2$, $\omega_i^m = 0.5$ to the fractional guard channel; and $T_i^m = 0$, $\omega_i^m = 0$ when no admission policy is used. Note that the WiMAX/WiFi system achieves very close blocking probabilities with and without resource sharing. Recall that the resource sharing is only for handovers. As expected, the use of admission control increases the probability of blocking new connections at the expense of a lower handover dropping probability. The highest blocking probabilities are achieved by cutoff priority because it rejects all new connections when the cell is in the reserved states for handovers, while fractional guard channel accept them with probability $\omega_i^m = 0.5$. Figure 7.6 shows the handover dropping probability at cell-level in the integrated WiMAX/WiFi system for the same increase in the arrival rates. This is the handover dropping probability for any type of handover (i.e., horizontal or vertical) within the dual-coverage area. The integrated WiMAX/WiFi network with resource sharing achieves lower handover dropping probabilities due to the additional capacity shared by the WiFi network in the dual-coverage areas. In this case, the highest dropping probabilities in both systems are achieved when no admission control is used, and the lowest when cutoff priority is used. By using reservation parameters $T_i^m = T_k^w = 2$, the handover dropping probability is reduced by 58 percent at an expense of a 30 percent increase in the new connection blocking probability. Note that the admission policy, and its parameters have to be selected carefully to meet the desired connection-level QoS.

7.4.3 Summary

For applications that require QoS guarantee (e.g., voice, real-time video, multimedia sessions), admission control is required to limit the number of connections in a network. A connection

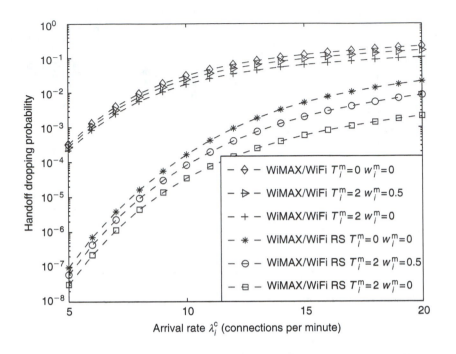

Figure 7.6 Handover dropping probability.

request will be blocked if the minimum bandwidth requirement cannot be satisfied. In this section, we have described an analytical model of an integrated WiMAX/WiFi system with admission control and resource sharing capabilities. Horizontal and vertical handovers due to mobility are considered in the traffic equations of the system. Finally, admission control is included by means of ω_i^m and ω_k^w in the WiMAX and WiFi networks, respectively. Note that by making those parameters state-dependent, other admission policies can be considered.

7.5 Conclusion

In this chapter, we have studied the major challenges of vertical handover in NG heterogeneous networks integrating WiMAX and WiFi technologies. Although horizontal handovers occur by necessity due to mobility, this is not necessarily the case with vertical handovers. For example, serveral WiFi hot spots may be overlaid by a WiMAX cell, so those handovers from a WiMAX cell to a WiFi hot spot would be by choice, whereas handovers from a hot spot to a cell could be by necessity. There are several additional issues that may be addressed by discretionary vertical handovers, such as optimization of user satisfaction and load balancing to alleviate congestion. The network's decision to proceed with a discretionary vertical handover should also take into consideration whether the QoS degradation that the connection may suffer during the handover process as well as the increase of signaling load in the networks due to the the vertical handover process are suitably compensated by the improvements in overall system performance or utility. We have reviewed different network selection and vertical handover decision schemes and proposed a location-based vertical handover decision algorithm that decides on a vertical handover into a hot spot on the basis of the expected duration that the MT will spend in the hot spot. We have further reviewed different vertical handover execution/management schemes and proposed a novel SIP-based handover scheme that provides seamless connection switching between the WiMAX and WiFi networks. Finally, we have modeled an integrated WiMAX/WiFi system and analytically evaluated two alternative admission control and resource sharing schemes that give priority to handover connections.

REFERENCES

1. P. Iyer, N. Natarajan, M. Venkatachalam, et al., All-IP network architecture for mobile WiMAX, IEEE Mobile WiMAX Symposium, Orlando, Florida, USA, pp. 54–59, March 2007.
2. E. Gustafsson and A. Jonsson, Always best connected, *IEEE Wireless Communications Magazine*, vol. 10, no. 1, pp. 49–55, February 2003.
3. M. Buddhikot, G. Chandranmenon, S. Han, Y. Lee, S. Miller, and L. Salgarelli, Integration of 802.11 and third-generation wireless data networks, Proceedings of IEEE Infocom 2003, San Francisco, California, USA, pp. 503–512, April 2003.
4. P. Neves, S. Sargento, and R. L. Aguiar, Support of real-time services over integrated 802.16 metropolitan and local area networks, IEEE Symposium on Computers and Communications, Pula-Cagliari, Sardinia, Italy, pp. 15–22, June 2006.
5. D. Niyato and E. Hossain, Integration of WiMAX and WiFi: Optimal pricing for bandwidth sharing, *IEEE Communications Magazine*, 140–146, May 2007.
6. Motorola and Intel White Paper, WiMAX and WiFi Together: Deployment Models and User Scenarios, 2007.
7. G. Pollini, Trends in handover design, *IEEE Communications Magazine*, 34(3), 82–90, March 1996.
8. C. Guo, Z. Guo, Q. Zhang, and W. Zhu, Seamless and proactive end-to-end mobility solution for roaming across heterogeneous wireless networks, *IEEE JSAC*, 22(5), 834–848, June 2004.
9. D. Chalmers and M. Sloman, A survey of quality of service in mobile computing environments, IEEE Communications Tutorials and Surveys, http://www.comsoc.org/pubs/surveys, Second Quarter 1999.
10. H. J. Wang, R. H. Katz, and J. Giese, Policy-enabled handoffs across heterogeneous wireless networks, Proceedings of IEEE Workshop on Mobile Computing Systems and Applications, New Orleans, Louisiana, USA, pp. 51–60, February 1999.
11. F. Zhu and J. MacNair, Optimizations for vertical handoff decision algorithms, Proceedings of IEEE WCNC'2005, New Orleans, Louisiana, USA, pp. 867–872, March 2005.
12. Q. Song and A. Jamalipour, A network selection mechanism for next generation networks, Proceedings of IEEE ICC'2005, Seoul, Korea, pp. 1418–1422, May 2005.
13. W. Chen and Y. Shu, Active application oriented vertical handoff in next-generation wireless networks, Proceedings of IEEE WCNC'2005, New Orleans, Louisiana, USA, pp. 1383–1388, March 2005.
14. A. Majlesi and B. H. Khalaj, An adaptive fuzzy logic based handoff algorithm for interworking between WLANs and mobile networks, Proceedings of IEEE PIMRC'2002, Lisbon, Portugal, pp. 2446–2451, September 2002.
15. Q. Guo, J. Zhu, and X. Xu, An adaptive multi-criteria vertical handoff decision algorithm for radio heterogeneous networks, Proceedings of IEEE ICC'2005, Seoul, Korea, vol. 4, pp. 2769–2773, May 2005.
16. W. Zhang, Handover decision using fuzzy MADM in heterogeneous networks, Proceedings of IEEE WCNC'2004, Atlanta, Georgia, USA, pp. 653–658, March 2004.
17. E. Stevens-Navarro and V. W. S. Wong, Comparison between vertical handoff decision algorithm for heterogeneous wireless networks, Proceedings of IEEE VTC'2006, Montréal, Canada, May 2006.
18. W. S. Soh and H. S. Kim, QoS provisioning in cellular networks based on mobility prediction techniques, *IEEE Communication Magazine*, 41(1), 86–92, January 2003.

19. J. Zhang, H. C. B. Chan, and V. C. M. Leung, A location-based vertical handoff decision algorithm for heterogeneous mobile networks, *IEEE Globecom*, San Francisco, California, USA, 1–5, December 2006.

20. W. Wang and I. F. Akyildiz, Intersystem location update and paging schemes for multitier wireless networks, Proceedings of ACM Mobicom, Boston, Massachusetts, pp. 99–109, 2000.

21. B. Liang and Z. J. Haas, Predictive distance-based mobility management for multidimensional PCS networks, *IEEE/ACM Transactions on Networking*, 11(5), 718–732, October 2003.

22. M. L. Puterman, *Markov Decision Processes: Discrete Stochastic Dynamic Programming*, John Wiley & Sons, New York, 1994.

23. J. Zhang, H. C. B. Chan, and V. C. M. Leung, A SIP-based handoff scheme for heterogeneous mobile networks, Proceedings of IEEE WCNC'2007, Hong Kong, China, pp. 3946–3950, March 2007.

24. I. F. Akyildiz, X. Jiang, and S. Mohanty, A survey of mobility management in next-generation all-IP-based wireless systems, *IEEE Wireless Communications Magazine*, 11(4), 16–28, August 2004.

25. N. Banerjee, W. Wu, and S. K. Das, Mobility support in wireless Internet, *IEEE Wireless Communications Magazine*, 10(5), 54–61, October 2003.

26. C. E. Perkins, IP mobility support for IPv4, RFC 3220, January 2002.

27. C. E. Perkins and D. B. Johnson, Route optimization in mobile IP, Internet draft, IETF, draft-ietf-mobileip-optim-11.txt, September 2001.

28. R. Hsieh, Z. G. Zhou, and A. Seneviratne, S-MIP: A seamless handoff architecture for mobile IP, Proceedings of IEEE INFOCOM'2003, San Francisco, California, USA, vol. 3, pp. 1774–1784, March–April 2003.

29. E. Gustaffson, A. Jonsson, and C. Perkins, Mobile IP Regional Registration, draft-ietf-mobileip-reg-tunnel-05.txt, September 2001.

30. A. Misra, S. Das, A. Dutta, A. Mcauley and S. K. Das, IDMP-based fast handoffs and paging in IP-based 4G mobile networks, *IEEE Communications Magazine*, 40(3), 138–145, March 2002.

31. H. Soliman, C. Castelluccia, K. El-Malki, and L. Bellier, Hierarchical Mobile IPv6 Mobility Management (HMIPv6), Internet draft, IETF, draft-ietf-mipshop-hmipv6-02.txt, Jun. 2004.

32. C. E. Perkins and K. Wang, Optimized smooth handoffs in mobile IP, IEEE Symposium on Computers and Communications, Sharm El Sheik, Red Sea, Egypt, pp. 340–346, July 1999.

33. A. C. Snoeren and H. Balakrishnan, An end-to-end approach to host mobility, Proceedings of Mobi-Com'2000, pp. 155–166, Boston, Massachusetts, September 2000.

34. Q. Zhang, C. Guo, Z. Guo, and W. Zhu, Efficient mobility management for vertical handoff between WWAN and WLAN, *IEEE Communications Magazine*, 41(11), 102–108, November 2003.

35. C. Guo, Z. Guo, Q. Zhang, and W. Zhu, A seamless and proactive end-to-end mobility solution for roaming across heterogeneous wireless networks, *IEEE JSAC*, 22(5), 834–848, June 2004.

36. L. Ma, F. Yu, and V. C. M. Leung, Performance improvements of mobile SCTP in integrated heterogeneous wireless networks, *IEEE Transactions on Wireless Communications*, 6(10), 3567–3577, 2007.

37. M. Handley, et al., SIP: Session Initiation Protocol, IETF RFC 2543, March 1999.

38. V. Koukoulidis and M. Shah, The IP multimedia domain in wireless networks: Concepts, architecture, protocols and applications, Proceedings of 6th IEEE ISMSE, Miami, Florida, USA, pp. 484–490, December 2004.

39. G. Camarillo and M. A. Garcia-Martin, *The 3G IP Multimedia Subsystem: Merging the Internet and the Cellular Worlds*, John Wiley & Sons Ltd., 2004.

40. W. Wu, N. Banerjee, K. Basu, and S. K. Das, SIP-based vertical handoff between WWANs and WLANs, *IEEE Wireless Communications Magazine*, 12(3), 66–72, June 2005.

41. D. Vali, S. Paskalis, A. Kaloxylos, and L. Merakos, An efficient micro-mobility solution for SIP networks, Proceedings of Globecom, San Francisco, California, USA, pp. 3088–3092, December 2003.

42. N. Banerjee, A. Acharya and S. K. Das, Seamless SIP-based mobility for multimedia applications, *IEEE Network Magazine*, 20(2), 6–13, March–April 2006.

43. R. Mahy and D. Petrie, The Session Initiation Protocol (SIP) "Join" Header, IETF RFC 3911, October 2004.

44. E. Stevens-Navarro and V. W. S. Wong, Resource sharing in an integrated wireless cellular/WLAN system, Proceedings of CCECE'07, Vancouver, Canada, April 2007.
45. Y. Fang and Y. Zhang, Call admission control schemes and performance analysis in wireless mobile networks, *IEEE Transactions on Vehicular Technology*, 51(2), 371–382, March 2002.
46. Y. Choi and S. Choi, Service charge and energy-aware vertical handoff in integrated IEEE 802.16e/802.11 networks, Proceedings of IEEE INFOCOM'2007, Anchorage, AK, May 2007.
47. D. Niyato and E. Hossain, A hierarchical model for bandwidth management and admission control in integrated IEEE 802.16/802.11 wireless networks, Proceedings of WCNC'2007, Hong Kong, China, March 2007.
48. A. Hasib and A. Fapojuwo, Performance analysis of common radio resource management scheme in multi-service heterogeneous wireless networks, Proceedings of WCNC'2007, Hong Kong, China, March 2007.
49. O. Falowo and H. Chan, Joint call admission control for next generation wireless networks, Proceedings of CCECE'2006, Ottawa, Canada, May 2006.
50. J. J. Roy, V. Vaidehi, and S. Srikanth, Always best-connected QoS integration model for the WLAN, WiMAX heterogeneous network, Proceedings of ICIIS'2006, Sri Lanka, August 2006.
51. S. Liu, V. Li, and P. Zhang, Joint radio resource management through vertical handoffs in 4G networks, Proceedings of IEEE GLOBECOM'2006, San Francisco, CA, November 2006.

Chapter 8

Piconet Interconnection Strategies in IEEE 802.15.3 Networks

Muhi A.I. Khair, Jelena Mišić, and Vojislav B. Mišić

CONTENTS

Wireless mesh networks (WMN) commonly refer to distribute, cooperative communication networks formed by many nodes with wireless communication capability, some of which may be mobile [1,2]. The main characteristic that distinguished WMNs from their predecessors, wireless

147

ad hoc networks, is the cooperative communication capability, which is facilitated by the fact that each node may function as data source, data destination (consumer), or router, as appropriate. On account of this capability, WMN technology has many potential applications with a huge consumer demand. Performance of a WMN is determined by many factors, not the least important of which is the Medium Access Control (MAC) protocol. Already a number of MAC protocols have been proposed for use in WMNs, but the IEEE 802.15.3 standard for high data rate wireless personal area networks (HR-WPANs) [3] is often singled out as a viable candidate. The IEEE 802.15.3 standard offers a combination of Carrier Sense Multiple Access with Collision Avoidance (CSMA/CA) and TDMA at the MAC layer, as well as a set of several physical (PHY) layer modulation techniques that allow operation at data rates up to 55 Mbps. Recently, the 802.15.3 MAC has even been coupled with ultra wideband (UWB) PHY layer technology to offer even higher data rates at reduced collision probability [4]. Similar to Bluetooth [5] 802.15.3 devices are organized in piconets controlled by a dedicated piconet coordinator (PNC). Unlike Bluetooth, however, devices in a 802.15.3 piconet can directly communicate with one another, which simplifies routing and improves throughput. Together, these features make the 802.15.3 standard a promising candidate for the implementation of WMNs.

In many cases, mesh networking requires coverage of larger physical areas in which distances can easily exceed the transmission range allowed by the power levels prescribed in the 802.15.3 standard. In such cases, the 802.15.3 technology can still be used, but the network must include two or more piconets interconnected through shared devices—bridges. Standard bridge configurations include the so-called master–slave bridge, where the coordinator of one piconet acts as the bridge during inactive periods, and slave–slave bridge, where the bridge node is an ordinary (i.e., non-coordinator) device in each of the piconets it visits. The main problems in multi-piconet networks are bridge and piconet scheduling, which are the focus of discussions in this chapter; although issues related to topology formation and maintenance are also important, they are beyond the scope of this chapter and will not be covered.

We begin this chapter with a brief overview of the MAC layer features and protocols as prescribed by the 802.15.3 standard. Then, we examine the problem of piconet interconnection, as well as bridge and piconet time scheduling, in the context of 802.15.3 networks. We present two basic strategies to interconnect the piconets to form a mesh network and discuss their pros and cons. Finally, we describe a simple piconet interconnection and scheduling protocol for IEEE 802.15.3-based mesh networks.

8.1 Basics of the IEEE 802.15.3 HR-WPAN Standard

The IEEE 802.15.3 standard [3] for HR-WPANs is designed to fulfill the requirements of high data rate suitable for multimedia applications while ensuring low end-to-end delay. It is also designed to provide easy reconfigurability and high resilience to interference, because it uses the unlicensed industrial, scientific, and medical (ISM) band at 2.4 GHz, which is shared with a number of other communication technologies such as WLAN (802.11b/g) and Bluetooth (802.15.1), among others.

Devices in 802.15.3 networks are organized in small networks called piconets, each of which is formed, controlled, and maintained by a single dedicated device referred to as the PNC. The network is formed in an ad hoc fashion: upon discovering a free channel, the PNC-capable device starts the piconet by simply transmitting period beacon frames; other devices that detect these frames then request admission, or association (as it is referred to in the 802.15.3 standard). The coordinator duties include transmission of periodic beacon frames for synchronization, admission

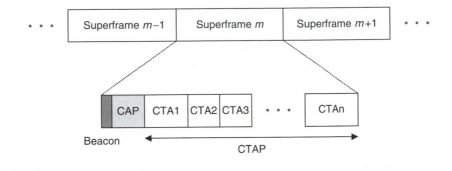

Figure 8.1 Superframe format in an IEEE 802.15.3 piconet. (Adapted from Standard for part 15.3: Wireless Medium Access Control (MAC) and Physical Layer (PHY) Specifications for High Rate Wireless Personal Area Networks (WPAN), IEEE, New York, IEEE Standard 802.15.3, 2003.)

of new devices to the piconet, as well as allocation of dedicated periods to allow unhindered packet transmission by the requesting device.

Time in an 802.15.3 piconet is structured in superframes delimited by successive beacon frame transmissions from the PNC. The structure of the superframe is shown in Figure 8.1. Each superframe contains three distinct parts: the beacon frame, the contention access period (CAP), and the channel time allocation period (CTAP). During the CAP, devices compete with each other for access; a form of CSMA/CA algorithm is used. This period is used to send requests for CTAs (defined below) and other administrative information, but also for smaller amounts of asynchronous data.

CTAP contains a number of individual subperiods (referred to as channel time allocation, or CTA), which are allocated by the PNC upon explicit requests by the devices that have data to transmit. Requests for CTAs are sent during the CAP; as such, they are subject to collision with similar requests from other devices. The decision to grant or not to grant the allocation request rests exclusively with the PNC, which must take into account the amount of resources available—most often, the traffic parameters of other devices in the network and the available time in the superframe. If a device is allocated a CTA, other devices may not use it, and contention-free access is guaranteed. CTA is announced in the next beacon frame; it may be temporary or may last until explicit deallocation by the PNC. Typically, CTAs are used to send commands and larger quantities of isochronous and asynchronous data.

Special CTAs known as management channel time allocation (MCTA) are used for communication and dissemination of administrative information from the PNC to the devices, and vice versa. There are three types of MCTA defined in the standard—association, open, and regular MCTA. Association and open MCTAs use the slotted aloha [6] medium access technique, while regular MCTAs use the TDMA mechanism.

Unlike other WPANs such as Bluetooth and 802.15.4, direct device-to-device communication is possible in an 802.15.3 piconet. In case the communicating devices are not within the transmission range of each other, the PNC (which, by default, must be able to communicate with both) may be involved as an intermediary, leading in effect to multi-hop intra-piconet communication. It is worth noting that problems of this nature may be alleviated not only by adjusting the transmission power but also by making use of the adaptive data rate facility provided by the 802.15.3 standard. Namely, if transmission at the full data rate of 55 Mbps suffers from too may errors because the signal-to-noise-plus-interference ratio (SINR) is too low, different modulation schemes with lower data rate may be used to give additional resilience. This problem and its solutions, however, are beyond the scope of this chapter.

Reliable data transfer in 802.15.3 networks is achieved by utilizing acknowledgments and retransmission of non-acknowledged packets. The standard defines three acknowledgment modes:

- No acknowledgment (No-ACK) is typically used for delay sensitive but loss tolerant traffic such as multimedia (typically transferred through UDP or some similar protocol)
- Immediate acknowledgment (Imm-ACK) means that each packet is immediately acknowledged with a small packet sent back to the sender of the original packet
- Delayed acknowledgment (Dly-ACK), where an acknowledgment packet is sent after successfully receiving a batch of successive data packets; obviously, this allows for higher throughput due to reduced acknowledgment overhead—but the application requirements must tolerate the delay incurred in this case, and some means of selective retransmission must be employed to maintain efficiency

8.2 Interconnecting IEEE 802.15.3 Piconets

The 802.15.3 standard contains provisions for the coexistence of multiple piconets in the same (or partially overlapping) physical space. Because the data rate is high, up to 55 Mbps, the channel width is large and there are, in fact, only five channels available in the ISM band for use of 802.15.3 networks. If 802.11-compatible WLAN (or, perhaps, several of them) are present in the vicinity, the number of available channels is reduced to only three to prevent excessive interference between the networks adhering to two standards. As a result, the formation of multiple piconets must utilize time division multiplexing, rather than the frequency division one, as is the case with Bluetooth. Namely, a piconet can allocate a special CTA during which another piconet can operate. There are two types of such piconets: a child piconet and a neighbor piconet.

A child piconet is the one in which the PNC is a member of the parent piconet. It is formed when a PNC-capable device that is a member of the parent piconet sends a request to the parent PNC asking for a special CTA known as a private CTA. Regular CTA requests include the device addresses of both the sender and the receiver; a request for a private CTA is distinguished by virtue of containing the same device address as both the sending and the receiving node. When the parent PNC allocates the required CTA, the child PNC may begin sending beacon frames of its own within that CTA, and thus may form another piconet that operates on the same channel as the parent piconet, but is independent from it. The private CTA is, effectively, the active portion of the superframe of the child piconet. The child superframe consists, then, of this private CTA that can be used for communication between child PNC and its devices (DEVs); the remainder of the parent superframe is reserved time—it cannot be used for communication in the child piconet.

8.2.1 Master–Slave Bridge

From the standpoint of piconet interconnection, the child piconet mechanism allows for simple implementation of the master–slave interconnection topology, because the two piconets are linked through the child PNC, which partakes in both of them, and thus can act as the bridge. Figure 8.2 schematically shows such topology in which Piconet 2 is the child of Piconet 1; the child piconet PNC is also acting as a master–slave bridge that links the piconets. The timing relationship of superframes in parent and child piconets is shown in Figure 8.3, where the top part corresponds

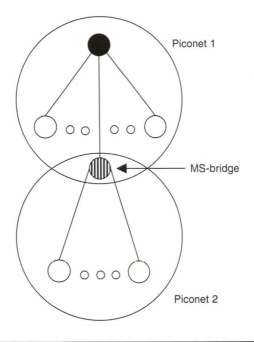

Figure 8.2   Piconet interconnection through a master–slave bridge.

to the parent piconet and the bottom part to the child piconet. Note that the distinction is logical rather than physical, because the piconets share the same RF channel.

A given piconet can have multiple child piconets, and a child piconet may have another child of its own. Obviously, the available channel time is shared between those piconets, at the expense of decreased throughput and increased delay; but the effective transmission range may be increased.

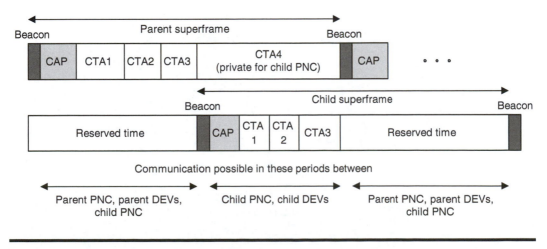

Figure 8.3   Superframe structure for parent and child piconets.

8.2.2 Slave–Slave Bridge

This interconnection topology may also be implemented, but in a slightly more complex manner. Namely, direct communication between the members of different piconets is not possible; the only shared device is the PNC of the child piconet. If an ordinary device wants to act as a bridge, it must explicitly associate with both parent and child piconets, and obtain a distinct device address in each of them. In this manner, multiple bridges may exist between the two piconets. The topology of two piconets interconnected through a slave–slave bridge is shown in Figure 8.4. Note that, in this case, the piconet may be linked through a parent–child relationship; but they could also use different RF channels, with a certain penalty because of the need for the bridge to synchronize with two independently running superframe structures.

8.2.3 Challenges

As can be seen from the discussion above, the main challenge in forming a multi-piconet network that uses the same RF channel—i.e., a complex network in which all piconets are related through parent–child relationships—is to develop a networkwide distributed scheduling algorithm that will allocate channel time to all devices in an efficient and fair manner. Because time division multiplexing among each parent–child piconet pair is used, we need not worry about the conflicts—collisions—between transmissions originating from different piconets in the networks: the transmissions during allocated CTAs are guaranteed to be conflict-free. The need to wait until the appropriate active portion of the superframe incurs some additional delays besides the usual transmission delay and access delay in the outbound queue of the source device; furthermore, the bridge device operates its own queues (one for each direction of the traffic) and these can also add delay to the total packet transmission time. As those queues are necessarily implemented with buffers of finite size, there exists nonzero probability that the buffer will overflow, in which case packets may be blocked from entering the queue; if reliable transfer is needed, the possibility of packet blocking necessitates the use of Imm-ACK or Dly-ACK acknowledgment policy. In addition, we must devise an efficient and fair algorithm to partition the available channel time between the piconets, taking into account the traffic intensity both within the piconets and between them.

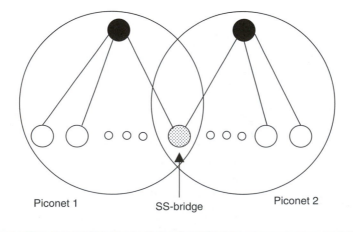

Piconet 1   SS-bridge   Piconet 2

Figure 8.4   Piconet interconnection through a slave–slave bridge.

8.2.4   Using Different RF Channels

Mesh networks can also be created using a different scenario, in which several multi-piconet networks operate in the same physical space but on different RF channels. Although physical conflicts between transmissions originating from different multi-piconet networks are still absent by virtue of frequency division multiplexing, scheduling conflicts between the piconets will be the main source of complexity, as the device that wants to act as a bridge must alternatively synchronize with piconets that operate according to entirely unrelated schedules. This precludes the use of master–slave bridges to interconnect such piconets. Namely, the master–slave bridges must not abstain from their duties as the PNCs in their respective piconets for prolonged periods of time. As a result, piconets operating on different RF channels favor interconnection through slave–slave bridges, i.e., devices that act as ordinary nodes in each of the piconets they belong to. As such devices have no coordinator duties, their absence from a given piconet will not cause any problems there. In fact, their absence might even go unnoticed if there happens to be no traffic directed to such devices during that time interval.

8.2.5   Neighbor Piconets

The 802.15.3 standard also provides the concept of the neighbor piconet, which is intended to enable an 802.15.3 piconet to coexist with another network that may or may not use the 802.15.3 communication protocols; for example, an 802.11 WLAN in which one of the devices is 802.15.3-capable. A PNC-capable device that wants to form a neighbor piconet will first associate with the parent piconet, but not as an ordinary piconet member; the parent PNC may reject the association request if it does not support neighbor piconets. If the request is granted, the device then requests a private CTA from the coordinator of the parent piconet. Once a private CTA is allocated, the neighbor piconet can begin to operate. The neighbor PNC may exchange commands with the parent PNC, but no data exchange is allowed. In other words, the neighbor piconet is simply a means to share the channel time between the two networks. Because, unlike the child piconet, data communications between the two piconets are not possible, this mechanism is unsuitable for the creation of multi-piconet networks, and, consequently, for mesh networking.

8.3   Implementing Mesh Networks with 802.15.3

In this section we will first explain the interconnection (bridging) mechanism, followed by our proposed scheduling algorithm for CTA in the mesh network. The superframe structure of our mesh MAC Protocol follows the IEEE 802.15.3 MAC superframe and the CTA is based on TDMA, during the guaranteed access period, and CSMA/CA, during the contention period.

Two common approaches, namely the master–slave bridge and the slave–slave bridge are used for piconet interconnection in different networks. In the case of a master–slave bridge, Figure 8.2, the bridge device is the PNC for Piconet 2 and a normal member of Piconet 1. In the case of a slave–slave bridge, Figure 8.4, the bridge device is an ordinary member (DEV) in both piconets. We can combine both types of bridges in the mesh environment to cover larger areas. The choice of the type of interconnection depends on location of the bridge device within the mesh network. The interconnection will be established through a master–slave bridge if a PNC-capable device is located such that it can easily control one piconet and participate in the other one. On the other hand, the slave–slave bridge can be used if no suitable PNC-capable device can be found, or if the two piconets operate on different RF channels, possibly because the traffic volume is too high to be serviced with half the available bandwidth.

8.3.1 Operation of the Master–Slave Bridge

The bridge establishes a connection between a parent and a child piconet where the bridge device acts as the PNC of the child piconet. The bridge device maintains two queues to temporarily store, and subsequently deliver, the traffic in both directions. As can be seen from Figure 8.3, the superframe duration is the same for both parent and child piconets; in fact, the child superframe is simply a private CTA from the parent superframe. The only setup operation needed in this case is for the child piconet PNC to request a private CTA as explained above. Once such a CTA is allocated by the parent piconet PNC, the child piconet PNC simply begins to send beacons at the beginning of the CTA, which is also the beginning of its own superframe. Devices that need to send data to the other piconet can simply request their own CTAs from their respective PNCs.

8.3.2 Operation of the Slave–Slave Bridge

A device that is already associated with a piconet can detect the presence of a new piconet by receiving a beacon sent by its PNC, or a control packet with a piconet identification number (PNCID) that is different from the existing one. Whenever a prospective bridge device detects the presence of two piconets within its transmission range, it initiates the connection establishment algorithm (Algorithm 1). First, the device waits for the MCTA period or CAP period to send a request command for bridging. Then, it will use the four-way handshake (RTS-CTS-DATA-ACK) to send the request command, piggybacking its current scheduling information to the neighbor PNC. The neighbor PNC adjusts its scheduling information based on the received scheduling information from its neighbor piconet. If the PNC is a master–slave bridge in its own right, it will request a private CTA from its parent PNC, trying to accommodate the demands of the bridge device. The bridge requirements are, simply, that the neighboring child piconets obtain channel time for transmission (i.e., private CTAs) without interfering with each other. A positive response from the parent PNC establishes the connection between the child piconets. After the connection establishment, the bridge device needs to maintain a table that keeps track of the scheduled times of activity in each piconet. The PNCID uniquely identifies each record in the table and helps the bridge device switch in a timely fashion between different piconets.

Algorithm 1:   The slave–slave bridge connection.

```
scan presence of overlapped coverage ;
if scan == positive then
  └ send join-request to neighbour PNC using four way handshaking protocol ;
  feedback from neighbour PNC ;
  update scheduling table with PNCID and received scheduling information ;
```

8.3.3 Channel Scheduling

The channel time scheduling of the network under consideration will be based on average queue size of the devices. The queue size of an ordinary device (i.e., not a PNC or a bridge device) primarily depends on packet arrival rate of that device. The queue size for a bridge device is based on the incoming packet queue sizes from neighbor piconets and outgoing packet queue sizes to the neighbor piconets. The bridge device will use the average of these two queues size to determine its channel time requirement. The devices send requests for channel time based on the average queue

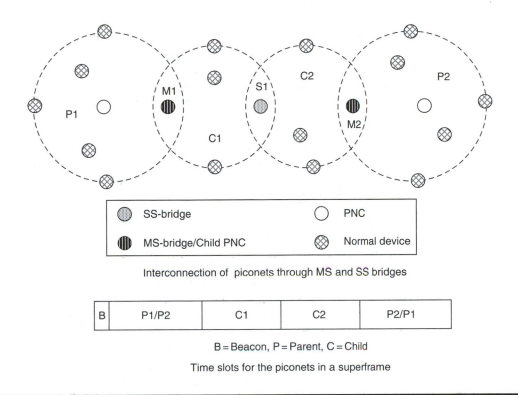

Interconnection of piconets through MS and SS bridges

B	P1/P2	C1	C2	P2/P1

B = Beacon, P = Parent, C = Child

Time slots for the piconets in a superframe

Figure 8.5 Multiple piconet interconnection with overlapped coverage and time slot management in the superframe.

size to their respective PNCs. The PNC uses Algorithm 2 based on the request from the bridge in question. In case of a request from a bridge device, the PNC schedules channel time and a private CTA (for the child piconet) such that there will be no overlap of channel time between the two adjacent piconets. A representative topology that employs both types of bridge interconnection is shown in Figure 8.5. In this network, the parent piconets P1 and P2 are located beyond each other's transmission ranges, and thus can operate on the same RF channel. However, the presence of two child piconets that can hear each other—they are, in fact, interconnected—presents a challenge for scheduling. To resolve this, the two parent piconets P1 and P2 will assign channel time for their children in different time slots, based on the scheduling information they exchanged during connection establishment. Let us consider time slots in the superframe in Figure 8.5. The time slots represented by P1/P2 (or P2/P1) imply that P1 and P2 can communicate at the same time. On the other hand, when a child piconet is operating, no other piconet in its range can talk. In this case, we can assume that a single superframe (actually two superframes from two different parent piconets) are divided into four time slots. Within each time slot, the devices will have guaranteed channel time and contention period. There are also MCTAs in each time slot during which a new node can join or a bridge can establish a connection. There is a chance of conflict during the MCTA period as the new devices do not have any knowledge of the current scheduling information resulting in the hidden terminal problem. We will use the Four-Way Handshaking Protocol to resolve this problem.

Algorithm 2: **Scheduling of channel time.**

```
if request command from master–slave bridge then
   assign private CTA anywhere in the superframe ;
if request command from slave–slave bridge then
     check piggyback data for neighbour scheduling information ;
     if no scheduling information then
        request for scheduling information ;
     scheduling information received ;
     calculate required channel time based on average queue size ;
     determine private CTA position ;
     assign private CTA and channel time ;
```

8.4 Related Work

The MAC protocols for WMNs are different from the traditional wireless MACs in terms of self-organization, distributed nature, multi-hop, and mobility. The WMNs can be designed with a single channel or multiple channels. For simplicity, we will focus on a single channel WMN in a parent–child interconnection.

WMNs have often been developed using the IEEE 802.11 DCF MAC protocol, and most of the research work in mesh networks has explored or modified the features of 802.11 MAC Protocol to improve network performance. Bicket et al. [7] have evaluated the performance of 802.11b mesh networks; their experiments have shown that an ad hoc mesh network implemented using 802.11b technology can achieve sustained throughput of around 630 Kbps, significantly below the supported data rate of 11 Mbps. By the same token, Yamada et al. [8] have identified two problems of 802.11b based mesh networks: limited throughput and degradation of fairness. To solve these problems they have introduced two new control packets, namely invite-to-send (ITS) and copied clear-to-send (CCTS). The use of ITS and CCTS leads to improvements in throughput, but at the cost of increased control overhead and delay. Also, the overhead due to ITS and CCTS packets and end-to-end packet delay will increase with the network load. However, in an 802.15.3 network, data communications are accomplished using dedicated periods; hence there is no need to introduce additional control packets such as ITS and CCTS.

MACA was developed to solve the hidden and exposed terminal problems of traditional CSMA [9] protocols. In MACA, the sender and receiver exchange RTS and CTS control packets before sending a data packet to avoid collisions. Fullmer and Garcia-Luna-Aceves [10] describe the scenario where MACA fails to avoid collisions due to hidden terminals. MACA may also make a device wait for a long period to access the medium because its use of the BEB* algorithm [11]. To overcome the problems of MACA, a new solution called Media Access protocol for wireless LANs (MACAW) was proposed by Bharghavan et al. [12]. Basically, MACAW is a modification of the BEB algorithm in MACA. It introduces acknowledgment and data-sending (DS) control packets producing the RTS-CTS-DS-DATA-ACK sequence for data transfer. The IEEE 802.11 standard [13] has been developed by adopting the CSMA and MACAW with further modifications to support wireless LANs. Both the IEEE 802.11 MAC and MACAW do not support real-time data transfer because of the absence of guaranteed time slots. Therefore, Lin and Gerla [14] proposed an

---

* In binary exponential backoff (BEB), a device doubles the size of its backoff window if a collision is detected.

enhanced version of MACA called MACA with piggybacked reservation (MACA/PR) to support real-time traffic.

The MACA/PR protocol is a contention-based protocol with a reservation mechanism. It has been designed to support multimedia data in multi-hop mobile wireless network providing guaranteed bandwidth through reservation. Every node keeps the reservation information of sending and receiving windows of all the neighbor nodes in a table, which is refreshed after every successful RTS-CTS-DATA-ACK (known as four-way handshaking protocol) cycle. The RTS and CTS packets are exchanged for the first packet in the transfer of a series of real-time data packets. The reservation information for the next data packet is piggybacked with the prior data packet and the receiver confirms this reservation in the acknowledgment control packet. The limitation of MACA/PR is that it requires help from the network layer routing protocol. However, MACA/PR has better performance in terms of latency, packet loss, mobility, and bandwidth share than both asynchronous packet radio network (PRNET*) and synchronous TDMA based MACs. The use of fixed reserved time slots in MACA/PR can result in wastage of bandwidth. Manoj and Murthy [15] have proposed a modification to the reservation mechanism of MACA/PR to prevent bandwidth wastage. In the modified scheme, the reserved slots can be placed at any position in the superframe and unused resources (channel time) are released after a successful transmission.

We note that the 802.15.3 MAC uses TDMA-based channel allocation to provide guaranteed time slots for data transfer. However, the piggybacked reservation information of MACA/PR can be employed together with the TDMA-based MAC to support real-time data transfer along with best-effort traffic in 802.15.3 based WMNs.

Xiao [16] has performed a detailed performance evaluation of the IEEE 802.15.3 [3] and IEEE 802.15.3a [17] standards through simulation and mathematical analysis. He has also done a throughput analysis of the 802.11 [13] protocol, which uses backoff with counter freezing during inactive portions of the superframe. The freezing and backoff techniques are essentially the same in the 802.11 and 802.15.3 MACs, except that different ways of calculating the backoff time are utilized. The backoff and freezing have an impact on the performance of the network; especially the backoff has a direct impact on the delay parameter. Large backoff windows can result in longer delays. On the other hand, small backoff windows may increase the probability of collisions. Xiao used the backoff procedures defined in the 802.11 and 802.15.3 MAC specifications; this work gives us performance of the protocol in terms of throughput over various payload sizes, but the performance of reliable transmission in error-prone wireless network during contention period needs more study.

8.5 Experimental Results

We have built a discrete event simulator of a two-piconet 802.15.3 network in a parent–child topology, using the Petri-net-based object-oriented simulation engine Artifex [18]. Different MAC parameters in our simulation have default values defined in the IEEE 802.15.3 standard, except where explicitly noted. In this master–slave architecture, the bridge device acts as the PNC for the child piconet, while acting as a normal member (DEV) of the parent piconet. The parent piconet has the PNC and nine devices, while the child piconet has the bridge (acting as PNC) and five

---

* In a PRNET, the devices use the same channel and share it dynamically.

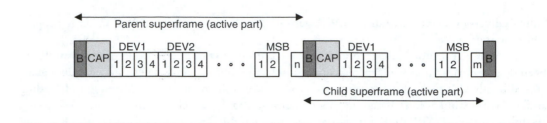

Figure 8.6 Superframe structure in the experimental setup.

devices. The bridge and the PNC of the parent piconet do not generate any packets: they simply receive and forward packets to their proper destinations. Ordinary devices generate packets according to a Poisson distribution for the overall data rate of 11 Mbps; another parameter, the locality probability P_l, determines the proportion of the packets that are sent to the destinations in the other piconet.

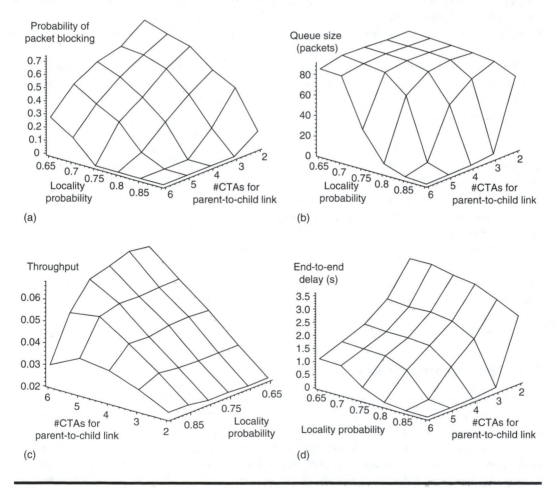

Figure 8.7 Performance of traffic sent from the parent piconet to the child piconet. (a) Packet blocking probability at the bridge; (b) mean queue size at the bridge; (c) bridge throughput relative to the rated capacity; and (d) end-to-end delay (in seconds).

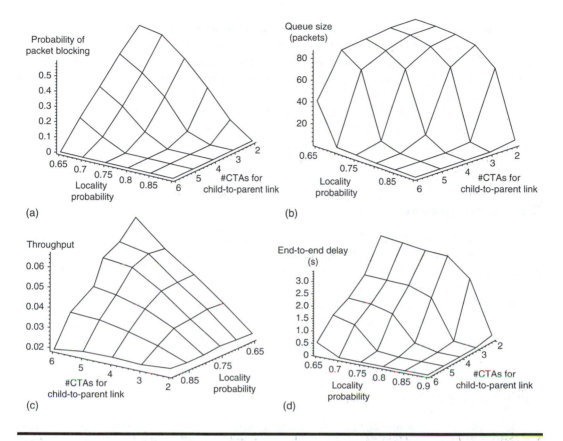

Figure 8.8 Performance of traffic sent from the child piconet to the parent piconet. (a) Packet blocking probability at the bridge; (b) mean queue size at the bridge; (c) bridge throughput relative to the rated capacity; and (d) end-to-end delay (in seconds).

The bridge maintains two queues, one for the uplink traffic (i.e., packets going from child piconet to the parent piconet) and the other for the downlink traffic (i.e., packets sent from the parent piconet to the child piconet). Each queue has a buffer of size 100 packets.

The beacon period is 100 μs, parent piconet CAP time is 3600 μs, and the child piconet CAP is 2000 μs long. The size of each CTA is 924 μs, which suffices for sending a single packet of the chosen packet size. Each of the ordinary devices gets four CTAs in each superframe, while the number of CTAs allocated to the bridge was made variable. Figure 8.6 shows the structure of the superframe.

Selected performance results for the downlink and uplink traffic are shown in Figures 8.7 and 8.8, respectively. In all the diagrams, independent variables were the locality probability and the number of CTAs allocated to the bridge for the appropriate traffic direction. For downlink traffic, Figure 8.7, the number of CTAs for uplink traffic was kept at 3. For uplink traffic, Figure 8.7, the number of CTAs for downlink traffic was kept at 4 (because the parent piconet has more devices than the child one).

As can be seen, the performance is critically dependent on the locality probability and the number of CTAs allocated to the traffic. In fact, the downlink (parent-to-child) traffic shows satisfactory performance only when the number of CTAs is above 5 and when less than 20 percent of the traffic is sent to the child piconet. If these conditions are not met, bridge buffer operates at high utilization ratio, above 80 percent, which leads to high blocking probability. The need to retransmit lost packets produces additional traffic and increases end-to-end packet delays. Similar observations hold for the uplink (child-to-parent) traffic, except that the region of normal traffic is slightly wider, due to the lower number of devices in the child piconet.

These results show that WMNs can efficiently be implemented using 802.15.3 high data rate WPAN technology, however, careful network design and the development of efficient scheduling algorithms are needed to achieve the full potential of this technology.

8.6 Conclusion

In this chapter, we have proposed interconnection schemes between 802.15.3 piconets and a networkwide scheduling policy to allocate channel time to the devices. We have discussed how the interconnection of piconets in mesh environment can affect scheduling when the bridge has overlapped coverage area with the same channel. Our proposed scheduling algorithm is simple and we expect that it will provide conflict-free communication and give good throughput and delay performance. Furthermore, proposed solution will help in the development of complex heterogeneous mesh network that can support mobility and dynamic topology change.

REFERENCES

1. I. F. Akyildiz, X. Wang, and W. Wang, Wireless mesh networks: A survey, *Computer Networks*, 47, 445–487, March 2005.
2. Wimedia Alliance: http://www.wimedia.org, 2006.
3. Standard for part 15.3: Wireless Medium Access Control (MAC) and Physical Layer (PHY) Specifications for High Rate Wireless Personal Area Networks (WPAN), IEEE, New York, IEEE Standard 802.15.3, 2003.
4. M.-G. Di Benedetto, T. Kaiser, A. F. Molisch, I. Oppermann, C. Politano, and D. Porcino, *UWB Communication Systems A Comprehensive Overview*. New York: Hindawi Publishing Corporation, 2006.
5. J. Mišic and V. B. Mišic, *Performance Modeling and Analysis of Bluetooth Networks; Polling, Scheduling, and Traffic Control*. Boca Raton, FL: Auerbach, 2005.
6. J. F. Kurose and K. W. Ross, *Computer Networking: A Top-Down Approach Featuring the Internet*. Boston, MA: Addison-Wesley Longman, 2000.
7. J. Bicket, D. Aguayo, S. Biswas, and R. Morris, Architecture and evaluation of an unplanned 802.11 b mesh network, 11th ACM International Conference on Mobile Computing and Networking, vol. 1, Cologne, Germany, September 2005, pp. 31–42.
8. A. Yamada, A. Fujiwara, and Y. Matsumoto, Enhancement of mesh network oriented IEEE 802.11MAC protocol, 10th Asia-Pacific Conference on Communications and 5th International Symposium on Multi-Dimensional Mobile Communications, vol. 1, Kanagawa, Japan, September 2004, pp. 142–146.
9. W. Stallings, *Data and Computer Communications*. Upper Saddle River, NJ: Prentice Hall, 2003.
10. C. L. Fullmer and J. J. Garcia-Luna-Aceves, Floor acquisition multiple access (FAMA) for packet-radio networks, ACM SIGCOMM '95, Cambridge, MA, September 1995, pp. 262–273.
11. C. S. Ram Murthy and B. Manoj, *Ad Hoc Wireless Networks, Architecture and Protocols*. Upper Saddle River, NJ: Prentice Hall, 2004.
12. V. Bharghavan, A. Demers, S. Shenker, and L. Zhang, MACAW: A media access protocol for wireless LAN's, ACM SIGCOMM '94, London, U.K., August 1994, pp. 212–225.

13. Standard for Wireless LAN Medium Access Control (MAC) and Physical Layer (PHY) Specifications, IEEE, New York, IEEE standard 802.11, May 1997.

14. C. R. Lin and M. Gerla, MACA/PR: An asynchronous multimedia multihop wireless network, IEEE INFOCOM '97, Kobe, Japan, March 1997, pp. 118–125.

15. B. S. Manoj and C. S. Ram Murthy, Real-time support for ad hoc wireless networks, IEEE ICON '02, Singapore, August 2002, pp. 335–340.

16. Y. Xiao, MAC layer issues and throughput analysis for the IEEE 802.15.3a UWB, *Dynamics of Continuous, Discrete and Impulsive Systems, Series B: Applications and Algorithms*, 12, 443–462, June 2005.

17. IEEE, IEEE 802.15 WPAN High Rate Alternative PHY Task Group 3a (TG3a). Available online at http://www.ieee802.org/15/pub/TG3a.html, August 2006.

18. RSoft Design, Inc., Artifex v.4.4.2, San Jose, CA, 2003.

Chapter 9

Quality of Service in Wireless Local and Metropolitan Area Networks

Haidar Safa and Mohamed K. Watfa

CONTENTS

Wireless technology has shown tremendous growth and acceptance as a solution for both wireless local area networks and wireless metropolitan area networks. The use of multimedia applications over IP with quality-of-service (QoS) support is now a reality in corporate networks and is rapidly expanding to the wireless networks. In this chapter, the state-of-the-art in supporting the QoS concepts in the IEEE 802.11-based wireless local area networks and the IEEE 802.16-based wireless metropolitan area networks is presented. The chapter is divided into two parts. The first part starts by describing the IEEE 802.11 standard that supports only best effort (BE) services before examining the new IEEE 802.11e that is introduced to support sophisticated services that guarantee QoS attributes such as bandwidth, delay, and jitter. The second part explores the QoS in the wireless metropolitan area networks as introduced in IEEE 802.16 standard and its IEEE 802.16e amendment. Open research issues pertaining to realizing QoS in these networks are identified and some of the solutions that are proposed to address these challenges are also presented.

9.1 Quality of Service in Wireless Local Area Networks

In this section we present the architectures, basic elements, and the QoS of wireless local area networks as introduced in IEEE 802.11 and IEEE 802.11e standards [1,2]. In this context, we describe two access mechanisms, the distributed coordination function (DCF) and the point coordination function (PCF), of the IEEE 802.11 medium access control (MAC) layer and the hybrid coordination function (HCF) access mechanism introduced in the IEEE 802.11e standard.

9.1.1 IEEE 802.11 Wireless Network Architectures

The IEEE 802.11 standard defines two basic architectures for wireless local area networks: Infrastructure-Based Architecture and Ad Hoc Architecture. In the Infrastructure-Based Architecture, the wireless network consists of access points (AP) and a set of mobile stations. The mobile stations and the AP that are within the same radio coverage form a basic service set (BSS) as shown in Figure 9.1a. If two stations in the same BSS want to communicate, then the communications flow from the source station to the AP and then from the AP to destination station. The BSS is the basic building block of IEEE 802.11 LAN. Several BSSs can be connected via a distribution system to form a single network called extended service set. The Ad Hoc Architecture contains no APs as shown in Figure 9.1b. In this architecture, mobile stations using the same frequency and situated in the transmission range of each other may form an independent BSS (IBSS) and communicate directly.

9.1.2 IEEE 802.11 MAC Layer

The IEEE 802.11 MAC layer defines two basic access mechanisms: the mandatory contention-based DCF, which offers an asynchronous data service, and the optional contention-free PCF that is built on top of the Carrier Sense Multiple Access with Collision Avoidance (CSMA/CA)-based

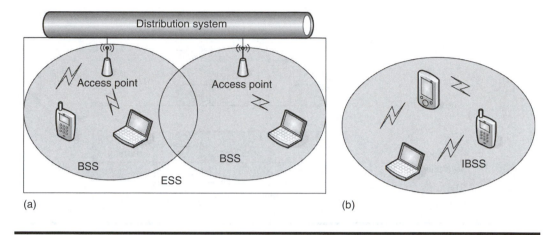

Figure 9.1 **(a) Architecture of 802.11 infrastructure-based network, (b) Architecture of 802.11 ad hoc network.**

DCF, as shown in Figure 9.2, to offer both asynchronous and time-bounded services. The DCF is based on the well-known CSMA/CA MAC access algorithm [1].

9.1.2.1 Carrier Sense Multiple Access with Collision Avoidance

The CSMA/CA Protocol is designed to reduce the collision probability between multiple stations accessing a shared medium [1]. A station wants to transmit senses if the medium is idle for a specific period. If it is, it starts transmitting. Otherwise, the station waits till the medium becomes idle again then resumes its operation as explained later.

The highest probability of a collision exists when the medium becomes idle following a busy medium because multiple stations could have been waiting for the medium to become idle again. This situation necessitates a random backoff procedure to resolve medium contention conflicts. Another type of collision may occur when two stations hidden from each other want to communicate with a third station. The two stations may sense an idle medium then transmit a frame that may cause a collision at the receiver. To avoid such collision, the CSMA defines two reservation information packets that announce the impending use of the medium prior to transmitting the actual data frame. These packets are the Request-to-Send (RTS) and the Clear-to-Send (CTS). The RTS and CTS frames contain a duration field that defines the period during which the medium is

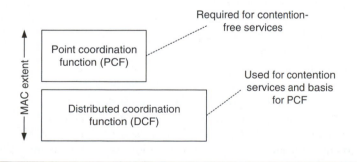

Figure 9.2 **Access mechanisms of MAC layer.**

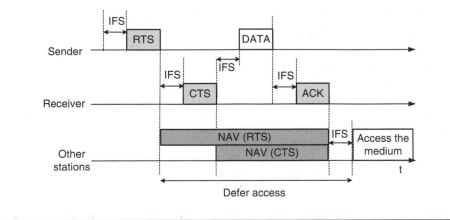

Figure 9.3 CSMA/CA.

to be reserved to transmit the actual data frame and the returning acknowledgment (ACK) frame. All stations within the reception range of either the station that transmits the RTS or the station that transmits the CTS learn of the medium reservation and set their network allocation vector (NAV) accordingly as shown in Figure 9.3. Thus, the NAV maintains a prediction of future traffic on the medium based on the duration information that is announced in the RTS/CTS frames.

In the example shown in Figure 9.3, a station that wants to transmit senses an idle channel for a period (inter-frame space [IFS] is explained in the next subsection) then transmits an RTS packet to the destination. All the nondestination neighboring stations that receive the transmission set their NAV to the duration announced in the RTS. The destination station replies to the RTS originator by transmitting a CTS packet after sensing an idle channel for a shorter period. After this transmission, the neighbors of the CTS sender will know about the time needed to complete the frame transmission and set their NAV accordingly. Then the data is transmitted and acked. All stations that have a NAV value larger than zero must refrain from using the medium. We may think about the NAV as a counter, which counts down to zero at a uniform rate. A NAV value of zero is an indication of an idle medium.

9.1.2.2 Inter-Frame Space

The minimum time interval between frames is called the IFS. A station determines that the medium is idle through the use of the carrier-sense function for the interval specified. Different IFSs are defined in IEEE 802.11 to provide priority levels for access to the wireless medium as shown in Figure 9.4; they are listed in order, from the shortest to the longest: short inter-frame space (SIFS), PCF inter-frame space (PIFS), and DCF inter-frame space (DIFS) [1].

The SIFS is used when a station has seized the medium and needs to keep it for the duration of the completion of the frame exchange sequence. Using the smallest gap between transmissions within the frame exchange sequence prevents other stations, which are required to wait for a longer gap for an idle medium, from attempting to use the medium, thus giving priority to complete the frame exchange sequence that is in progress. SIFS are mostly used for an ACK frame, a CTS frame, the second or subsequent fragment burst, and by a station responding to any polling by the PCF.

The PIFS is used only by stations operating under the PCF to gain priority access to the medium at the start of the contention-free period (CFP).

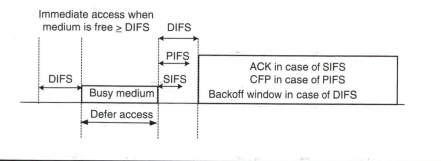

Figure 9.4 IFS types.

The DIFS is used by stations operating under the DCF to transmit data and management frames. A station using the DCF is allowed to transmit if its carrier-sense mechanism determines that the medium is idle for a DIFS period and its backoff time has expired.

9.1.2.3 Distributed Coordination Function

The DCF is the fundamental access method of the IEEE 802.11 MAC layer. It is implemented in all stations, for use within both infrastructure and ad hoc architectures [1]. The DCF access technique employs a contention window (CW)-based channel access function and uses the CSMA/CA Access Protocol to avoid collision in the the event of two or more stations attempting to transmit simultaneously. Under DCF, a station that intends to transmit must sense an idle medium for a DIFS period then selects a backoff timer (time slot) within a backoff window. The backoff timer is decreased only when the medium is idle; it is frozen when another station is transmitting, as shown in Figure 9.5. Each time the medium becomes idle, the station waits for a DIFS period then starts continuously decrementing the backoff timer. As soon as the backoff timer expires, the station is authorized to access the medium and transmit. The backoff timer is derived from a uniform distribution over the interval [0, CW], where CW, the contention window size, is a value between [CW_{min}, CW_{max}].

The set of CW values shall be sequentially ascending integer powers of 2, minus 1. At the very first transmission attempt, CW value is equal to the initial backoff window size CW_{min}. For every unsuccessful transmission, the value of CW   1 is doubled until CW_{max} is reached. After transmitting a frame, the station expects to receive an ACK from the destination station following SIFS time period. If the acknowledgment is not received, the sender assumes that the transmitted frame was collided, so it schedules a retransmission and enters the backoff process again. After every successful transmission, the CW is reset to CW_{min}.

The DCF employs a discrete time backoff scale. The time that immediately follows the DIFS is slotted and the station is permitted to transmit only at the beginning of each slot. The length of the slot is set equal to the time needed at any station to detect the transmission of a packet from any other station. More precisely, a slot time is defined in the standard as aCCATime   aRxTxTurnaroundTime   aAirPropagationTime   aMACProcessingDelay where aCCATime is the minimum time the clear channel assessment mechanism has available to assess the medium within every time slot to determine whether the medium is busy or idle; aRxTxTurnaroundTime is the maximum time the physical layer requires to change from receiving to transmitting the start of the first symbol; aAirPropagationTime is the anticipated time it takes a transmitted signal to go

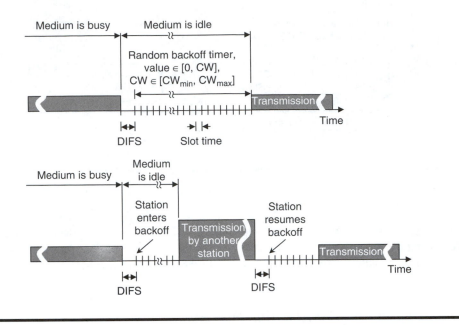

Figure 9.5 IEEE 802.11 DCF channel access.

from the transmitting station to the receiving station; aMACProcessingDelay is the nominal time that the MAC layer uses to process a frame and prepare a response to the frame.

9.1.2.4 Point Coordination Function

The PCF provides a time-bounded service and is especially utilized for asynchronous data, voice, and mixed applications (voice, data, video) [1]. It is a polling-based contention-free MAC access mechanism, used in a wireless local area network that operates in an infrastructure mode where APs are used as point coordinators (PC). The PCF is built on top of the CSMA/CA-based DCF access mechanism. It controls the frame transfers during a CFP. The CFP should alternate with the contention period (CP) in which the DCF controls the frame transfers as shown in Figure 9.6 [1]. Thus, PC allows contention and contention-free mechanisms to coexist. Each CFP should begin with a Beacon frame. The CFPs occur at a defined repetition rate that is synchronized with the beacon interval. The contention-free repetition rate (CFPRate) is defined as a number of delivery traffic indication message (DTIM) intervals where DTIM interval itself is a number of beacon intervals.

At the nominal beginning of each CFP, the PC senses the medium and gains control of it by waiting a PIFS period between transmissions. When the medium is determined to be idle for one PIFS period, the PC transmits a Beacon frame, which specifies the maximum time needed starting from the transmission of this beacon to the end of this CFP. All stations in the BSS set their NAVs to the duration value of the CFP. This prevents contention by preventing transmissions by other stations. The PC transmits a contention-free end frame at the end of each CFP. Stations that receive this frame reset their NAVs.

After the initial Beacon frame, the PC waits for a SIFS period, and then transmits a Data frame, a Polling frame, a contention-free end frame, or a combination of these frames. The Contention-Free Transfer Protocol is based on a polling scheme controlled by a PC operating at the AP of the

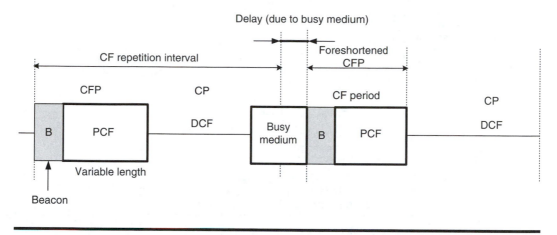

Figure 9.6   IEEE 802.11 PCF channel access.

BSS. During a CFP, the PC maintains a list of registered stations and polls them accordingly. A station can start transmitting only after it is polled. The size of each Data frame is bounded by the maximum MAC frame size. Stations receiving directed, error-free frames from the PC are expected to acknowledge the frame after a SIFS period. With PCF, stations are allowed to transmit even if the frame transmission cannot finish before the start of the next CF repetition interval. The duration of the beacon to be sent defers the transmission of data frames during the next CFP, as shown in Figure 9.6.

9.1.3   QoS in IEEE 802.11e MAC Layer

The new IEEE 802.11e standard provides an HCF through the services of DCF as shown in Figure 9.7 [2]. The HCF combines and enhances aspects of the access methods to provide QoS-stations (QSTAs) with prioritized and parameterized QoS access to the wireless medium, while continuing to support non-QoS stations for BE transfer. The HCF is compatible with both DCF and PCF. It defines two medium access mechanisms: a contention-based channel access referred to as enhanced distributed channel access (EDCA), and controlled channel access referred to as HCF controlled channel access (HCCA).

9.1.3.1   Enhanced Distributed Channel Access

The IEEE 802.11 DCF access mechanism can support only the BE services [3,4]. In DCF mode, all the stations compete for the resources and channel with the same priorities. There is no differentiation mechanism to guarantee bandwidth, packet delay, and jitter for high-priority traffic or multimedia flows [5]. The EDCA is proposed in IEEE 802.11e to support prioritized QoS services [2]. It provides differentiated, distributed access to the wireless medium for QoS stations using eight different user priorities (UPs) that are shown in Table 9.1. The user priority is a value assigned to a data packet in the layers above the MAC layer to indicate how the packet is to be handled. At the MAC layer, EDCA introduces four different first-in first-out queues, called access categories (ACs), and multiple independent backoff entities that are shown in Figure 9.8 [2]. The eight traffic priorities are mapped into four queues (ACs) as shown in Table 9.1. Thus, four backoff entities exist in every 802.11e station. Each AC queue works as an independent DCF station and uses its own

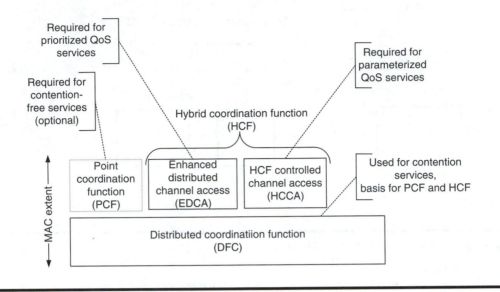

Figure 9.7   MAC architecture in 802.11e.

contention parameters such as CW_{min}, CW_{max}, and arbitrary inter-frame space (AIFS), as shown in Figure 9.8. The AIFS is introduced in EDCA in place of DIFS in DCF. Each AIFS has arbitration length that is computed as follows: AIFS[AC]   SIFS   AIFSN[AC] × slot_time where AIFSN[AC] is called the AIFS number.

Similar to a DCF station, each AC starts a backoff timer after detecting an idle channel for a time interval equal to an AFIS length. The backoff value is chosen to be a random number between 1, CW   1, where CW is initially set to CW_{min} and increases whenever collision occurs up to CW_{max}, CW increases in accordance with the following equation [5]:

$$CW_{new}[AC]   (CW_{current}[AC]   1) × 2 - 1$$

Table 9.1   User Traffic Priorities Mapped to ACs

User Priority from Lowest to Highest	Designation	Access Category
1	BK (Background)	AC_BK
2	BK (Background)	AC_BK
0	BE (Best-effort)	AC_BE
3	EE (Video/excellent-effort)	AC_BE
4	CL (Video/controlled load)	AC_VI
5	VI (Video)	AC_VI
6	VO (Voice)	AC_VO
7	NC (Network control)	AC_VO

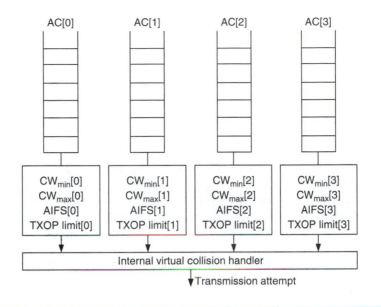

Figure 9.8 AC queues.

Whenever a station seizes the channel, it can transmit for a transmission opportunity time interval (TXOP). A TXOP is defined by its starting time and duration. The duration of TXOP is limited by a parameter referred to as TXOPlimit. In case of successful transmission, the CW value of the AC queue is reset to CW_{min}.

Additionally, the purpose of using different contention parameters for different queues is to give the low-priority traffic the longer waiting time than a high-priority trafic. Thus, the high-priority traffic has values for AIFS, CW_{max}, and CW_{min} smaller than those of the low-priority traffic as shown in Table 9.2. Consequently, the higher priority traffic will enter the CP and access the wireless medium earlier than the lower priority traffic.

Note that the backoff timers of different ACs in one QoS station are randomly generated and may reach zero simultaneously. This can cause an internal collision. In such a case, a virtual scheduler inside every QoS station, as shown in Figure 9.8, allows only the highest-priority AC to transmit frames [4].

Table 9.2 User Traffic Priorities Mapped to ACs

Access Category	AC VO	AC VI	AC BE	AC BO
AIFS	2	2	3	7
CW_{min}	7	15	31	31
CW_{max}	15	31	1023	1023
TXOPLimit	3.264	6.016	0	0

9.1.3.2 HCF Controlled Channel Access

The PCF mode of the IEEE 802.11 standard has some major problems that lead to poor QoS performance [3,4,6]. Indeed, the PCF defines only a single-class round-robin scheduling algorithm, which cannot handle the various QoS requirements of different types of traffic. In PCF, stations are allowed to transmit even if the frame transmission cannot finish before the start of the next CF repetition interval. This delays the following data frames. A polled station is allowed to send a frame of any length between 0 and 2304 bytes, which may introduce a variable transmission time.

The IEEE 802.11e HCCA was proposed as the contention-free part of HCF to overcome the limitations of the PCF [2]. Unlike PCF, HCCA stations are not allowed to transmit packets if the frame transmission cannot finish before the next beacon starts [2]. In addition, HCCA uses a $TXOP_{Limit}$ parameter to bound the transmission time of polled QoS stations. HCCA provides parameterized QoS support. It uses a QoS-aware centralized coordinator, called a hybrid coordinator (HC) that is collocated with the QoS AP and has a higher priority of access to the wireless medium. This allows HCCA to initiate frame exchange sequences and allocate transmission opportunities (TXOPs) to itself and other QoS stations.

The most important enhancement of the HCCA is the ability to provide a limited-duration CAP for contention-free transfer of QoS data during the CP. When the HC needs to access the wireless medium to start a CFP or a CAP in the CP, it should sense an idle wireless medium for one PIFS period. After an idle medium for PIFS period, the HC transmits the first permitted frame, with the duration value set to cover CFP or the CAP. The first permitted frame in a CFP is the Beacon frame, as shown in Figure 9.9. The HC can start a CAP during the CP by sending a poll or data frame after sensing an idle medium for a PIFS period. Because the PIFS length is shorter than the AIFS length (used by EDCA), the HC is able to interrupt the contention operation and generate a CAPs at almost any moment (with at most one packet length delay). The HC can start several CAPs after detecting a medium being idle for a PIFS period. To leave enough space for EDCA, the maximum duration for HCCA is limited by a $T_{CAPLimit}$ parameter. After the TXOP, the HC may reclaim the channel if the channel remains idle for a duration of PIFS. A CAP ends when the HC does not reclaim the channel after the end of a transmission opportunity.

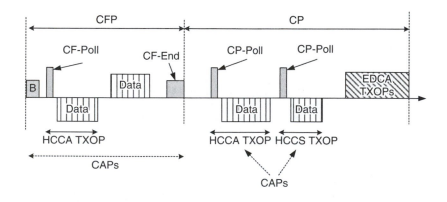

Figure 9.9 Controlled access phases in HCCA.

9.1.4 Current Challenges and Enhancements

The two access mechanisms of the HCF IEEE 802.11e have several drawbacks. The main drawback of the EDCA mechanism is that the values of CW_{min}, CW_{max}, and backoff time of each queue are static and do not take into account wireless channel conditions. The static reset method has been proved to be ineffective in maximizing channel utilization whenever the demand for channel access (e.g., traffic load) increases [6–8]. Indeed, the probability of collisions in a busy wireless channel is high and will likely cause CW to approach CW_{max} almost every time before the station succeeds in transmitting its data. After a successful transmission, and the next time the station wants to transmit a frame, CW will start at CW_{min} and will subsequently increase (double) at each unsuccessful step until the station succeeds in transmitting its packet. As a result, a station may have to try several times before it succeeds and hence, significant time is wasted due to resetting CW to CW_{min} and consequently, the channel becomes underused.

The HCCA of the 802.11e standard does not specify the scheduling discipline that determines when the controlled access phase are generated and leaves it to system developers to devise such a scheme [9]. In addition, the HCCA allocates transmission opportunities (TXOPs) to itself and to other QoS stations using the reported mean transmission rates. The HCCA scheduler allocates a fixed TXOP to each QoS station based on its mean rate requirements. When the transmitted flow is of variable bit rate (VBR) and its rate is larger than the mean transmission rate, the packets will be queued causing a delay increase. If the peak transmission rate of VBR applications are reported to the HC and used to calculate the TXOPs, the TXOPs will be large enough for delivering packets. However, the channel will be underutilized when a smaller number of VBR flows are admitted and the gaps between the peak and mean rates are considerable [5].

Several techniques have been proposed to enhance the performance of IEEE 802.11e by adapting CW to the network state [7,9–12]. In Ref. [12], a scheme is proposed for the purpose of improving the IEEE 802.11e performance under different load rates and increasing the service differentiation in EDCA-based networks. The scheme uses a dynamic procedure to change the CW value after a collision or a successful transmission by resetting CW to adaptive values that are different from the CW_{min}, taking into account their current sizes and the average collision rate. Furthermore, the scheme suggests changing the mechanism of doubling the CW when a collision occurs. In Ref. [7], an approach was proposed to replace the EDCA technique and is based on adapting the AIFS in response to network conditions. The rationale behind adapting the AIFS is to reduce the waiting time for the high-priority applications and increasing it for the low-priority ones. When the network is congested, the AIFS of the high-priority traffic is decreased while the AIFS of the BE traffic is increased. Conversely, if the network is in normal conditions, high-priority AIFS is increased while BE AIFS is decreased. It follows that this technique favors high-priority streams when the network is overloaded and tries to serve all traffic when the network is in normal conditions. A link adaptation strategy that provides differentiation not only at the MAC layer but also at the PHY layer was proposed in Ref. [10]. This strategy exploits the positive ACK procedure to evaluate the quality of the link. If the transmitter does not receive an ACK frame it concludes that the last transmission was failed. For each active link, a transmitter maintains two counters, a success counter and a failure counter. If a frame is successfully transmitted, then the success counter is incremented by one and the failure counter is reset to zero. If the transmission fails then the failure counter is incremented by one and the success counter is reset to zero. These two counters are used to determine whether the quality of the link is good or not. The transmission rate is adjusted with respect to the link quality. A low-complexity adaptive EDCA algorithm that adapts the CW to channel conditions and adjusts it depending on the network utilization and performance was proposed in Ref. [11]. The proposed technique outperforms IEEE 802.11e and is comparable to the other enhancement schemes while maintaining relatively low-complexity requirements and providing a faster adaptation to the network state. A new access scheduling framework designed to improve the HCCA access mechanism was proposed in Ref. [9]. This framework is capable of providing per-session QoS guarantees for interactive voice and video applications over WLAN. It provides guaranteed services to flows that make reservation with the WLAN AP by means of the available MAC signaling methods, while at the same time, allowing the normal contention-based access to take place using the remaining capacity of the channel. This approach is different from the existing polling mechanisms in which long alternating contention-free and CPs are generated, resulting in uncontrolled delay bounds and an inefficient operation.

9.2 Quality of Service in Wireless Metropolitan Area Networks

The WiMAX technology based on the IEEE 802.16 standard plays a key role in fixed broadband wireless metropolitan area networks [13]. It has proven to be a cost-effective wireless alternative to cabled access networks (i.e., fiber optic links, digital subscriber line [DSL]). This section gives an overview of the WiMAX networks and its QoS requirements as presented in the IEEE 802.16 standard and its IEEE802.16e amendment [13,14].

9.2.1 *WiMAX Network: Entities, Architecture, and Operation Modes*

To give an understanding of how QoS can be achieved in WiMAX networks, we first present the architecture of these networks, the main entities, and the operation modes. We then describe the data and scheduling services supported by the MAC layer and how resources are allocated.

9.2.1.1 *Mesh Mode versus Point-to-Multipoint Mode*

The basic architecture of WiMAX consists of two fixed stations: base station (BS) and subscriber station (SS). The BS is a central equipment set providing connectivity, management, and control of several SSs situated at varying distances. An SS can represent a building equipped with a conventional wireless or wired local area network. The IEEE 802.11 standard defines two operation modes: an optional mesh mode and a mandatory point-to-multipoint (PMP) mode [13]. The mesh mode supports direct communications between SSs without the need for a BS. Access coordination is distributed among the SSs. In PMP mode, a controlling BS connects multiple SSs to various public networks as shown in Figure 9.10.

9.2.1.1.1 PMP Mode

In PMP, the communication path is bidirectional, downlink (from the BS to the SS) and uplink (from the SS to the BS). The uplink and downlink data transmissions are duplexed using frequency division duplex (FDD) or time division duplex (TDD). In FDD, the uplink and downlink subframes occurs simultaneously on separate frequencies while in TDD they occur at different times but usually share the same frequency as shown in Figure 9.11. To schedule the uplink and downlink grants to meet the negotiated QoS requirements, the BS starts the downlink subframe with a downlink map (DL-MAP) followed by an uplink map (UL-MAP). The downlink MAP contains the

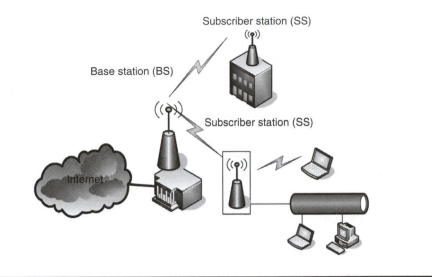

Figure 9.10   WiMAX in PMP mode.

timetable for downlink grants in the forthcoming downlink subframe. Downlink grants directed to SSs with the same downlink interval usage code (DIUC) are advertised in the DL-MAP as a single burst. The uplink MAP tells each SS about the boundaries of its allocated bandwidth within

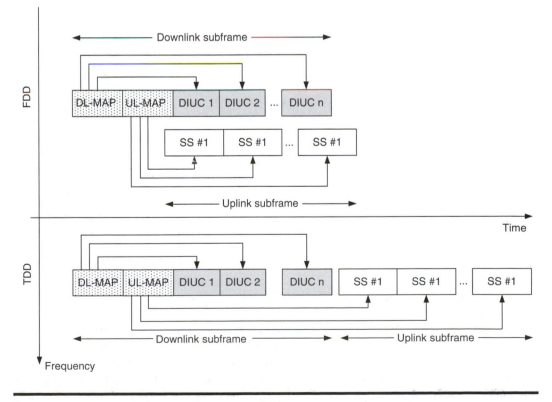

Figure 9.11   Frame structure with FDD and TDD.

the current uplink subframe. The SSs transmit in their assigned allocations using the burst profile specified in the uplink MAP entry granting them bandwidth. The downlink is generally broadcast. In cases where the DL-MAP does not explicitly indicate that a portion of the downlink subframe is for a specific SS, all SSs capable of listening to that portion of the downlink subframe shall listen. However, SSs check the received subframe and retain only the parts addressed to them. In addition to messages that are individually addressed, messages may also be sent on multicast connections (control messages and video distribution are examples of multicast applications) as well as broadcast to all stations. SSs share the uplink to the BS on a demand basis. Depending on the class of service utilized, the SS may be issued continuing rights to transmit, or the right to transmit may be granted by the BS after receipt of a request from the user.

9.2.1.1.2   Mesh Mode

In mesh mode, the traffic can be routed through other SSs and can occur directly between SSs. A system that has a direct connection to backhaul services outside the mesh network is termed a mesh BS. All the other systems of a mesh network are termed mesh SS. Within mesh context, the uplink and downlink are defined as a traffic in the direction of the mesh BS and traffic away from the mesh BS, respectively. The other three important terms of mesh systems are neighbor, neighborhood, and extended neighborhood. The stations with which a node has direct links are called neighbors. Neighbors of a node form a neighborhood. Nodes, neighbors are considered to be one hop away from the node. An extended neighborhood contains, additionally, all the neighbors of the neighborhood. Using distributed scheduling, all the nodes including the mesh BS coordinate their transmissions in their two-hop neighborhood to ensure that the resulting transmissions do not cause collisions with the data and control traffic scheduled by any other node. Using centralized scheduling, resources are granted in a more centralized manner. The mesh BS gathers resource requests from all the mesh SSs within a certain hop range. It then determines the amount of granted resources for each link in the network and communicates these grants to all the mesh SSs within the hop range. All the communications are in the context of a link, which is established between two nodes. One link is used for all the data transmissions between the two nodes.

The rest of this chapter focuses on PMP mode because it is anticipated that providers will use it to connect customers to the Internet [15].

9.2.1.2   *Addressing and Connections*

Each SS has a 48-bit universal MAC address that uniquely defines its air interface and serves mainly as an equipment identifier [14]. This address is used during the initial ranging process to establish the appropriate connections for an SS and is also used in the authentication process between the BS and the SS.

The MAC layer of the IEEE 802.16 is a connection-oriented layer. All data services are in the context of a connection. A connection is defined as a unidirectional mapping between the MAC peers of the BS and the SS. The MAC defines two kinds of connections: management connections that are used for the purpose of transporting management messages or standard-based messages and transport connections that are used to transport user data. Connections are identified by a 16-bit connection identifier (CID). At the SS initialization, two pairs of management connections (uplink and downlink) are established between the SS and the BS and a third pair may be optionally generated. These three pairs of management connections reflect the fact that there are inherently three different levels of QoS for managing traffic between an SS and the BS. These connections are (1) the basic connection that is used by the BS MAC and the SS MAC to exchange short, time-urgent MAC management messages; (2) the primary management connection that is used by the BS MAC and the SS MAC to exchange longer, more delay-tolerant MAC management messages such as those used for authentication and connection setup; (3) the optional secondary management connection that is used by the BS and the SS to transfer delay tolerant, standard-based (Dynamic Host Configuration Protocol [DHCP], Trivial File Transfer Protocol [TFTP], SNMP, etc.) messages. In addition, the IEEE 802.16 standard defines another two management connections: the broadcast connection that is used by the BS to send MAC management messages on a downlink to all SSs and the initial ranging connection that is used by the SS and the BS during the initial ranging process. The initial ranging connection is identified by a well-known constant value within the protocol because an SS has no addressing information available until the initial ranging process is complete.

9.2.1.3 Data and Scheduling Services

Scheduling services represent the data handling mechanisms supported by the MAC scheduler for data transport on a connection. Each connection is associated with a single scheduling service. A scheduling service is determined by a set of QoS parameters that quantify aspects of its behavior. These parameters are managed using dynamic service messages that allow the BS and the SS to add, modify, or delete the characteristics of a service flow. The key QoS parameters are [13]

- Traffic priority: specifies the priority assigned to a service flow. Given two service flows identical in all QoS parameters besides priority, the higher priority service flow should be given lower delay and higher buffering preference.
- Maximum sustained traffic: defines the peak information rate of the service and is expressed in bits per second. Data units deemed to exceed the maximum sustained traffic rate may be delayed or dropped.
- Maximum traffic burst: describes the maximum continuous burst that the system should accommodate for the service.
- Minimum reserved traffic rate: specifies the minimum amount of data to be transported on behalf of the service flow when averaged over time. It is expressed in bits per second. The specified rate is only honored when sufficient data is available for scheduling. The BS and the SS are able to transport traffic up to its minimum reserved traffic rate. If less than the minimum reserved traffic rate is available for a service flow, the BS and the SS may reallocate the excess reserved bandwidth for other purposes.
- Tolerated jitter: defines the maximum delay variation (jitter) for the connection.
- Maximum latency: specifies the maximum latency between the reception of a packet by the BS or SS on its network interface and the forwarding of the packet to its radio frequency Interface.

Well-known scheduling services can be implemented by specifying a specific set of QoS parameters. Four scheduling services are supported [14]:

1. Scheduling service to support real-time data streams consisting of fixed-size data packets issued at periodic intervals, such as Voice-over-IP (VoIP) without silence suppression. The key QoS parameters are the maximum sustained traffic rate, the maximum latency, and the tolerated jitter.

2. Scheduling service to support real-time data streams consisting of variable-sized data packets that are issued at periodic intervals, such as moving pictures experts group (MPEG) video. The key QoS service flow parameters for this scheduling service are the minimum reserved traffic rate, the maximum sustained traffic rate, and the maximum latency.

3. Scheduling service to support delay-tolerant data streams consisting of variable-sized data packets for which a minimum data rate is required, such as FTP. The key QoS service flow parameters for this scheduling service are the minimum reserved traffic rate, the maximum sustained traffic rate, and the traffic priority.

4. Scheduling service to support data streams for which no minimum service level is required and therefore may be handled on a space-available basis. The key QoS service flow parameters for this scheduling service are the maximum sustained traffic rate and the traffic priority.

9.2.1.3.1 Uplink Request/Grant Scheduling

The uplink request/grant scheduling is performed by the BS with the intent of providing each subordinate SS with bandwidth for uplink transmissions or opportunities to request bandwidth. By specifying a scheduling type and its associated QoS parameters, the BS scheduler can anticipate the throughput and latency needs of the uplink traffic and provide polls or grants at the appropriate time. There are five uplink scheduling algorithms [14]:

1. Unsolicited grant service (UGS) scheduling algorithm is designed to support real-time uplink service flows that transport fixed-size data packets on a periodic basis such as VoIP. The BS provides data grant burst to the SS at periodic intervals based upon the maximum sustained traffic rate of the service flow. The size of these grants should be sufficient to hold the fixed-length data associated with the service flow but may be larger at the discretion of the BS scheduler. The grant size and period are negotiated in the session initialization process. This eliminates the overhead and latency of SS bandwidth requests (BW-REQ) and assures that grants are available to meet the flow's real-time needs.

2. Real-time polling service (rtPS) scheduling algorithm is designed to support real-time uplink service flows that transport variable size data packets on a periodic basis such as MPEG video. The service offers real-time, periodic, unicast BW-REQ opportunities, which allow the SS to specify the size of the desired grant. Thus, this service requires more request overhead than UGS, but supports variable grant sizes for optimum data transport efficiency. The BS provides unicast request opportunities. For this service to work correctly, the SS is prohibited from using any contention request opportunities for that connection. The BS may issue unicast request opportunities as prescribed by this service even if prior requests are currently unfulfilled. This results in the SS using only unicast request opportunities and data transmission opportunities to obtain uplink transmission opportunities.

3. Extended rtPS (ertPS) scheduling algorithm is designed to support real-time service flows that generate variable size data packets on a periodic basis such as VoIP. This scheduling mechanism, recently proposed in IEEE 802.16e, builds on the efficiency of both UGS and rtPS. The BS provides unicast grants in an unsolicited manner like in UGS, thus saving the latency of a BW-REQ. However, whereas UGS allocations are fixed in size, ertPS allocations are dynamic. The SS requests the bandwidth using extended piggyback request (PBR) bits of the grant management subheader. The BS provides periodic uplink allocations according to the requested size until the SS requests a different size of bandwidth. When the SS data rate increases, SS requests the bandwidth using BR (BW-REQ) bits of the bandwidth request

header. The BS assigns uplink bandwidth according to the requested size periodically. The BS does not change the size of uplink allocations until receiving another bandwidth change request from the SS.

4. Non-real-time polling (nrtPS) uplink scheduling is designed to support delay-tolerant data streams consisting of variable-sized data packets for which a minimum data rate is required such as FTP. The nrtPS offers unicast polls on a regular basis, which assures that the uplink service flow receives request opportunities even during network congestion. The BS typically polls nrtPS CIDs on an interval on the order of one second or less. The BS provides timely unicast request opportunities. For this service to work correctly, the SS is allowed to use contention request opportunities as well as unicast request opportunities and data transmission opportunities.

5. BE scheduling is designed to support data streams for which no minimum service level is required. The intent of the BE grant scheduling type is to provide an efficient service for BE-traffic in the uplink. For this service to work correctly, the SS is allowed to use contention request opportunities as well as unicast request opportunities and data transmission opportunities.

9.2.1.4   Bandwidth Allocation and Request Mechanisms

In IEEE 802.16, all packets from the application layer in the SS are classified by the connection classifier based on the CID and are forwarded to the appropriate queue as shown in Figure 9.12. According to the incoming traffic service flow, the SS sends BW-REQ messages that report the current queue size of each SS connection to the BS uplink bandwidth allocation scheduler, which controls all the uplink packet transmissions. The BS schedules the requests in the different service flow queues according to their QoS requirements and generates a MAP message to the SS scheduler. The MAP message contains the information element (IE) parameter, which includes the time slotsin which the SS can transmit during the uplink subframe. The three management connec-

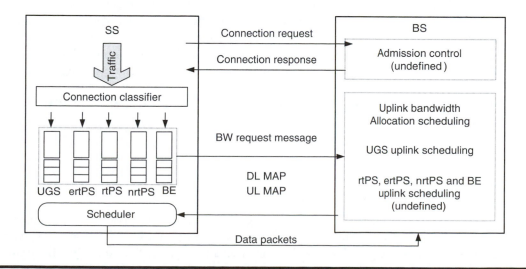

Figure 9.12   QoS architecture of the IEEE 802.16.

tions that are assigned to the SS during the initialization process are used for sending and receiving BW-REQ messages and other control messages. These connections allow differentiated levels of QoS to be applied to the different connections carrying MAC management traffic. Increasing (or decreasing) bandwidth requirements is necessary for all services except the incompressible constant bit rate UGS connections. There are numerous methods by which the SS can get the BW-REQ message to the BS.

Requests: Requests refer to the mechanism that SSs use to indicate to the BS that they need uplink bandwidth allocations. A request may come as a standalone BW-REQ header or may come as a PBR. BW-REQs are made in terms of the number of bytes needed to carry the MAC header and the payload. The BW-REQ message may be transmitted during any uplink allocation, except during the initial ranging interval. SS bandwidth requests reference individual connections. They may be incremental or aggregate. When the BS receives an incremental BW-REQ, it adds the quantity of the requested bandwidth to its current perception of the connection bandwidth needs. When the BS receives an aggregate BW-REQ, it replaces its perception of the connection bandwidth needs with the quantity of the requested bandwidth. Capability of incremental BW-REQs is optional for the SS and mandatory for the BS. Capability of aggregate BW-REQs is mandatory for SS and BS.

Grants: The BS grants bandwidth resources to the SS, not to individual CIDs. When the SS receives a grant shorter than expected (scheduler decision, request message lost, etc.), no explicit reason is given. On the basis of the latest information received from the BS and the status of the request, the SS may decide to perform backoff and request again or to discard the transmission.

Polling: Polling is the process by which the BS allocates bandwidth to the SSs for making BW-REQs. These allocations may be to individual SSs (unicast polling) or to groups of SSs (multicast polling). The allocations are not in the form of an explicit message, but are contained as a series of IEs within the uplink MAP. When an SS is polled, no explicit message is transmitted to poll the SS. Rather, the SS is allocated in the uplink MAP, bandwidth sufficient to respond with a BW-REQ. If insufficient bandwidth is available to individually poll many inactive SSs, some SSs may be polled in multicast groups or a broadcast poll may be issued. Certain CIDs are reserved for multicast groups and for broadcast messages. An SS belonging to the polled group may request bandwidth during any request interval allocated to that CID in the UL-MAP. To reduce the likelihood of collision with multicast and broadcast polling, only SSs needing bandwidth reply; they should apply the contention resolution algorithm to select the slot in which the initial BW-REQ is to be transmitted.

Note that polling is done on SS basis. Bandwidth is always requested on a CID basis and is allocated on an SS basis.

9.2.2 IEEE 802.16 QoS Architecture

The IEEE 802.16 standard defines several QoS related mechanisms: (1) service flow QoS scheduling; (2) dynamic service establishment; and (3) two-phase activation model. These concepts are used to support QoS for both uplink and downlink traffic through the SS and the BS. The principal mechanism for providing QoS is to associate packets traversing the MAC interface into a service flow as identified by the transport connection.

9.2.2.1 Service Flow QoS Scheduling

A service flow is a MAC transport service that provides unidirectional transport of packets either to uplink packets transmitted by the SS or to downlink packets transmitted by the BS. A service flow is partially characterized by several attributes that include details of how the SS requests uplink bandwidth allocations and the expected behavior of the BS uplink scheduler. Some of these attributes are

- Service flow ID (SFID) that is assigned to each existing service flow to serve as the principal identifier for the service flow between a BS and an SS.
- Connection ID (CID) of the transport connection exists only when the service flow is admitted or active. The relationship between the SFID and the transport CID is unique. An SFID is associated with only one transport CID, and a transport CID is associated with only one SFID.
- Admitted QoS parameters that define QoS parameters for which the BS (and possibly the SS) are reserving resources.
- Active QoS parameters that specify a set of QoS parameters defining the service actually being provided to the service flow. Only an active service flow may forward packets.

It is useful to think of three types of service flows:

1. Provisioned: A service flow may be provisioned but not immediately activated (sometimes called deferred).
2. Admitted: This type of service flow has resources reserved by the BS for its admitted QoS parameters, but these parameters are not active. Admitted service flows may have been provisioned or may have been signaled by some other mechanisms.
3. Active: This type of service flow has resources committed by the BS for its active QoS parameters.

Service flows are first admitted, then activated. An authorization module in the BS approves or rejects a request regarding a service flow. The authorization module can activate a service flow immediately or defer activation to a later time. Every change to the service flow QoS parameters should be approved by the authorization module. This includes every Dynamic Service message to activate, update, or delete an existing service flow.

9.2.2.2 Dynamic Service Flow Establishment

Creation of service flows may be initiated by either the SS (optional capability) or the BS (mandatory capability). In the SS-initiated protocol, an SS wishing to create either an uplink or downlink service flow sends a request to the BS using a Dynamic Service Addition Request (DSA-REQ) message (Figure 9.13a). The BS checks the integrity of the message and, if the message is intact, sends a Dynamic Service Received (DSX-RVD) response message to the SS. The BS checks the SSs authorization for the requested service and whether the QoS requirements can be supported, generating an appropriate response using a DSA Response (DSA-RSP) message indicating acceptance or rejection. The SS concludes the transaction with a DSA Acknowledgment (DSA-ACK) message.

In the BS-initiated protocol, a BS wishing to establish either an uplink or a downlink dynamic service flow with an SS checks first the authorization of the destination SS for the requested service

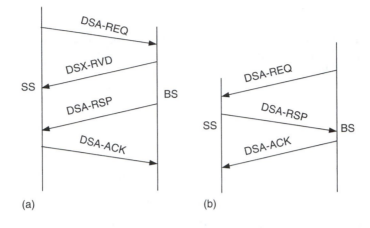

Figure 9.13 (a) SS-initiated protocol; (b) BS-initiated protocol.

flow and determines whether the QoS requirements can be supported. If the service can be supported, the BS generates a new SFID with the required class of service and informs the SS using a DSA-REQ message (see Figure 9.13b). If the SS checks that it can support the service, it responds using a DSA-RSP message. The transaction completes with the BS sending a DSA-ACK message.

In addition to the methods for creating service flows, protocols are defined for modifying and deleting service flows. The Dynamic Service Change (DSC) set of messages is used to modify the flow parameters associated with a service flow. Specifically, DSC can modify the service flow specification. Implementation of the DSC initiated by BS is mandatory while it is optional by SS.

A single DSC message exchange can modify the parameters of either one downlink service flow or one uplink service flow. The BS controls the uplink scheduling, the downlink scheduling, and the downlink transmit behavior. The BS always changes scheduling on receipt of a DSC-REQ (SS-initiated transaction) or DSC-RSP (BS-initiated transaction). The timing of scheduling changes is independent of direction and whether it is an increase or decrease in bandwidth. The change in the downlink transmit behavior is always coincident with the change in downlink scheduling as BS controls both.

The SS controls the uplink transmit behavior. The timing of SS transmit behavior changes is a function of which device initiated the transaction and whether the change is an increase or decrease in bandwidth. If an uplink service flows bandwidth is being reduced, the SS reduces its payload bandwidth first and then the BS reduces the bandwidth scheduled for the service flow. If an uplink service flows bandwidth is being increased, the BS increases the bandwidth scheduled for the service flow first and then the SS increases its payload bandwidth.

Any service flow can be deleted with the the Dynamic Service Delete (DSD) messages. When a service flow is deleted, all resources associated with it are released.

9.2.2.3 Activation Model

The IEEE 802.16 standard supports a two-phase activation model that is often utilized in telephony applications. In the two-phase activation model, the resources for a call are first admitted, and then once the end-to-end negotiation is completed the resources are activated. The two-phase model serves the following purposes:

- Conserving network resources until a complete end-to-end connection has been established.
- Performing policy checks and admission control on resources as quickly as possible, and in particular, before informing the far end of a connection request.
- Preventing several potential theft-of-service scenarios.

9.2.3 Undefined QoS Requirements, Challenges, and Enhancements

Three components are necessary to manage QoS in the IEEE 802.16 standard. These are (1) admission control, which determines whether a new request for a connection can be granted or not according to the remaining free bandwidth; (2) scheduling, which determines which packet will be served first to guarantee QoS requirements; and (3) buffer management, which controls buffer size and decides which packets to drop. IEEE 802.16 defines the signaling mechanism for information exchange between the BS and the SS such as connection setup, BW-REQ, and MAP messages. It defines also the UGS uplink scheduling to support real-time data streams consisting of fixed-size data packets issued at periodic intervals. However, IEEE 802.16 does not define the uplink rtPS and ertPS scheduling to support real-time uplink service flows that transport variable size data packets on a periodic basis (see Figure 9.12). IEEE 802.16 does not define also the nrtPS and BE uplink scheduling. In addition, the admission control in the BS is left undefined as well.

In IEEE 802.16, service data units deemed to exceed the maximum sustained traffic rate may be delayed or dropped. However, the standard does not define or recommend any algorithm for measuring whether a flow exceeds its maximum sustained traffic rate.

The QoS mechanisms of the IEEE 802.11-based networks are difficult to apply on the IEEE 802.16 networks due to the difference in the nature of these two technologies. Indeed, the IEEE 802.11 MAC is a connectionless and a contention-based technology in which the MAC uses acknowledgments and timeouts, which may cause overhead and delays. However, the IEEE 802.16 is a connection-oriented protocol that uses service flows. In IEEE 802.16, overhead and delays between users are eliminated because of its grant-based nature that does not require the use of acknowledgment and timers like SIF, PIFS, DIFS, and AIFS of the IEEE 802.11. This allows a better QoS handling. In addition, IEEE 802.11 has a fixed channel size while the channel size is changeable in IEEE 802.16.

Several architectures were recently proposed [16–22] to support QoS in WiMAX. Most of these proposals aim to complete the missing parts of the IEEE 802.16 QoS architecture. In Ref. [16], a two-layer scheduling structure of the bandwidth allocation was proposed to support all types of service flows. In the first layer, the deficit fair priority queue (DFPQ) was used to distribute total bandwidth among flow services in different queues. Six queues were defined according to their direction (uplink or downlink) and service classes. In the second layer scheduling, packets within each of the six queues will be served according to a certain scheduling algorithm. For rtPS connections, packets with earliest deadline will be scheduled first. The information module determines the packets' deadline that is calculated by its arrival time and maximum latency. For nrtPS connections, packets are scheduled based on their weight, which is defined as ratio between a connection's nrtPS minimum reserved traffic rate and the total sum of the minimum reserved traffic rate of all nrtPS connections [23]. For BE connections, the remaining bandwidth is allocated to each BE connection using round robin [24]. Because UGS will be allocated fixed bandwidth in transmission, their bandwidths will be directly cut before each scheduling. In Ref. [17], a preemption-based variation of the scheduling algorithm presented in Ref. [16] was proposed. The proposed scheme focuses on giving rtPS service flow packets more chances to meet their deadline and decrease their delay to better guarantee the QoS requirements of this class. Indeed, in addition to checking if the available bandwidth is enough for granting requests, information related to the rtPS service flows that are admitted are tracked by maintaining a table that is used to approximate the expected delay of each rtPS connection. These delays are used later in the scheduling algorithm.

9.3 Summary

This chapter presented the QoS concepts in wireless local area networks (i.e, WiFi/IEEE 802.11) and wireless metropolitan area networks (i.e., WiMAX/IEEE 802.16). It described the BE services IEEE 802.11 standard and the IEEE 802.11e standard that was introduced to support sophisticated services that guarantee QoS attributes such as bandwidth, delay, and jitter. In this context, the architectures, basic elements, the DCF MAC access mechanism, the PCF access mechanism, and the HCF access mechanism were described. The chapter also examined the QoS in the wireless metropolitan area networks as introduced in the IEEE 802.16 standard and its IEEE 802.16e amendment. In this context, the WiMAX Architecture, operation modes, scheduling services and algorithms, resource allocation, and QoS requirements were explored. Throughout the chapter, the QoS aspects of the wireless local and metropolitan area networks were addressed. Current challenges and drawbacks of the IEEE 802.11 and IEEE 802.16 were highlighted and some proposed enhancements were surveyed.

REFERENCES

1. IEEE Std 802.11, Wireless LAN Medium Access Control (MAC) and Physical Layer (PHY) Specification, 1999 edition
2. IEEE Std 802.11e-2005 IEEE Standard for Information Technology—Telecommunications and Information Exchange between Systems—LAN/MAN Specific Requirements Part 11: Wireless LAN Medium Access Control (MAC) and Physical Layer (PHY) Specifications Amendment 8: Medium Access Control (MAC) Quality of Service Enhancements.
3. S. Mangold, S. Choi, P. May, G. Hiertz, O. Klein, and B. Walke, Analysis of IEEE 802.11e for QoS support in wireless LAN, *IEEE Wireless Communications*, December 2003.
4. Q. Ni, L. Romdhani, and T. Turletti, A survey of QoS enhancements for IEEE 802.11 wireless LAN, *Wireless Communications and Mobile Computing*, 4, 2004.
5. D. Gu and J. Zhang, QoS enhancement in IEEE 802.11 wireless area networks, IEEE *Communications Magazine*, 41(6), June 2003.
6. Q. Ni, Performance analysis and enhancements for IEEE 802.11e wireless networks, *IEEE Network*, 19(4), 21–27, July–August 2005.
7. A. Ksentini, M. Naimi, A. Nafss, and M. Gueroui, Adaptive service differentiation for QoS provisioning in IEEE 802.11 wireless ad hoc networks, Proceedings of the 1st ACM International Workshop on Performance Evaluation of Wireless Ad Hoc, Sensor, and Ubiquitous Networks, Venezia, Italy, 2004.
8. J. Naoum-Sawaya, B. Ghaddar, S. Khawam, H. Safa, H. Artail, and Z. Dawy, Adaptive approach for QoS support in IEEE 802.11e wireless LAN, IEEE International Conference on Wireless and Mobile Computing, Networking and Communications, 2005 (WiMob'2005), Montreal, Canada, August 22–24, 2005.
9. Y. Fallah and H. Alnuweiri, Hybrid polling and contention access scheduling in IEEE 802.11e WLANs, *Journal of Parallel and Distributed Computing*, 67, 2007.
10. M. Bandinelli, F. Chifi, R. Fantacci, D. Tarchi, and G. Vannuccini, A link adaptation strategy for QoS support in IEEE 802.11e-based WLANs, IEEE Wireless Communications and Networking Conference, New Orleans, Louisiana March 2005.

11. H. Artail, H. Safa, J. Naoum-Sawaya, B. Ghaddar, and S. Khawam, A simple recursive scheme for adjusting the contention window size in IEEE 802.11e wireless ad hoc networks, *Computer Communications*, 29(18) 18, November 2006.

12. L. Romdhani, Q. Ni, and T. Turletti, Adaptive EDCF: Enhanced service differentiation for IEEE 802.11 wireless ad-hoc networks, IEEE Wireless Communications and Networking Conference (WCNC'03), New Orleans, Louisiana, 2003.

13. IEEE Std 802.16-2004, IEEE Standard for Local and Metropolitan Area Networks Part 16: Air Interface for Fixed Broadband Wireless Access Systems, 2004.

14. IEEE Std 802.16e-2005, IEEE Standard for Local and Metropolitan Area Networks Part 16: Air Interface for Fixed Broadband Wireless Access Systems, Amendment 2: Physical and Medium Access Control Layers for Combined Fixed and Mobile Operation in Licensed Bands, 2006.

15. A. Ghosh, D. Walter, J. Andrews, and R. Chen, Broadband wireless access with WiMax/802.16: Current performance benchmarks and future potential, *IEEE Communications Magazine*, February 2005.

16. J. Chen, W. Jiao, and H. Wang, A service flow management strategy for IEEE 802.16 broadband wireless access systems in TDD mode, Proceedings IEEE International Conference on Communications (ICC 2005), Seoul, Korea, May 2005.

17. H. Safa, H. Artail, M. Karam, R. Soudah, and S. Khayat A new scheduling architecture for IEEE 802.16 wireless metropolitan area network, Proceedings of the 5th ACS/IEEE International Conference on Computer Systems and Applications, AICCSA '2007, Amman, Jordan, May 2007.

18. H. Alavi, M. Mojdeh, and N. Yazdani, A QoS architecture for IEEE 802.16 standards, Proceedings of IEEE Asia Pacific Conference on Communications, Perth, Australia, October 2005.

19. J. Chen, W. Jiao, and Q. Guo, An integrated QoS control architecture for IEEE 802.16 broadband wireless access systems, Proceedings of IEEE Global Telecommunications Conference (GLOBECOM'05), St. Louis, Missouri, 2005.

20. D.-H Cho, J.-H Song, M.-S Kim, and K.-J Han An architecture for efficient QoS support in the IEEE 802.16 broadband wireless access network, Proceedings of 4th International Conference on Networking, April 2005.

21. G. Chu, D. Wang, and S. Mei, A QoS architecture for the MAC protocol of the IEEE 802.16 BWA system, Proceedings of IEEE Conference on Communications, Circuits, and Systems, St. Petersburg, Russia, 2002.

22. Y. Shang and S. Cheng, An enhanced packet scheduling algorithm for QoS support in IEEE 802.16 wireless network, Proceedings of 3rd International Conference on Networking and Mobile Computing, August 2005.

23. A. Demers, S. Keshav, and S. Shenker Analysis and simulation of a fair queuing algorithm, ACMSIG-COMM19, Austin, Texas, 1989.

24. E. L. Hahne and R. G. Gallager, Round robin scheduling for fair flow control in data communication networks, International Conference on Communications, June 1986.

Chapter 10

Fast MAC Layer Handoff Schemes in WLANs

Li Jun Zhang and Samuel Pierre

CONTENTS

10.1   Introduction

The explosion of lightweight handheld devices with built-in wireless network cards and the significant benefits of ubiquitous Internet access have driven the deployment of wireless local area networks (WLANs). On the basis of the IEEE 802.11 standard series, WLANs offer users an array of benefits such as user-friendly operations, low cost, large bandwidth, high throughput, etc., so that many multimedia application such as Voice-over-IP (VoIP), media-streaming services, etc., tend to run on top of WLANs. However, when a mobile station moves outside the radio range of its current access point (AP), handoff takes place to ensure the transfer of ongoing calls or data sessions. This handoff procedure involves a series of message exchanges between a mobile station and APs that results in unacceptable delays and packet lost. Fast Media Access Control (MAC) layer handoff schemes are thus required for IEEE 802.11 WLANs.

This chapter introduces state-of-the-art concepts to improve MAC layer handoff performance. Firstly, the components of the MAC layer handoff procedure in WLANs are introduced to demonstrate the necessity to enhance performance. Then, current approaches to reduce handoff latencies and packet loss rates are outlined in detail, along with their strengths and weaknesses. Open research issues pertaining to handoff management are exposed. Finally, potential solutions to address such challenges are proposed.

10.2   IEEE 802.11 Handoff Process

The IEEE 802.11 standard defines two operation modes: infrastructure and ad hoc. In the infrastructure mode, an AP comprises a basic service set (BSS) and provides network connectivity to its associated mobile stations. One or more APs comprise an extended service set (ESS) to cover a larger area. In the ad hoc mode, two or more mobile stations form a peer-to-peer wireless network without deploying any APs. Note that this chapter focuses only on the infrastructure mode.

An ideal WLAN can provide successive radio signal coverage for mobile stations in its service area. A mobile station may decide to handoff from one AP to another for mobility reasons, AP load balancing state, or signal fading. The legacy MAC layer (L2) handoff process specified in the IEEE 802.11 standard [1] comprises three phases: scanning, authentication, and reassociation. The latter two subprocesses are also referred as re-authentication. The following subsections investigate these three subprocesses in detail.

10.2.1   Scanning

Mobile stations can operate either in a passive or active scanning mode depending on their configuration parameters.

During passive scanning, mobile stations listen for periodic Beacon frames generated by APs, which announce their presence on each channel and wait for at least a full beacon interval to ascertain beacons receipt from as many APs as possible and at most a ChannelTime, on each channel. While scanning, mobile stations cannot transmit frames but they do rather listen for Beacon frames on each channel.

Active scanning involves generating Probe Request frames and the subsequent processing of received Probe Response frames. For each channel to be scanned, a mobile station uses active scanning to perform the following [1]:

1. Wait until the ProbeDelay time has expired or reception of an indication from the physical layer.
2. Perform wireless medium control using any normal channel access procedure, i.e., Carrier Sense Multiple Access with Collision Avoidance (CSMA/CA).
3. Broadcast a Probe Request frame with the broadcast destination, Service Set Identifier (SSID), and broadcast basic Service Set Identifier (BSSID).
4. Clear and start a ProbeTimer.
5. Clear network allocation vector (NAV) and scan the next channel if the wireless medium is idle before the ProbeTimer reaches MinChannelTime, otherwise, continue accepting Probe Responses sent periodically by APs within radio range until MaxChannelTime and process all received Probe Responses.
6. Clear NAV and scan the next channel.

As indicated above, the passive scanning delay is determined by the number of scanned channels, ChannelTime, and Beacon interval. The probe delay bound, T_p, can be expressed as the following:

$$N * \text{BeaconInterval} \leq T_p \leq N * \text{ChannelTime}$$

where ChannelTime refers to the maximal time during which a mobile station listens on each channel and N represents the number of channels available ($N = 32$ for 802.11a [2], $N = 11$ for 802.11b [3], and $N = 11$ for 802.11g [4]).

In the same vein, the active scanning delay depends on the number of probed channels, MinChannelTime and MaxChannelTime. The probe delay bound, T_a, can be expressed as follows:

$$N * \text{MinChannelTime} \leq T_a \leq N * \text{MaxChannelTime}$$

where N depicts the number of channels available; MinChannelTime and MaxChannelTime show the minimal and maximal probe-waiting time on each channel.

Generally, scanning ends with a set of potential BSSs. Furthermore, passive scanning delays are much longer than those generated by active scanning, because mobile stations are mandated to iterate on all available channels for beacons from APs in range at a set rate (default: 100 ms per beacon). In addition, mobile stations must dwell on each channel for at least a beacon interval to discover as many APs as possible.

10.2.2 Authentication

Authentication aims to identify a mobile station to become a member of a specific BSS, as well as to authorize this mobile station to communicate with other stations in the same BSS. Authentication occurs after a target AP is found. Two authentication methods have been specified for the IEEE 802.11 standard: open system and shared key authentication [1]. Open system authentication involves a pair of frames, an Authentication Request as well as an Authentication Response, which are exchanged between a mobile station and the target AP. Generally, all mobile stations can be authenticated.

Shared key authentication is an optional four-step process that uses the Wired Equivalent Privacy (WEP) key. A mobile station launches the authentication process by transmitting an Authentication Request to the target AP. Upon receiving this request, the target AP generates a challenge text using a WEP key, and sends an Authentication Response with this challenge text to the mobile station. The latter then encrypts the received challenge text with a shared WEP key and returns an Authentication Request along with the encrypted challenge text to the target AP. Later, the target AP decrypts this request with the shared key and compares the original challenge text with the decrypted one. When they are identical, the target AP transmits an Authentication Response, which confirms a successful authentication.

Regardless of the authentication method used, the IEEE 802.11 standard requires mutually acceptable, successful authentication. Furthermore, authentication is required before an association can be established. Owing to the current security flaws in open system and shared key authentication, the authentication methods specified in IEEE 802.11 are superseded by IEEE 802.11i [5]. However, considering compatibility, IEEE 802.11i allows open system authentication and exchanges authentication messages after the reassociation phase [5,6].

10.2.3 Reassociation and Association

Association consists of establishing AP and mobile station mapping and enabling station invocation of the distribution system services whereas reassociation enables an established association to be transferred from one AP to another [1].

Reassociation is an important component of L2 handoff after a successful authentication. Because the IEEE 802.11 standard specifies that each mobile station must be associated with a single AP at any given time [1] and a mobile station must issue a Reassociation Request to the new AP during handoff, this request frame contains the previous BSSID, the mobile stations MAC address, etc. and triggers the Inter-Access Point Protocol (IAPP) [7] to deliver relevant context information.

Upon receiving this frame, the new AP sends an Access-Request message to the RADIUS (Remote Authentication Dial-In User Service [8]) server, which then looks up the IP address of the previous AP and verifies the BSSID, before returning an Access-Accept message to the new AP. This message contains the previous AP's IP address along with security block items required to establish a secure communication channel between APs. After exchanging security elements through Send-Security-Block and ACK-Security-Block packets, both APs have obtained sufficient information to encrypt all further packets. Afterward, the new AP sends an encrypted MOVE-Notify packet to the prior AP asking for the context of the concerned itinerant station. Upon verifying the mobile station's association, the old AP removes the mobile station from its association table and replies an encrypted MOVE-Response packet to the new AP, including the concerned Context Block. Then, the new AP adds the mobile station into its association table and broadcasts a Layer 2 Update frame to inform any layer 2 devices, such as bridges and switches, so that they can update their forwarding table for the specific mobile station. At last, the new AP sends a Reassociation Response to the mobile station [7,9,10]. The overall handoff process is completed when this response is received.

Briefly, the IAPP allows an AP to communicate with other APs in a common ESS, while minimizing opportunities for the transmission of mobile stations' security information over the air. However, context transfer using IAPP results in additional delays during handoff. Figure 10.1 illustrates the overall handoff process for WLANs.

Numerous studies have been conducted to improve MAC layer handoff performance in terms of handoff delays (time required to complete scanning, authentication, and reassociation) and packet loss rates for mobile hosts roaming in IEEE 802.11 networks. This chapter builds on previous work in Refs. [6,11], focusing on the investigation of typical fast handoff schemes in recent literature. To simplify the analysis, two categories are investigated, namely those that reduce delays pertaining to probing and re-authenticating.

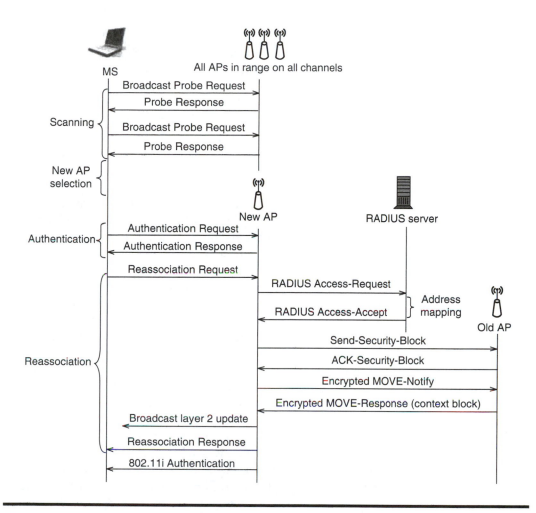

Figure 10.1 Handoff process for WLANs.

10.3 Fast Handoff Schemes to Reduce Probe Delays

Probe delays consist of the main contributor to the overall MAC layer handoff latency [12] and, most recently proposed handoff schemes aim to reduce this lengthy delay. These schemes can be further classified into fast scanning, bypass scanning, and cross-layer design.

10.3.1 Fast Scanning

Fast scanning methods rely on reducing the number of probed channels, the time taken on each channel, scanning-related timers, such as MinChannelTime and MaxChannelTime for active scanning, ChannelTime and Beacon Interval for passive scanning, etc. Such methods can be further classified into full scanning and selective scanning.

10.3.1.1 Full Scanning

Full scanning means that all available channels are probed while the values of MinChannelTime, MaxChannelTime, probe-waiting time, or Beacon Interval are optimized. Usually, full scanning is based on the assumption that mobile stations have no pre-knowledge of existing neighboring APs within their range when the handoff occurs and, as a result, all available channels must be searched consecutively. There are several full scanning methods. Here are some examples:

■ Tuning technique [13] aims to find an optimal value for MinChannelTime and MaxChannel-Time to reduce active scanning delays. Through rigorous calculations, the authors conclude that MinChannelTime and MaxChannelTime should be set to 1.024 and 10.24 ms, respectively. Furthermore, link-layer handoff detection delays can be reduced upon the loss of three consecutive frames.

■ Intelligent channel scanning [14] is designed to minimize the probe-waiting time on each channel. Instead of waiting for MaxChannelTime on a busy channel, mobile stations stop searching immediately once they have collected all available Probe Responses. Continuous monitoring techniques are thus used by mobile stations to evaluate the number of APs (N) on a probed channel. Accordingly, mobile stations stop scanning after collecting a maximum of N responses on a specific channel.

■ SyncScan [15] replaces the active scanning procedure with passive channel monitoring on nearby APs. Furthermore, a continuous tracking technique is devised by synchronizing short listening periods at mobile stations with regulated periodic beacon transmission from APs. As a result, mobile stations can passively scan by switching channels at the exact moment a beacon is about to arrive. To do so, a staggered periodic schedule of beacon periods is created and spread across channels. For example, all APs operating on Channel 1 are forced to broadcast Beacons at time T, APs on Channel 2 broadcast Beacons at time $(T + d)$, APs on Channel 3 broadcast Beacons at time $(T + 2d)$, and so on. Therefore, if a mobile station connected to an AP on Channel c receives Beacons from Channel c at time T_c, it can receive Beacons from APs on Channel $(c + 1)$ at time $(T_c + d)$. In a nutshell, SyncScan reduces the cost of continuous scanning and yields better handoff decisions. However, time synchronization is a critical issue among all neighboring APs. On the other hand, multiple APs operating on the same channel attempt to generate and broadcast Beacons simultaneously, hence bringing about beacon conflicts. Consequently, more collisions take place on wireless mediums. This side effect reduces productivity on wireless links and system throughput. In addition, more packets are lost while mobile stations explore other channels.

■ Smooth handoff schemes [16] focus on smooth channel scanning by classifying channels into different groups. Rather than scanning all channels consecutively, once a group of channels is scanned, the mobile station pauses before switching back to normal data transmission mode. This scheme works as a scheduled full scan.

10.3.1.2 Selective Scanning

Instead of probing all available channels individually, selective scanning reduces the number of channels required to discover APs. Thus, probe delays are significantly minimized compared to the full scanning method. A number of selective scanning approaches have been proposed in the literature. Here are some examples:

■ Channel mask scheme [17] allows mobile stations to selectively scan channels with a mask built in previous scans. See Section 10.3.4 below for details.

■ Neighbor graph (NG) method [18,19] allows mobile stations to scan only channels that neighbor APs operate on, which drastically reduces the number of probed channels. Together with the NG method, several schemes are proposed to minimize probe-waiting time on each neighboring channel, i.e., NG-pruning approach [18], unicast Probe Request with fast switching between each probed channel [19]. See Section 10.3.5 below for details regarding the NG and NG-pruning methods in Ref. [18]. A modified NG solution [19] is developed with a unicast Probe Request to a neighbor AP previously selected by an NG server. And to shorten probe-waiting time, mobile stations switch to another channel when they receive a Probe Response from the specific AP [19].

■ Handover assisted by geolocation information [20] aims to predict the next AP and the associated subnet using the mobile station's position and the topology information of domain APs. Section 10.3.6 below provides further details.

■ Sensor network-assisted handoff [21] is designed to reduce scanning delays by limiting the number of probed channels during handoff. Before the actual handoff, the sensor network is deployed and overlaid with a WLAN. Sensors collect the parameters of neighboring APs. Consequently, before a handoff decision, a mobile station broadcasts an AP List Request. Equipped with this request, sensors located within range act as neighboring relay nodes and reply with an AP List Response that contains all required information about surrounding APs. Accordingly, at the moment when the handoff occurs, mobile stations solely scan channels indicated on the list they have received.

10.3.2 Bypass Scanning

As scanning is the most time-consuming component of the overall link-layer handoff, certain solutions focus on bypassing scanning to eliminate the probe latency. For example, the caching technique or multiple radio interface deployed either at the AP or at the mobile station uses to decouple scanning with handoff so that mobile stations can search proactively for alternate APs while being associated with an AP and interleaving data communication. Here are some examples:

■ Caching technique aims to buffer neighboring APs' information and exploit this information to accelerate the scanning procedure. Usually, this approach implies trivial modifications at the mobile station and the size of caching tables are either fixed [17] or dynamic [22]. AP caching [17], neighbor graph caching (NGC) [23], and adjacent APs [24] represent typical examples.

■ Multiple radio interfaces at mobile station approach [25–28] is a physical layer approach designed to completely eradicate scanning. During handoff, one wireless interface is used for normal data communication with the associated AP, while the other is used to search surrounding APs and find a candidate one to reassociate with. MultiScan [25] consists of a relevant example of such a strategy.

■ Multiple radio interfaces at AP approach [29–31] also consists of a physical layer method conceived to eradicate scanning. Additional radio interfaces eavesdrop on neighboring channels to rapidly detect mobile station movements [29,30]. Moreover, this additional radio transceiver can be used to search neighboring stations located within range and control their handoff operations [31].

■ Pre-scanning methods allow mobile stations to scan neighboring APs while they are associated with an AP. They interleave scanning with data communication, such as pre-scanning with selective channel mask [22], periodic scan [24], proactive scan with smart triggers [32], pre-active scan [33], anticipated handover [34], anticipated scanning [35], continuous monitoring with smart triggers [36], etc.

■ Location-based fast handoff method [37] allows mobile stations to select potential APs by predicting the path of their movement. By doing so, a location server is deployed to provide APs' information to mobile stations so that they can reassociate with the new AP directly, without scanning channels. However, this method relies on precise localization methods.

10.3.3 Cross-Layer Design

It may be beneficial that mobile stations maintain IP connectivity during their movements: this brings about additional requirements for efficient mobility management in wireless LANs. This new research objective provides support for seamless handoff and real-time multimedia applications in WLANs. Generally, user mobility is managed using Mobile IP version 6 (MIPv6) [38] or Mobile IP version 4 (MIPv4) [39]. The typical handoff procedure comprises movement detection, new address configuration and registration, etc. and turns in unacceptable delays for real-time services. Therefore, several handoff schemes have been proposed to improve handoff performance using cross-layer design strategies. Some of them are briefly introduced as follows:

■ Beacon with sufficient IP layer information [40,41] allows an enhanced AP to assist and handle fast new address configuration by inserting IP layer information, such as Router Advertisement (RA) [40] and network prefix information [41] into Beacon frames. This approach makes it possible to drastically decrease overall handoff latencies (both at MAC layer and IP Layer).

■ IP-IAPP scheme [42,43] enhances APs with advanced routing functionalities so that they act as mobility agents for mobile stations. They are also responsible for IP mobility management.

■ Link-layer triggers and topology information-aided fast handoff [44] use pre-handoff triggers to discover agents or address configuration prior to IP layer handoffs. In addition, post-handoff triggers are applied to eliminate movement detection delays.

Sections 10.3.4 through 10.3.6 delve into three typical fast handoff schemes.

10.3.4 Channel Masks and AP Caching Schemes

Shin et al. [17] propose channel masks and AP caching schemes to reduce probe delays to a level where VoIP communication becomes seamless. These schemes focus on reducing the probing time of non-existing channels through selective scanning, as well as the frequency of selective scanning using caching techniques.

Selective scanning is performed using channel masks that are built when drivers are first loaded on mobile stations. A full scan is conducted through broadcasting Probe Requests on all available channels. When a Probe Response arrives, a channel mask is set for the examined channel. In addition, channel masks are set by default for Channels 1, 6, and 11 as they are most likely to be used in well-configured wireless networks. Once all channels are scanned, the mobile station selects the best AP, based on the received signal strength. Then, it performs authentication and reassociation with the newly selected AP. Accordingly, channel masks are updated by removing the currently associated AP from the channel mask. When the ensuing handoff occurs, only channels equipped with a mask are probed. If no APs are found, the channel mask is inverted and a new selective scan is required. If no APs are found, a full scan is conducted to build new channel masks.

AP caching scheme consists of a cache table where the current AP's MAC address is indexed as a key. The list composed of the adjacent APs discovered during the scanning phase corresponds to

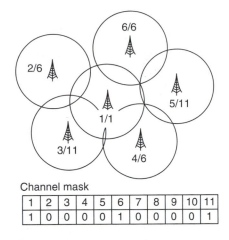

Cache Table 1

Key	1st AP	2nd AP
1/1	3/11	4/6
...	...	...
6/6	1/1	5/11

Cache Table 2

Key	1st AP	2nd AP
1/1	...	...
2/6	1/1	3/11
...	...	...
6/6	1/1	5/11

Channel mask

1	2	3	4	5	6	7	8	9	10	11
1	0	0	0	0	1	0	0	0	0	1

Figure 10.2 Channel masks and AP caching schemes.

the key. When a mobile station becomes associated with an AP, the latter is entered as a key into the cache. Cache entries are checked when a handoff is launched. If no entries are found (this case is called cache miss), the mobile station performs selective scanning and inserts two APs with the highest received signal strength into the cache. If an entry is found in the cache (this case is called cache hit), the station initially attempts to connect to the first AP. Once reassociation is done with success, the handoff is over; otherwise, it will try to associate with the second AP in the cache. When reassociation with this AP is done successfully, the handoff is over; otherwise, selective scanning is necessary to build new channel masks and find new APs for further reassociations.

Figure 10.2 illustrates channel masks and AP caching schemes. The symbol m/n denotes AP_m operating on Channel n. Each mobile station maintains a channel mask built after a full scan. During ensuing handoffs, mobile stations only scan mask-wearing channels. In Figure 10.2, a mobile station currently connected to AP_1 only scans Channels 1, 6, and 11 during the handoff. In addition, Cache Table 1 allows this mobile to reassociate with AP_3 and AP_4 without scanning. However, Cache Table 2 results in a cache miss during handoff.

In short, the channel masks scheme presents a fast selective scanning technique as mobile stations solely need to scan channels endowed with masks after a full scan whereas the AP caching scheme reflects a bypassing scanning approach. These schemes result in enhanced handoff performance. However, as caching tables are built from previous scanning results, mobile stations are likely to select an incorrect AP during handoff, thus triggering false handovers. Furthermore, cache misses hinder network performance.

10.3.5 NG and NG-Pruning Schemes

NG and NG-pruning schemes are proposed to enhance MAC layer handoff performance when mobile stations roam in WLANs [18]. This newly discovery method aims to reduce the total number of probed channels as well as the probe-waiting time on each channel. NGs dynamically capture the mobility topology of wireless networks [45,46] to assist mobile stations in making decisions regarding whether or not a channel needs to be scanned. Meanwhile, using non-overlapping graphs, mobile stations can find out whether to wait longer for Probe Responses on an examined channel, before the MaxChannelTime expires.

Before generating an NG, a mobility graph is defined to aggregate stations mobility traces in WLANs. Using adaptive estimation techniques, an NG is created by abstracting handoff relationships between adjacent APs. A nonoverlapping graph is created by abstracting nonoverlapping relationships among APs. Three methods are used to implement NGs:

- Centralized method: an NG server is used to restore the NGs and provide mobile stations with an NG as they join the network.
- Distributed method: each AP stores its local NG and mobile stations retrieve this graph from the AP after reassociation [45,46].
- Personal or user-oriented method: each mobile station keeps track of its mobility patterns to create its own NG.

An NG can be defined as: $G = (V, E)$ where $V = \{AP_i | i = 1, 2, \ldots, n\}$ (represents the set of all APs of a wireless system), $E = \{(AP_i, AP_j) | i \neq j\}$ (at least one mobile station handoffs from AP_i to AP_j). A nonoverlapping graph can be defined as: $G = (V, E)$ where $V = \{AP_i | i = 1, 2, \ldots, n\}$ (depicts the set of all APs of a wireless system), $E = \{(AP_i, AP_j) | i \neq j\}$ (mobile stations cannot communicate with AP_i and AP_j simultaneously with acceptable link quality).

An example of APs location map and the corresponding NG is illustrated in Figure 10.3. The symbol m/n denotes AP_m operating on Channel n. In this figure, mobile stations associated with AP_1 can handoff to AP_2 and AP_5, AP_2 to AP_1, and AP_3 to AP_1 and AP_4. Using the NG scheme,

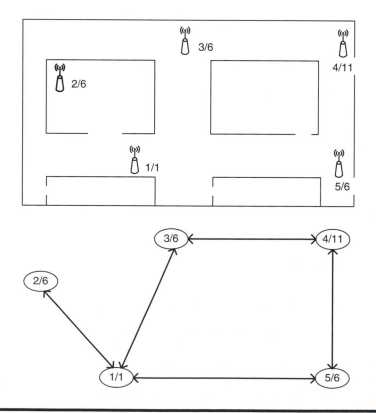

Figure 10.3   APs location map and corresponding NG.

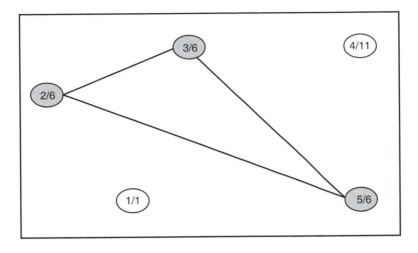

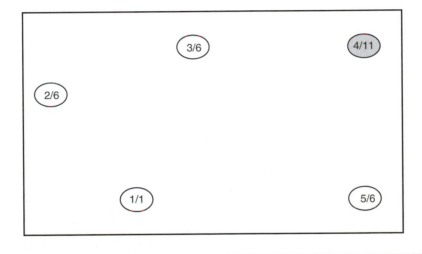

Figure 10.4   Nonoverlapping graphs for Channel 6 (up) and Channel 11 (down).

mobile stations covering by AP_1 only scan Channel 6 during the handoff. The number of probed channels is thus drastically reduced using the NG.

Nonoverlapping graphs for the aforementioned NG are depicted in Figure 10.4. On the basis of these nonoverlapping graphs, the NG-pruning technique makes it possible for a mobile station scanning Channel 6 to stop searching the same channel upon receiving a Probe Response from AP_2 or AP_3 or AP_5, as they do not overlap. NG-pruning scheme can thus drastically reduce probe-waiting time on each probed channel. For example, mobile stations probing Channel 11 stop scanning when they receive a Probe Response from AP_4, because a single AP operates on Channel 11.

Briefly, NG and NG-pruning schemes allow mobile stations to scan only a subset of all available channels and spend less waiting time on each probed channel, compared with the conventional handoff process defined in IEEE 802.11. However, mobile stations require global knowledge of the wireless environments before the actual handoff. Moreover, the quality of the NG scheme can be impaired by important topology changes in cases where APs are added or removed [18]. Furthermore, building such graphs involves a considerable amount of time and maintaining mobility graphs happens to be a complex task.

10.3.6 Handoff Assisted by Geolocation Information

A fast handoff method using geolocation information provided by a GPS system is proposed to reduce MAC and IP layer handoff latencies [20]. All mobile stations are equipped with a GPS receiver, which estimates the station's position and reports the obtained measurements to the station every second. The station calculates the distance, (d_1), between its current and previous locations using a Haversine formula. If $(d_1 > 1\ m)$, the mobile station sends a Location Update (LU) message that includes its coordinates to a GPS server. Note that this server keeps a list of domain APs' topology information and their relative parameters, such as $< AP\_ID, (x, y), Channel, SSID, IPv6\_Prefix >$. Upon receiving the LU message, the GPS server assesses the distance, (d_2), between the mobile station and its current AP. If $(d_2 > G)$, G denotes a predefined threshold, the GPS server selects the AP closest to the mobile station as the new AP and sends a Handover Initiate (HI) message to the mobile station. This message contains the target AP's ID, its operating channel, and IPv6 prefix. After receiving this message, the mobile station launches a handover by sending out a Probe Request over the new AP's channel and waits for a Probe Response from the target AP. When the mobile station receives a Probe Response, it launches authentication and reassociation processes. Simultaneously, the mobile station looks up the new AP's IPv6 prefix. If this prefix differs from the previous one owned by the station, the mobile station starts IP layer handover immediately, without waiting for a RA. Then, the mobile station configures a new IPv6 address and sends a Binding Update (BU) to its home agent (HA), which replies with a Binding Acknowledge (BA) message to the mobile station, thus completing the handoff when the BA is received.

In summary, GPS-assisted handoff scheme consists of a fast, selective scanning approach because mobile stations only need to scan a single channel during the MAC layer (L2) handoff. It also embodies a cross-layer design method as mobile stations can find out their exact new subnet prefix before completing the L2 handoff. Moreover, it contains a new fast IP layer (L3) movement detection method without the support of RAs. However, as this handoff scheme uses a GPS server, it introduces a centralized system and the server becomes a traffic flow bottleneck. In addition, those preconfigured parameters have a significant impact on system performance, which could be hindered by certain values. Moreover, in cases of fast movements performed by mobile stations, it becomes unreliable for the GPS server to select a new AP as it is likely to choose an inappropriate AP, leading to wrong handoff decisions.

10.4 Fast Handoff Schemes to Reduce Re-Authentication Delays

IEEE 802.11 defines open system and shared key authentication methods. Open system authentication admits all stations in the distribution system while shared key authentication relies on WEP to demonstrate the knowledge of a WEP encryption key. However, given that numerous security flaws [6,47] render both methods vulnerable to attacks, they have been replaced by IEEE 802.11i [5], a standard designed to enhance 802.11 security aspects, by introducing key management and establishment mechanisms, along with encryption and authentication improvements [47]. IEEE 802.11i incorporates IEEE 802.1X [48] as its authentication enhancements. As IEEE 802.1X is commonly deployed in many IEEE 802 series standards and uses a RADIUS [8] server to manage authentication, authorization, and accounting (AAA)-related activities, re-authentication (including authentication and reassociation) processes result in important delays. Therefore, fast handoff schemes to reduce this lengthy latency have been proposed in the literature. An overview of fast authentication methods is provided in Ref. [6], most of which are designed for intra-ESS handoffs. Some typical fast authentication methods are introduced in the following subsections.

10.4.1 IEEE 802.11i Pre-Authentication

Specified in IEEE 802.11i and designed for mobile stations already associated with an AP in the ESS, pre-authentication is launched by mobile stations that act as IEEE 802.1X supplicant [48]. It allows mobile stations to authenticate multiple APs at once [5], rendering authentication independent from roaming.

The roaming station sends an EAPOL-Start (Extensible Authentication Protocol Over LANs-Start) message to the new AP via its associated AP. Then, the new AP acts as authenticator to initiate an IEEE 802.1X authentication process by transmitting an EAP-Request/Identity to the mobile station. Following that, the mobile station returns an EAP-Response/Identity message to the new AP. Subsequently, the new AP forwards a RADIUS-Access-Request message to the authentication server (AS), which replies with a RADIUS-Access-Challenge. Thereafter, the new AP forwards this challenge text to the mobile station in an EAP-Request/Auth message. The mobile station then encrypts the challenge text using a shared secret with the AS, and transmits an EAP-Response/Auth to the new AP, which forwards a RADIUS-Access-Request containing the encrypted challenge text to the AS. This server decrypts the challenge, which is then compared to the original. When identical, the server returns a RADIUS-Access-Accept message to the new AP, which, in turn, forwards an EAP-Success message to the mobile station [6,47]. Upon a successful 802.1X authentication, a shared secret is created and cached between the new AP and the mobile station.

Briefly, pre-authentication allows mobile stations to prevent reassociation during re-authentication, thus leading to significantly shorter delays. However, pre-authentication also introduces new opportunities for denial-of-service (DoS) attacks and unnecessary burden on the AS [6].

10.4.2 Proactive Key Distribution Schemes

Proactive key distribution schemes, including the proactive neighbor caching (PNC) [46] and the selective neighbor caching (SNC) schemes [49], are designed to reduce authentication latency by pre-distributing key material one hop ahead of mobile stations' movement. NGs are used to dynamically capture the mobility topology of a WLAN for pre-distributing mobile stations' contexts to all neighbor APs [46]. The PNC scheme consists of pre-positioning mobile stations' contexts to all neighboring APs while SNC scheme pre-distributes the context to a set of selective APs, based on handoff probabilities.

Figure 10.5 illustrates the PNC scheme. The symbol m/n denotes AP_m operating on Channel n. When a mobile station associates with AP_2, its context information is propagated to all neighboring APs, $\{AP_1, AP_3, AP_4\}$. When this mobile station handoffs to AP_4, no additional authentication is required because AP_4 has received and cached the mobile station's credentials. Simultaneously, the mobile station's contexts are removed from other non-neighboring APs, i.e., AP_1. When reassociation occurs, the context information is broadcasted to all neighbors of the AP_4.

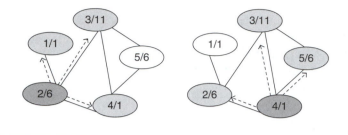

Figure 10.5 PNC scheme.

Figure 10.6 illustrates the SNC scheme. The symbol m/n denotes AP_m operating on Channel n. Suppose that the predefined handoff probability threshold is 0.3. The SNC scheme allows mobile stations' contexts to be transferred to surrounding APs, whose handoff probabilities are ≥ 0.3. When a mobile station associates with AP_2, its context information is propagated to a set of selective neighboring APs, $\{AP_3, AP_4\}$, according to handoff probabilities. As a result, when this mobile station handoffs to AP_4, normal authentication is not required because AP_4 has received and cached the mobile station's credentials. Simultaneously, the mobile station's contexts are removed from other non-neighboring APs. When reassociation occurs, the context information is broadcasted to AP_4's selected neighbors $\{AP_2, AP_5\}$. If this mobile station continues roaming and handoffs to AP_3, it must perform legacy re-authentication as AP_3 has removed the related context.

In a few words, proactive key distribution schemes can reduce re-authentication delays because mobile stations' contexts are transferred to neighboring APs during reassociation. However, these schemes result in high signaling overhead, especially in WLANs that contain a very dense population of mobile users. To reduce context transfer signaling costs in the PNC scheme, the SNC scheme adds neighbor weights to all edges of the NG. Neighbors' weight translates into handoff probabilities for each neighboring AP and is generated by monitoring mobile stations' handoff patterns among APs. Using the SNC scheme, mobile stations contexts are solely transferred to neighboring APs with higher handoff probabilities. The SNC scheme provides similar handoff performance when mobile stations handoff to a carefully selected AP. However, performance degradation is noticed when stations handoff to an AP without cached context. In addition, as handoff probabilities consist of

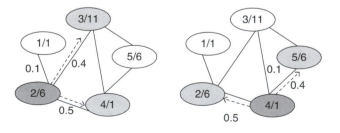

Figure 10.6 SNC scheme.

variable factors, the SNC scheme introduces additional complexity compared to the PNC scheme.

10.4.3 Predictive Authentication Scheme

A predictive authentication scheme is proposed to reduce re-authentication delays using frequent handoff regions (FHRs) [50–52]. This scheme is referred to as the FHR scheme in Ref. [11]. Mobile stations' authentication information is proactively distributed to multiple APs chosen by an FHR selection algorithm that considers the mobile stations' mobility patterns, service classes, etc. The FHR comprises a subset of adjacent APs most likely to be visited by mobile stations in the near future. The essence of the approach is that when a mobile station enters an AP's covered area, it performs authentication procedures for multiple APs in the FHR. As a result, when it handoffs to an AP in the specific FHR, re-authentication delays are eliminated as the mobile has already registered and authenticated this AP.

Figure 10.7 depicts the operation of the FHR scheme. A mobile station is associated with AP_2 and its corresponding FHR comprises $\{AP_1, AP_2, AP_4\}$. The mobile station's authentication information is propagated to these three APs. As a result, when the mobile handoffs to AP_4, a further authentication procedure is not required, as the AS has already dispatched the station's context to AP_4. When connected to AP_4, the mobile station generates a new FHR $\{AP_3, AP_4, AP_5\}$, and its pertaining authentication material is propagated to the APs in the new FHR by the AS.

The FHR scheme is based on mobility predictions to dispatch mobile stations' authentication information to a set of APs in FHRs. Authentication delays are eliminated in cases where mobile stations handover to an AP within the FHR. However, as the FHR is based on centralized authentication, the AS creates bottleneck problems.

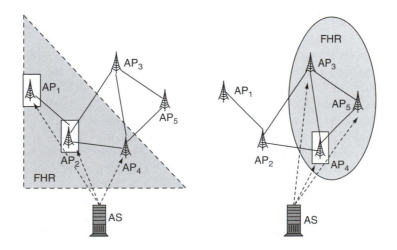

Figure 10.7   FHR scheme.

10.5 Handoff-Related Open Research Challenges

Even though a collection of fast handoff schemes has been proposed for mobile stations roaming in WLANs, seamless mobility support remains an important and challenging issue. On the other hand, real-time service support for roaming users consists of another complex issue. So far, several IEEE working groups have pooled their efforts to improve the conventional IEEE 802.11 standard: 802.11e strives to enhance quality of service (QoS) [53], 802.11i to reinforce security [5], 802.11f to upgrade the IAPP [7], 802.11k [54] to manage radio resources, 802.11r [55] to support fast roaming (transit), 802.21 [56] to improve handoff between heterogeneous wireless networks, etc.

Regarding fast handoff support, the IEEE 802.11r working group is drafting a protocol to facilitate the deployment of IP-based telephony over IEEE 802.11-enabled phones by speeding up handoffs between APs or cells in a WLAN. Moreover, as mobile devices with multiple interfaces emerge [57], a set of technical management issues needs to be taken into consideration to provide seamless connectivity from one interface to another. In this context, the IEEE 802.21 working group is standardizing a general interface to manage network interface cards as well as extending their work to hide network heterogeneity from end users.

Supporting seamless mobility and users' demands on multimedia applications imply that future WLAN handoff solutions must meet the following requirements:

- Backward compatibility: As a number of WLANs have been deployed using IEEE 802.11b/g, new handoff solutions must be compatible with the current legacy WLAN systems [11].
- Application diversity: Various applications will be proposed for WLANs. Future handoff schemes must be designed to meet users' demands for value-added services.
- Integration with other heterogeneous networks: To expand the territory of wireless services, new fast handoff solutions will be designed to provide broader radio coverage and seamless service for WLAN colocated with cellular networks, wireless LAN-based mesh networks, etc.
- Network-controlled handoff: The inherited IEEE 802.11 standard requires handoffs to be managed, autonomously and independently, by each mobile station without pre-knowledge of the wireless environment. However, most current solutions pertain to mobile station-controlled handoffs. As wireless networks themselves are endowed with the capacity of leveraging considerable information regarding their topology and station proximity, network-controlled handoff schemes will tend to be significantly more present in future proposals.
- Continuous signal monitoring capacity: Because most cellular networks provide facilities with continuously monitoring signal quality between mobile stations and all neighboring APs, future WLANs should develop these types of functionalities for mobile stations, using multimode radio interfaces.
- Load balancing: As overloaded APs cannot provide services to newly handoff mobile stations, new fast handoff scheme must introduce novel techniques to leverage the work-load in wireless networks. Several load balancing solutions have been reported in the literature, such as cell-breathing [58,59], yet further progress is required for future handoff designs.
- Handoff in a mixed architecture: Given that the inherited IEEE standard defines two architectural components—infrastructure and ad hoc mode—it is possible for mobile stations to operate with both modes simultaneously, using double radio interfaces. Thus, new fast handoff solutions will be valuable for handoffs in mixed architectures, such as ad hoc-assisted handoff scheme [60].

10.6   Conclusion

This chapter presents link-layer fast handoff schemes for mobile hosts roaming in WLANs. As shown above, providing seamless mobility and supporting real-time applications represent challenging issues for WLANs. IEEE 802.11 standard working groups are currently striving to enhance and standardize fast handoff management. Novel handoff approaches that meet the aforementioned design requirements represent a promising field of research in the near future.

REFERENCES

1. IEEE 802.11, Part 11: Wireless LAN Medium Access Control (MAC) and Physical Layer (PHY) Specifications, IEEE, 1999.
2. IEEE 802.11a, Part 11: Wireless LAN Medium Access Control (MAC) and Physical Layer (PHY) Specifications: High-Speed Physical Layer in the 5 GHz Band, IEEE, 1999.
3. IEEE 802.11b, Part 11: Wireless LAN Medium Access Control (MAC) and Physical Layer (PHY) Specifications: Higher-Speed Physical Layer Extension in the 2.4 GHz Band, IEEE, September 1999.
4. IEEE 802.11g, Part 11: Wireless LAN Medium Access Control (MAC) and Physical Layer (PHY) Specifications, Amendment 4: Further Higher Data Rate Extension in the 2.4 GHz Band, IEEE, June 2003.
5. IEEE 802.11i, Part 11: Wireless LAN Medium Access Control (MAC) and Physical Layer (PHY) Specifications, Amendment 6: Medium Access Control (MAC) Security Enhancements, IEEE, July 2004.
6. M. S. Bargh, R. J. Hulsebosch, E. H. Eertink, A. Prasad, H. Wang, and P. Schoo, Fast authentication methods for handovers between IEEE 802.11 wireless LANs, the 2nd ACM International Workshop on Wireless Mobile Applications and Services on WLAN Hotspots (WMASH 2004), Philadelphia, PA, October 1, 2004. pp. 51–60.
7. IEEE 802.11F, IEEE Trial-Use Recommended Practice for Multi-Vendor Access Point Interoperability via an Inter-Access Point Protocol Across Distribution Systems Supporting IEEE 802.11 Operation, IEEE, July 2003.
8. C. Rigney, S. Willens, A. Rubens, and W. Simpson, Remote Authentication Dial In User Service (RADIUS), IETF RFC-2865, June 2000.
9. P.-J. Huang, Y.-C. Tseng, and K.-C. Tsai, A fast handoff mechanism for IEEE 802.11 and IAPP networks, the 63rd IEEE Vehicular Technology Conference (VTC 2006-Spring), Melbourne, Australia, May 7–10, 2006. pp. 966–970.
10. C.-T. Chou and K. G. Shin, An enhanced Inter-Access Point Protocol for uniform intra and intersubnet handoffs, *IEEE Transactions on Mobile Computing*, 4(4), 321–334, July/August 2005.
11. S. Pack, J. Choi, T. Kwon, and Y. Choi, Fast handoff support in IEEE 802.11 wireless networks, *IEEE Communications Surveys and Tutorials*, 9(1), 2–12, January 2007.
12. A. Mishra, M. Shin, and W. Arbaugh, An empirical analysis of the IEEE 802.11 MAC layer handoff process, *ACM SIGCOMM Computer Communication Review*, 33(2), 93102, April 2003.
13. H. Velayos and G. Karlsson, Techniques to reduce IEEE 802.11b MAC layer handover time, the IEEE International Conference on Communications (ICC 2004), Paris, France, June 20–24, 2004. pp. 3844–3848.
14. K. Kwon and C. Lee, A fast handoff algorithm using intelligent channel scan for IEEE 802.11 WLANs, the 6th International Conference on Advanced Communication Technology (ICACT 2004), Phoenix Park, Republic of Korea, February 9–11, 2004. pp. 46–50.
15. I. Ramani and S. Savage, SyncScan: Practical fast handoff for 802.11 infrastructure networks, the IEEE 24th Annual Conference on Computer Communications (INFO-COM 2005), Miami, FL, March 13–17, 2005. pp. 675–684.

16. Y. Liao and L. Gao, Practical schemes for smooth MAC layer handoff in 802.11 wireless networks, the IEEE International Symposium on a World of Wireless, Mobile and Multimedia Networks (WoW-MoM06), Niagara-Falls/Buffalo, NY, June 26–29, 2006. pp. 181–190.

17. S. Shin, A. G. Forte, A. S. Rawat, and H. Schulzrinne, Reducing MAC layer handoff latency in IEEE 802.11 wireless LANs, the 2nd ACM International Workshop on Mobility Management and Wireless Access Protocols (MobiWac 2004), Philadelphia, PA, September 26–October 1, 2004. pp. 19–26.

18. M. Shin, A. Mishral, and W. A. Arbaugh, Improving the latency of 802.11 handoffs using neighbor graphs, the 2nd International Conference on Mobile Systems, Applications, and Services (MobiSys 2004), Boston, MA, June 6–9, 2004. pp. 70–83.

19. H.-S. Kim, S.-H. Park, C.-S. Park, J.-W. Kim, and S.-J. Ko, Selective channel scanning for fast handoff in wireless LAN using neighbor graph, the 2004 International Technical Conference on Circuits/Systems, Computers and Communications (ITC-CSCC2004), Sendai/Matsushima, Japan, July 6–8, 2004. 7F2P-29-1 to 7F2P-29-4.

20. J. Montavont and T. Noel, IEEE 802.11 handovers assisted by GPS information, the IEEE 2nd International Conference on Wireless and Mobile Computing, Networking and Communications (WiMob06), Montreal, Canada, June 19–21, 2006. pp. 166–172.

21. S. Waharte, K. Ritzenthaler, and R. Boutaba, Selective active scanning for fast handoff in WLAN using sensor networks, the 6th IFIP/IEEE International Conference on Mobile and Wireless Communication Networks (MWCN 2004), Paris, France, October 25–27, 2004. pp. 59–70.

22. N. Mustafa, W. Mahmood, A. A. Chaudhry, and C. M. Ibrahim, Pre-scanning and dynamic caching for fast handoff at MAC layer in IEEE 802.11 wireless LANs, the IEEE 2nd International Conference on Mobile Ad Hoc and Sensor Systems (MASS 2005), Washington, DC, November 7–10, 2005.

23. C.-S. Li, Y.-C. Tseng, and H.-C. Chao, A neighbor caching mechanism for handoff in IEEE 802.11 wireless networks, 2007 International Conference on Multimedia and Ubiquitous Engineering (MUE 2007), Seoul, Korea, April 26–28, 2007. pp. 48–53.

24. J. Montavont, N. Montavont, and T. Noel, Enhanced schemes for L2 handover in IEEE 802.11 networks and their evaluations, the IEEE 16th Annual International Symposium on Personal Indoor and Mobile Radio Communications (PIMRC05), Berlin Germany, September 11–14, 2005. pp. 1429–1434.

25. V. Brik, A. Mishra, and S. Banerjee, Eliminating handoff latencies in 802.11 WLANs using multiple radios: Applications, experience, and evaluation, the Internet Measurement Conference 2005 (IMC 05), Berkeley, CA, October 19–21, 2005. pp. 299–304.

26. M. Ohta, Smooth handover over IEEE 802.11 wireless LAN, IETF Draft, June 2002. draft-ohta-smooth-handover-wlan-00.txt.

27. K. Ramachandran, S. Rangarajan, and J. C. Lin, Make-before-break MAC layer handoff in 802.11 wireless networks, the IEEE International Conference on Communications (ICC 2006), Istanbul, Turkey, June 11–15, 2006. pp. 4818–4823.

28. S. Shin, A. G. Forte, and H. Schulzrinne, Seamless layer-2 handoff using two radios in IEEE 802.11 wireless networks, Columbia University Technical Report CUCS-018-06, New York, April 2006.

29. H.-S. Kim, S.-H. Park, C.-S. Park, J.-W. Kim, and S.-J. Ko, Fast handoff scheme for seamless multimedia service in wireless LAN, the 5th International IFIP-TC6 Networking Conference (Networking 2006), Coimbra, Portugal, May 15–19, 2006. pp. 942–953.

30. C.-S. Park, H.-S. Kim, S.-H. Park, K.-H. Jang, and S.-J. Ko, Fast handoff algorithm using access points with dual RF modules, the 3rd European Conference on Universal Multiservice Networks (ECUMN 2004), Porto, Portugal, October 25–27, 2004. pp. 20–28.

31. T. Manodham and T. Miki, A novel AP for improving the performance of wireless LANs supporting VoIP, *Journal of Networks*, 1(4), 41–48, August 2006.

32. H. Wu, K. Tan, Y. Zhang, and Q. Zhang, Proactive scan: Fast handoff with smart triggers for 802.11 wireless LAN, the 26th Annual IEEE Conference on Computer Communications (INFOCOM 2007), Anchorage, AK, May 6–12, 2007. pp. 749–757.

33. T. Manodham, L. Loyola, and T. Miki, Pre-active scan phase for the latency handover time issues in IEEE 802.11 wireless LANs, the International Conference on Communication and Broadband Networking (ICBN 2004), Kobe, Japan, April 7–9, 2004.

34. N. Montavont and T. Noel, Anticipated handover over IEEE 802.11 networks, the IEEE 1st International Conference on Wireless and Mobile Computing, Networking and Communications (WiMob05), Montreal, Canada, August 22–24, 2005. pp. 64–71.

35. N. Montavont and T. Nol, Fast movement detection in IEEE 802.11 networks, *Wireless Communications and Mobile Computing*, 6(5), 651–671, July 2006.

36. V. Mhatre and K. Papagiannaki, Using smart triggers for improved user performance in 802.11 wireless networks, the 4th International Conference on Mobile Systems, Applications and Services (MobiSys 2006), Uppsala, Sweden, June 19–22, 2006. pp. 246–259.

37. C.-C. Tseng, K.-H. Chi, M.-D. Hsieh, and H.-H. Chang, Location-based fast handoff for 802.11 networks, *IEEE Communications Letters*, 9(4), 304–306, April 2005.

38. D. Johnson, C. Perkins, and J. Arkko, Mobility support in IPv6, IETF RFC-3775, June 2004.

39. C. Perkins, IP Mobility Support for IPv4, IETF RFC-3344, August 2002.

40. B. Park, Y.-H. Han, and H. Latchman, EAP: New fast handover scheme based on enhanced access point in mobile IPv6 networks, *International Journal of Computer Science and Network Security*, 6(9), 69–75, September 2006.

41. N. Jordan, A. Poropatich, and R. Fleck, Link-layer support for fast mobile IPv6 handover in wireless LAN based networks, the 13th IEEE Workshop on Local and Metropolitan Area Networks (LANMAN 2004), San Francisco Bay Area, CA, April 25–28, 2004. pp. 139–143.

42. I. Samprakou, C. Bouras, and T. Karoubalis, Fast IP handoff support for VoIP and multimedia applications in 802.11 WLANs, the 6th IEEE International Symposium on a World of Wireless Mobile and Multimedia Networks (WoWMoM 2005), Taormina, Italy, June 13–16, 2005. pp. 332–337.

43. I. Samprakou, C. J. Bouras, and T. Karoubalis, Improvements on 'IP-IAPP': A Fast IP Handoff Protocol for IEEE 802.11 wireless and mobile clients, *Wireless Networks (WINET) Journal*, Special Issue on Broadband Wireless Multimedia, 13(4), 497–510, August 2007.

44. C.-C. Tseng, L.-H. Yen, H.-H. Chang, and K.-C. Hsu, Topology-aided cross-layer fast handoff designs for IEEE 802.11/mobile IP environments, *IEEE Communications Magazine*, 43(12), 156–163, December 2005.

45. A. Mishra, M. Shin, and W. A. Arbaugh, Context caching using neighbor graphs for fast handoffs in a wireless network, the IEEE 23rd Conference on Computer Communications (INFOCOM 2004), Hong Kong, March 7–11, 2004. pp. 351–361.

46. A. Mishra, M. Shin, N. Petroni, T. C. Clancy, and W. A. Arbaugh, Proactive key distribution using neighbor graphs, *IEEE Wireless Communications*, 11(1), 26–36, February 2004.

47. J.-C. Chen, M.-C. Jiang, and Y.-W. Liu, Wireless LAN security and IEEE 802.11i, *IEEE Wireless Communications*, 12(1), 27–36, February 2005.

48. IEEE 802.1X, IEEE Standards for Local and Metropolitan Area Networks, Port-Based Network Access Control, IEEE, December 2004.

49. S. Pack, H. Jung, T. Kwon, and Y. Choi, SNC: A selective neighbor caching scheme for fast handoff in IEEE 802.11 wireless networks, *ACM SIGMOBILE Mobile Computing and Communications Review*, 9(4), 39–49, October 2005.

50. S. Pack and Y. Choi, Fast handoff scheme based on mobility prediction in public wireless LAN systems, *IEE Proceedings Communications*, 151(5), 489–495, October 2004.

51. S. Pack and Y. Choi, Fast inter-AP handoff using predictive authentication scheme in a public wireless LAN, the IEEE Joint International Conference on Wireless LANs and Home Networks and Networking (NETWORKS 2002), Atlanta, GA, August 26–29, 2002. pp. 15–26.

52. S. Pack and Y. Choi, Pre-authenticated fast handoff in a public wireless LAN based on IEEE 802.1x model, the IFIP TC6/WG6.8 Working Conference on Personal Wireless Communications (PWC2002), Singapore, October 23–25, 2002. 175–182.

53. IEEE 802.11e, Part 11: Wireless LAN Medium Access Control (MAC) and Physical Layer (PHY) Specifications: Amendment 8: Medium Access Control (MAC) Quality of Service Enhancements, IEEE, November 2005.

54. D. Simone, 802.11k Makes WLANs Measure Up, *Network World*, March 2004.

55. P. Calhoun and B. O'Hara, 802.11r Strengthens Wireless Voice, *Network World*, August 2005.

56. IEEE 802.21 Working Group Web site, http://www.ieee802.org/21/.

57. F. Cacace and L. Vollero, Managing mobility and adaptation in upcoming 802.21-enabled devices, the 4th ACM International Workshop on Wireless Mobile Applications and Services on WLAN Hotspots (WMASH 2006), Los Angeles, CA, September 29, 2006. 1–10.

58. Y. Bejerano and S.-J. Han, Cell breathing techniques for load balancing in wireless LANs, the IEEE 25th International Conference on Computer Communications (INFOCOM 2006), Barcelona, Spain, April 23–29, 2006. pp. 1–13.

59. H. Velayos, V. Aleo, and G. Karlsson, Load balancing in overlapping wireless LAN cells, the IEEE International Conference on Communications (ICC 2004), Paris, France, June 20–24, 2004. pp. 3833–3836.

60. M. He, T. D. Todd, D. Zhao, and V. Kezys, Ad hoc assisted handoff for real-time voice in IEEE 802.11 infrastructure WLANs, the IEEE Wireless Communications and Networking Conference (2004 WCNC), Atlanta, GA, March 21–25, 2004. pp. 201–206.

Chapter 11

Security in Wireless LANs

Mohamed K. Watfa and Haidar Safa

CONTENTS

The rapid deployment of wireless LANs is testimony to the inherent benefits of this technology. Unfortunately, most wireless deployments are, at this time, fundamentally insecure. This chapter provides an overview of security issues, explains how security works in Wi-Fi networks, and explores various security and authentication protocols. The chapter starts with a coverage of the basic elements of IEEE 802.11. It describes the types of messages that are exchanged and explains how a portable device can find, select, and connect to an access point (AP). This chapter contains a moderate amount of detail to highlight some of the security risks between Wi-Fi components. We will see how the Wi-Fi LAN fits into a stack of layers between the operating system and the wireless medium. The security mechanisms are tied up with the process of making connections and passing data. This chapter also outlines why Wi-Fi networks are vulnerable to attacks and what type of attackers one might encounter. By understanding the motivations and resources of attackers, an efficient security policy can be established. After introducing the overall message exchange between the components, the chapter digs deeply and laboriously into the security protocols for Wi-Fi. It describes the original Wi-Fi security approach, Wired Equivalent Privacy (WEP), and explains why this method is no longer considered secure. It then covers the new approaches of Wi-Fi Protected Access (WPA) and 802.11i Robust Security Networks (RSNs) both of which are scalable from small networks of few devices up to international corporations. This part also describes several methods that can be used in conjunction with RSN and WPA Wi-Fi networks. It introduces some of the protocols that are central to the new security solutions starting with access control, which are built around the IEEE 802.1X standard. This section also looks at the upper-level authentication protocols covering the way that Transport Layer Security and Kerberos V5 work and how they can be applied to Wi-Fi security.

At the end of this chapter, we focus on practical security issues in the cellular/Wi-Fi world. In the past, mobile phone networks and infrastructure have been quite separate from the Internet technologies used by the computer industry. Therefore, a way is needed to bridge the gap between mobile phone infrastructure and Internet infrastructure. This part will look at interesting methods such as Protected EAP (PEAP) and GSM-SIM, which allow Wi-Fi systems to be authenticated by cellular phone infrastructure. This part also reviews one of the new security protocols that was developed specifically for use with existing Wi-Fi equipment, Temporal Key Integrity Protocol (TKIP). We start with an overview of TKIP, one of the options for implementing encryption and message authentication under RSN, and work through each of its functionalities comparing it with Advanced Encryption Standard (AES), a block ciphersuite considered as the default mode of IEEE 802.11i. We conclude this chapter by highlighting some security-related open issues.

11.1   IEEE 802.11 Basic Elements

In this section, a brief overview of IEEE 802.11 is provided. To understand the security system of IEEE 802.11, you need a basic knowledge of how IEEE 802.11 operates. This section starts with a coverage of the basic elements of IEEE 802.11. It describes the types of messages that are exchanged and explains how a portable device can find, select, and connect to an AP.

11.1.1   IEEE 802.11 Introduction

The complete IEEE 802.11 [1] architecture is shown in Figure 11.1. The basic service set (BSS) is the basic building block of an IEEE 802.11 LAN and this consists of devices referred to as stations (STA). Basically, the set of STAs that can talk to each other can form a BSS. Multiple BSSs are interconnected through an architectural component, called distribution system (DS), to form an extended service set (ESS). An AP is a STA that provides access to the DS by providing DS services. IEEE 802.11 specifies the services that are to be provided by STAs and DS. It does not specify how a DS is to be implemented; it can be either distributed or central. A service provided by a station

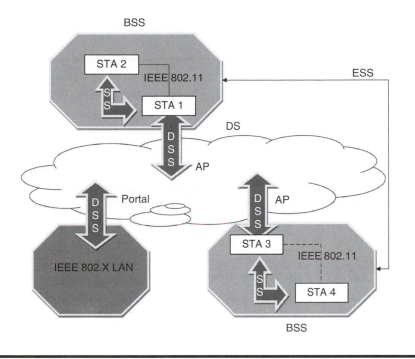

Figure 11.1   IEEE 802.11 architecture.

is called station service (SS) and a service by a DS is called distributed system service (DSS). SSs include the following four services:

1. Authentication: This service is used by all stations to establish their identity to stations with which they will communicate.
2. Deauthentication: This is invoked when an existing authentication is to be removed.
3. Privacy: The standard specifies optional WEP, which provides a security level equivalent to a wired LAN.
4. MSDU delivery: This is for the delivery of Medium Access Control (MAC) service data unit.

A station willing to join an ESS first chooses an AP from all available APs or can start a new BSS by assuming the role of an AP. The available APs are found by looking for the periodic beacons they broadcast over the medium. A station upon choosing an AP will authenticate itself to this AP and then associates with it. This association basically provides this STA to AP mapping information to the DS, which is utilized in routing the messages in the ESS. The IEEE 802.11 standard places specifications on the parameters of both the physical (PHY) and the MAC layers of the network. The PHY layer, which actually handles the transmission of data between nodes, can use either direct sequence spread spectrum, frequency-hopping spread spectrum, or infrared (IR) pulse position modulation. The IEEE 802.11 makes provisions for data rates of either 1 or 2 Mbps, and calls for operation in the 2.4–2.4835 GHz frequency band (in the case of spread-spectrum transmission), which is an unlicensed band for industrial, scientific, and medical (ISM) applications, and 300–428,000 GHz for IR transmission [2]. The MAC layer is a set of protocols responsible for maintaining order in the use of a shared medium. The 802.11 standard specifies a Carrier Sense Multiple Access with Collision Avoidance (CSMA/CA) protocol. Whenever a packet

Preamble	PLCP	MAC	User data	CRC

Figure 11.2   IEEE 802.11 frame format.

is to be transmitted, the transmitting node first sends out a short ready-to-send (RTS) packet containing information on the length of the packet. If the receiving node hears the RTS, it responds with a short Clear-to-Send (CTS) packet. After this exchange, the transmitting node sends its packet. When the packet is received successfully, as determined by a cyclic redundancy check (CRC), the receiving node transmits an acknowledgment (ACK) packet. This back-and-forth exchange is necessary to avoid the famous hidden node problem. Next, we focus on some features of the main IEEE 802.11 protocol that the security protocols depend on.

11.1.2   IEEE 802.11 Frame Formats

Every transmission over the wireless medium has a similar form as shown in Figure 11.2. A special pattern called the preamble is first sent out, which is used for identification purposes. By the end of the preamble, all the receivers in range should have adjusted themselves to interpret the data that is to follow. The preamble is followed by the Physical Layer Convergence Protocol header (PLCP header), which contains information relevant to the receiver logic (data rate, packet length, etc.). Following the PLCP header is the MAC header. The MAC header comes in three basic types depending on whether the frame is a control frame, a management frame, or a data frame. It contains the source and destination addresses (DAs) which are 6 bytes long [1]. The DA could be unicast (deliver to one device), multicast (deliver to several devices), or broadcast (deliver to all devices). The MAC header can have four addresses: transmitter address (TA), receiver Address (RA), source address (SA), and DA. In an ad hoc network, only two addresses are contained in the MAC header as the device that created the message is the one that sent it. In an infrastructure network, three addresses are needed because the AP plays as an intermediate node. The next section looks at the technical characteristics that make Wi-Fi LANs vulnerable and provides an overview of the different types of attacks that a wireless LAN must defend against.

11.2   Wi-Fi Security Risks

Wireless LANs use radio propagation that make them very vulnerable to attacks in every direction. We first look at the types of attackers and discuss some of their motivations [3].

11.2.1   Different Types of Attackers

Attackers fall into different categories as depicted in Figure 11.3. At the bottom of the pyramid are "script kiddies" with some weak tools. As you move up the pyramid, the number of attackers decreases but the complexity of their tools increases. Halfway up the pyramid, one might encounter some cryptographic attack tools that seek into breaking into secure systems rather than just searching for systems where the security is turned off. At the top of the pyramid are the most sophisticated attackers called ego hackers.

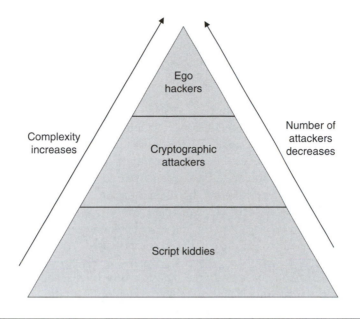

Figure 11.3 Pyramid structure of the different motivations behind attacks.

In many situations, the wireless LAN user is in an unsecure region. One response is to handle LAN users in the same way that you handle remote users by using a virtual private network (VPN). The alternative is to make the wireless LAN itself secure and therefore impenetrable and thus can be regarded as secure physical wiring. This thinking led to the original security system of IEEE 802.11 called WEP [4]. In the next section, we provide an overview of the different types of attacks that a wireless LAN must defend against.

11.2.2 Different Types of Attacks

The wireless networks based on 802.11 have been plagued by some security failings. The following is a list of some of the different types of attacks a wireless LAN might encounter [3]:

- Jamming: DoS (denial-of-service) attacks are easily applied to the wireless world, where legitimate information cannot reach the clients or APs, mainly because the legitimate traffic overwhelms the frequencies.
- Insertion attacks: These occur when you place unauthorized devices on the wireless network without going through a security process and review. This type of attack can happen when an attacker tries to connect a wireless client to an AP without authorization.
- Interception and monitoring: As in wired networks, it is possible to intercept and monitor the network traffic across a wireless LAN.
- Misconfiguration: Many APs ship in an unsecured configuration so that they can be handled and deployed easily.
- Client-to-client attacks: Two wireless clients can communicate with each other, bypassing the AP. Therefore, there is a need for users to defend the clients not just against an external attack but also against each other.

The attacker can use some of these methods with or without having access to your secret network keys although having access to the secret keys will move you to a different category of danger. It is sometimes possible to breach the network security without ever having access to the network keys. Some common attacks are

■ Traffic analysis attack: Attackers can use passive network monitoring extensively to capture possible useful information. Through passive monitoring, an attacker can gain a thorough understanding of the network's topology: the services that are available, the operating systems that are in use, and the vulnerabilities that may be taken advantage of on the network. The attacker cannot see the content of the message, but combined with other techniques the attacker might gain some extra knowledge of the content.

■ Man-in-the-middle (MITM) attack: An MITM attack is an attack in which an attacker is able to read, insert, and modify at will messages between two parties without either party knowing that the link between them has been compromised.

As a brief overview of how the security mechanism works is what follows. A ciphertext (after encryption) is created by processing plaintext (before encryption) with the cipher (algorithm) using the keys (secret) as illustrated in Figure 11.4. So, the attacker has a copy of the ciphertext that was snooped and also one of the modern rules of cryptography is that one should assume that the attacker also knows the algorithm used for encryption. There are also many ways an attacker can get a sample of the plain text: In IEEE 802.11, the MAC header is not encrypted, but the rest of the message is. The header always occurs at the start of the packet and some fields have fixed values that can easily be found. Also, some IP messages are of known format, such as DHCP discover messages used in assigning network addresses. These are encrypted but can be identified from their length. An attacker might correctly guess the entire plain text. We will assume that they have all three components (the cipher text, the plain text, and the cipher) and their ultimate goal is to get the keys. The following are some traditional techniques:

■ Brute force attack: The most basic attack used to decrypt packets sent on an encrypted network is the brute force attack. This attack is performed by first intercepting communications occurring on the network. Once the data has been intercepted, it is a matter of guessing keys until the attack finds the right one. The attacker does this by generating keys from different initial strings and then attempts to decrypt the data he or she collected with these keys. The feasibility of this attack depends entirely on the length of the key used. Clearly, a 104-bit

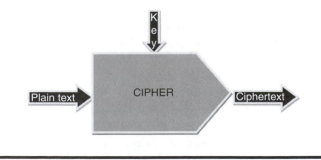

Figure 11.4 Encryption technique illustration.

key is more secure than a 40-bit key. Both key sizes represent a large key space, but with the growing speed of computers and the parallel computational properties of this attack, it is susceptible to a distributed computing solution. A 40-bit key can be cracked in about 200 days with a single computer doing about 60,000 guesses per second. If the attacker had access to a large collection of computers, he or she could split up the attack and could discover a key in a matter of hours or even days. However, a 104-bit key would take a single computer about 10^{19} years to crack doing the same 60,000 guesses per second. Even with a collection of computers on which to run this attack, it would still take far too long to discover a key.

■ Algorithmic attacks: One approach the attacker might use would be to try to find a flaw in the encryption algorithm. A weakness in the algorithm allows 1 byte of the key to be attacked at a time. The total time needed is proportional to the number of bytes and therefore it is slightly more difficult to crack a 104-bit key than it is to crack a 40-bit key.

11.2.3   *Wired Equivalence Privacy*

In this section, we focus on the original security method used in Wi-Fi LANs and which has now been discredited due to its numerous security vulnerabilities. We first look at the general design of WEP [4–7] and then analyze the reasons behind its failure.

11.2.4   *WEP Overview and Pitfalls*

The IEEE 802.11 defines a mechanism for encrypting the contents of 802.11 data frames. This scheme uses five elements directly relevant to its analysis:

■ A key shared between all the members of the BSS (there are really four shared keys, but this is irrelevant to the analysis).
■ An encryption algorithm: For WEP, this is the RC4 stream cipher, used to generate a key stream, which is XORed against the plaintext to produce the ciphertext.
■ A decryption algorithm: For WEP, this is the same as the encryption algorithm. RC4 is used to generate a key stream, which is XORed against the ciphertext to reproduce the plaintext.
■ A 24-bit initialization vector (IV): WEP appends the IV to the shared key; WEP uses this combined key and IV to generate the RC4 key schedule. WEP selects a new IV for every packet.
■ An encapsulation: The encapsulation transports the IV and the ciphertext from the sender (encryptor) to the receiver (decryptor).

WEP also uses a CRC of the frame payload plaintext in its encapsulation. The CRC is computed over the data payload and then appended to the payload before encryption. WEP encrypts the CRC with the rest of the data payload. The operation of WEP is very simple to describe. First, each member of the BSS is initialized with the shared key via an unspecified, implementation-specific, out-of-band mechanism. To send a WEP-encapsulated frame, the sender calculates the CRC of the frame payload and appends it to the frame. It then selects a new IV, appends this to the shared key to form a per-packet key, and uses the result to generate a RC4 key schedule. The sender then uses RC4 to generate a key stream equal to the length of the frame payload plus the CRC. The sender XORs the generated key stream against the plaintext payload data and CRC. The sender also inserts the IV into the appropriate field in the frame header, and sets a bit indicating this is a WEP-encrypted packet. At this point, the WEP encapsulation is complete, and the frame can be sent to the peer. To process a WEP frame, the receiver checks the encrypted bit in the arriving frame. If it is set, the receiver extracts the IV from the frame, appends it to the BSS shared key, and generates the per-packet RC4 key schedule. RC4 is applied to the key schedule to produce a key stream. The receiver then XORs this key stream against the encrypted payload to extract the plaintext. Finally, the receiver verifies the CRC of the decrypted payload data to verify that the data frame has been correctly decrypted. Some possible attacks on WEP are summarized in the next subsection.

11.2.5   Attacks on WEP

Some possible attacks on WEP include the following:

- Passive attack to decrypt traffic: A passive eavesdropper can intercept all wireless traffic, until an IV collision occurs. By XORing two packets that use the same IV, the attacker obtains the XOR of the two plaintext messages. The resulting XOR can be used to infer data about the contents of the two messages. IP traffic is often very predictable and includes a lot of redundancy. This redundancy can be used to eliminate many possibilities for the contents of messages. Further educated guesses about the contents of one or both of the messages can be used to statistically reduce the space of possible messages, and in some cases it is possible to determine the exact contents. Once it is possible to recover the entire plaintext for one of the messages, the plaintext for all other messages with the same IV follows directly, because all the pairwise XORs are known.
- Active attack to inject traffic: Suppose an attacker knows the exact plaintext for one encrypted message. He or she can use this knowledge to construct correct encrypted packets. The procedure involves constructing a new message, calculating the CRC-32, and performing bit flips on the original encrypted message to change the plaintext to the new message. The basic property is that $RC4(X) \oplus X \oplus Y = RC4(Y)$. This packet can now be sent to the AP or mobile station (MS), and it will be accepted as a valid packet. A slight modification to this attack makes it much more insidious. Even without complete knowledge of the packet, it is possible to flip selected bits in a message and successfully adjust the encrypted CRC to obtain a correct encrypted version of a modified packet. If the attacker has partial knowledge of the contents of a packet, he or she can intercept it and perform a selective modification on it.
- Active attack from both ends: In this case, the attacker makes a guess not about the contents but rather about the headers of a packet. Armed with this knowledge, the attacker can flip appropriate bits to transform the destination IP address to send the packet to a machine he or she controls, somewhere in the Internet, and transmit it using a rogue MS. Most wireless installations have Internet connectivity; the packet will be successfully decrypted by the AP and forwarded unencrypted through appropriate gateways and routers to the attacker's machine, revealing the plaintext. If a guess can be made about the TCP headers of the packet, it may even be possible to change the destination port on the packet to be port 80, which will allow it to be forwarded through most firewalls.
- Table-based attack: The small space of possible initialization vectors allows an attacker to build a decryption table. Once the attacker learns the plaintext for some packet, he or she can compute the RC4 key stream generated by the IV used. This key stream can be used to decrypt all other packets.

11.3   Wi-Fi Protected Access and RSNs

In this section, we introduce the new security protocols that replace WEP and provide real security. We first define some terms and explain the process under which these protocols were developed.

11.3.1   IEEE 802.11i, WPA, and RSN

The long-anticipated 802.11i [8,9] specification for wireless LAN security was finally ratified by the IEEE in June 2004. It had been in the works for years. Unlike 802.11a, b, and g specifications, all of which define physical layer issues, 802.11i defines a security mechanism that operates between the MAC sublayer and the network layer. The new spec offers significant improvements over the old standard, WEP. The specifications were developed by the IEEE's TGI task group, headed by David Halasz of Cisco. The standard IEEE 802.11i is designed to provide secured communication of wireless LAN as defined by all the IEEE 802.11 specifications. IEEE 802.11i defines a new type of wireless network called an RSN. In a true RSN, the AP allows only RSN-capable mobile devices to connect and because many people using pre-RSN devices will want to upgrade, IEEE 802.11i defines a transitional security network (TSN) in which RSN and WEP systems can operate in parallel.

Wi-Fi manufacturers found that customers were not willing to throw away all their existing equipment to switch to RSN; however, they would want to upgrade their products through software. This lead to the definition of the TKIP [9]. The Wi-Fi alliance adopted a new security approach based on RSN but only specifying TKIP, which was called WPA. WPA [10,11] was created by the Wi-Fi Alliance in 2002. To avoid multiple standards and conflicts later on, WPA was designed from the get-go to be compatible with 802.11i and was based on its early draft specifications. WPA provides several security advantages. First, it uses a stronger key management scheme, by implementing the TKIP. TKIP creates encryption values that are mathematically derived from a master key, and changes these encryption keys and IV values automatically and transparently to the user so to prevent key stream reuse. This is important because WEP keys have to be changed manually. TKIP also uses a Message Integrity Code called Michael that uses a 64-bit key. The integrity checker is designed to block forged messages. There are two methods for generating the master key, and WPA operates in two different modes, depending on whether pre-shared keys are used or a central authentication server is available. For home users, WPA offers easy setup since one big problem with WEP was that many users found it too difficult or confusing to set up and manage. Authentication is based on the Extensible Authentication Protocol (EAP) [12,13] and can use pre-shared keys that make it simple to configure on the WAP and clients in small network settings. At the large network level, operating in Enterprise mode, WPA supports Remote Authentication Dial-In User Service (RADIUS) [13] so that users can be authenticated through a centralized server. WPA 802.1x [14] authentication methods include EAP over Transport Layer Security (EAP-TLS), EAP-Tunneled TLS (EAP-TTLS), EAP-LEAP, EAP-PEAP, and other implementations of EAP. WPA uses the same encryption algorithm for encrypting data that WEP uses: the RC-4 cipher stream algorithm. However, TKIP uses a 48-bit initialization vector, as opposed to the weaker 24-bit IV used by WEP.

On the other hand, RSN dynamically negotiates the authentication and encryption algorithms to be used for communications between WAPs and wireless clients. This means that as new threats are discovered, new algorithms can be added. RSN uses the AES [15], along with 802.1x and EAP. The security protocol that RSN builds on AES is called the Counter Mode CBC MAC Protocol (CCMP). AES supports key lengths up to 256 bits, but is not compatible with older hardware. However, there is a specification designed to allow RSN and WEP to coexist on the same wireless LAN; it is called TSN.

11.3.2 Authentication Process

In this section, we introduce another entity, the AP, also generally known as a network access server (NAS), and the protocol that clients use to communicate with APs during the authentication process, the EAP. The authentication process of WPA and IEEE 802.11i adopted the three-entity model of IEEE 802.1x. IEEE 802.1x is the port-based access control protocol that was originally designed for the Point-to Point Protocol (PPP), such as modem connections and wired LANs [16]. The three entities are the client, the AS, and the NAS. The AS resides in the network, and the client, who initially does not have access to the network, is connected to the NAS. The NAS is the entity that initially blocks the client's access to the network and also serves as a broker between the client and the AS during the authentication process. Thus, the NAS acts as a security guard for the network, allowing only those who are successfully authenticated by the AS, which makes the access decisions. In WLANs, the AP and wireless links replace the NAS and modem connections.

Although there have been some changes made in 802.1x to allow the WLANs to adopt port-based access control, the relationship among the three entities remains the same. The three entities use EAP to communicate during the authentication process. EAP has four message types: Request, Respond, Success, and Failure. EAP can encapsulate other authentication protocols, such as TLS and SRP, in its Request and Respond messages. The AS uses the Success or Failure message to notify the AP whether the client authentication was successful. The way EAP messages are used portrays the AP's role in WPA and 802.11i. Figure 11.5 shows the EAP message flow.

The AP cares only about the AS's decision whether to grant the client the access to the network. Thus, the AP simply passes along the EAP Request messages from the AS to the client, and EAP

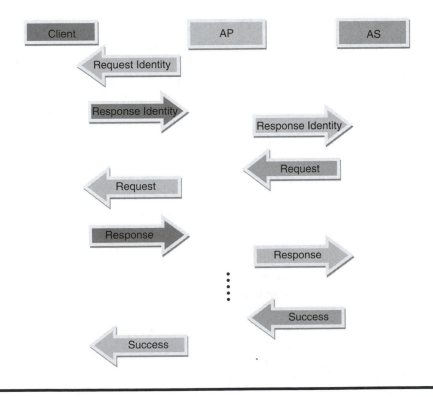

Figure 11.5   EAP message flow.

Response messages from the client to the AS. The contents of these messages are not important to the AP. Meanwhile, the AP listens for the EAP Success or Failure message from the AS. If the AS sends the Success message, the AP admits the client into the network; if the AS sends the Failure message, the AP leaves the client disconnected from the network. An important role of the authentication process is to establish a temporary secret that the client and the AP can use for message protection. The message protection process starts only when the authentication process finishes with the EAP message Success, which includes the session key from the AS to use for message protection. The client also computes the same session key from the information exchanged in the authentication protocol. The AP and the client use this session key for message protection.

In summary, the authentication process of WPA and IEEE 802.11i involves three entities: the client, the AP, and the AS. The client seeks access to the network. The AP guards the access to the network, allowing only the clients that the AS has authenticated. Finally, the AS decides whether the client is eligible to gain access to the network. They use EAP, which provides a way for the three entities to embed other authentication protocols.

11.3.3 *Authentication Protocols*

In this section, we survey many recent WLAN authentication protocols. We group the protocols into three categories: secret-key methods, public-key methods, and tunneled methods [17].

11.3.3.1 *Secret-Key Approach*

In secret-key authentication methods, the AS and the client have the same secret and establish trust by providing to each other the knowledge of the shared secret key. Secret-key authentication protocols are efficient and require little computational power. This advantage is especially important in WLANs because many wireless devices, such as PDAs and mobile Voice-over-IP [VoIP] phones, have little computational power. However, secret-key authentication methods have several drawbacks. Because most secret-key authentication protocols derive the shared secret from the user's password, and because most users choose bad passwords, it is easy for the attacker to gather enough encrypted messages and extract the secret key from them, using dictionary attacks [18,19]. Although some secret-key authentication methods such as EAP-SRP protect the client's password from dictionary attacks, these methods require much greater computational power than other secret-key methods. Moreover, it is hard to securely distribute the shared secret to both parties. We discuss and compare three secret-key authentication protocols that are being used for WLAN authentication: Lightweight Extensible Authentication Protocol (LEAP), Kerberos, and EAP over Secure Remote Password (EAP-SRP).

LEAP: LEAP [20] was developed by Cisco to address WEP's weaknesses. Because LEAP uses WEP for the message protection process, LEAP does not meet the level of security that WPA and 802.11i RSN provide. LEAP's authentication process, however, is noteworthy because it is the first commercial use of IEEE 802.1x and EAP for WLANs [3]. LEAP's authentication protocol also includes mutual authentication and temporary session keys, two of many missing features of WEP authentication. Figure 11.6 shows the LEAP authentication message flow. In the figure, the notation $\{X\}Y$ denotes that X is encrypted with Y. Initially, the client and the AS share a secret Microsoft version of the challenge-handshake authentication protocol (MS-CHAP) key to accomplish mutual authentication. First, the client sends a random challenge to the AS, and the AS responds to the challenge by encrypting it with the secret key that is shared between the AS and client. The client authenticates the AS by decrypting the response from the AS and comparing it to the challenge. If the decrypted

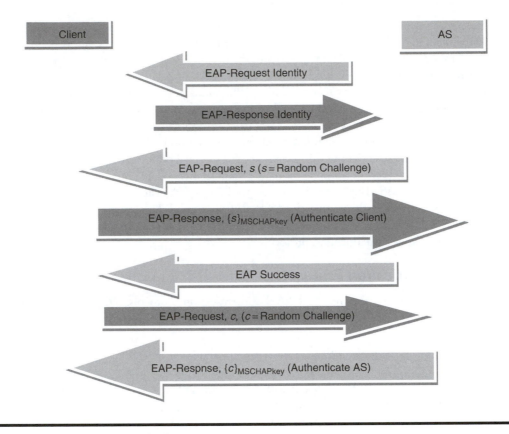

Figure 11.6 Message flow for leap.

response matches the challenge, the AS is authenticated. Similarly, the AS authenticates the client with a challenge. If the mutual authentication is successful, the client and the authentication server derive a temporary session key from the information exchanged during the authentication process. The client and the AP use this session key for the message protection process. Although LEAP supports mutual authentication and session key derivation, LEAP has some flaws. LEAP does not protect the client's identity because the EAP identity messages are sent in plaintext. Moreover, because an eavesdropper can easily sniff the challenge–response pair sent between the client and the AS during the MSCHAP authentication, LEAP is vulnerable to dictionary attacks [21]. LEAP also does not consider other desired properties such as delegation and fast reconnect.

Kerberos: Developed at MIT in 1993, Kerberos is an authentication protocol designed for TCP/IP [22]. Many places that use wired LAN already use Kerberos for authenticating its clients for services, such as accessing the e-mail server. In the Kerberos model, every service requires a ticket and a session key. The tickets and the session keys are issued by the Kerberos AS and its ticket granting server (TGS). Each ticket contains enough information about the client so that the server controlling the service can verify that the client is indeed the one to whom the ticket is issued. This information may include the server's name, the client's name, the network address of the client, the beginning and ending validity time for a ticket, and the session key. Tickets are encrypted with the secret shared between the TGS and the server controlling the service so that any tampered information can be detected. There are two kinds of tickets: ticket granting tickets (TGTs) and service tickets. Service tickets are the tickets that grant access to services to the client. A TGT allows the client to gain service tickets, so the client must first obtain a TGT to obtain service tickets. Kerberos also uses authenticators to prevent replay attacks. An authenticator may include the client's name, the client's network address, and a timestamp. The client presents an authenticator with a fresh timestamp along with the ticket and encrypts them with the session key of the ticket so that an attacker who sniffs the authenticator and the ticket cannot replay them. The AS issues a TGT and a corresponding session key to the client. This session key is encrypted with the secret shared between the client and the AS. Thus, if an imposter pretending to be the client receives the encrypted session key, he or she will not be able to decrypt the encrypted session key and gain the service ticket because the server can check that the session key value in the ticket does not match the session key that the imposter used to encrypt the authenticator. Moreover, an imposter pretending to be the AS will not be able to generate a correctly encrypted ticket and session key to the client. Thus, in the Kerberos model, the usage of the encrypted session key allows the client and the AS to mutually authenticate each other.

One way of implementing Kerberos in WLAN authentication is to treat the AP as a service that passes data packets to and from the network (Figure 11.7). In this model, authentication occurs when the client obtains the TGT and the AP service ticket. This idea is consistent with the port control model of IEEE 802.1x because the AP acts as the server that provides access to the wireless network. But because the Kerberos AS resides within the IP network, which an

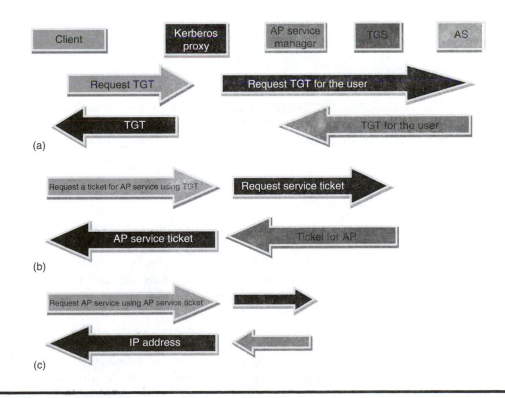

Figure 11.7 Kerberos model for WLAN. (a) Requesting the TGT. (b) Requesting the AP service ticket. (c) Requesting the AP service.

unauthenticated client would not be able to access, the client would have no way to obtain the TGT and the AP service ticket from the Kerberos AS. To solve this problem, Kerberos authentication in WLANs requires the AP to have a proxy Kerberos application server, which is specifically designed to help connect the AS and the TGS with clients who do not have access to the AS and the TGS [23]. Moreover, the Kerberos authentication protocol can support delegation. To support delegation, Kerberos TGTs also include a field AUTHORIZATION-DATA, which specifies application restrictions on the delegate, and flags such as FORWARDABLE (the receiver can further delegate TGTs to another client), FORWARDED (this TGT was obtained through delegation), and PROXYABLE (the receiver can delegate service tickets to another client). Kerberos can also provide fast reconnect for handoffs by having all the APs share the same secret with the AS. Then, after the client obtains an AP service ticket for a particular AP, it can reuse the ticket at any AP until the ticket expires. Kerberos has a few disadvantages. It is vulnerable to dictionary attacks, because an eavesdropper can sniff the encrypted ticket that contains the ticket's timestamp, which the eavesdropper can easily guess [18]. Moreover, because EAP identity messages are not protected, Kerberos does not protect the client's identity from an eavesdropper.

EAP-Secure Remote Password (EAP-SRP): The secret-key methods so far described are vulnerable to dictionary attacks. The SRP protocol, proposed by Wu, is one example of secret-key protocols known as strong password protocols, which resist dictionary attacks [19]. Strong password protocols include strong password encrypted key exchange (SPEKE) and augmented encrypted key exchange (AEKE). Informally speaking, SRP is a Diffie–Hellman variant. The Diffie–Hellman key exchange is a way for two entities to agree on a secret key by using asymmetric keys [24]. In SRP, the client and the AS compute these asymmetric keys from their shared secret. They exchange these keys and compute a session key based on them. Rather than checking to see if they share the same master secret (the secret they used to compute the asymmetric keys), they check to see if they agree on the value of the computed session key.

Even though LEAP and Kerberos are well understood and widely deployed, both are vulnerable to dictionary attacks. EAP-SRP overcomes this vulnerability to dictionary attacks using temporary asymmetric keys that are based on the shared symmetric key. EAP-SRP, however, lacks implementation in WLANs.

11.3.3.2 Public-Key Approaches

Unlike the secret-key approach, the public-key approach uses a mathematically connected key pair, a public key, and a private key. If a message is encrypted with the public key, it can be decrypted only with the corresponding private key. To ensure that a client's public key is legitimate and to prevent an imposter from advertising his or her public key as a legitimate client's key, the AS and the client need to establish trust, typically through certification authorities (CAs) trusted independent third parties that issue certificates. CAs sign their certificates using their private keys so that one can verify the validity of the certificate using their public keys. Clients are assumed to have, in advance, a copy of the CA's public key to use for validating certificates. The requirement of well-implemented CAs makes most public-key methods considerably more complicated to deploy than the secret-key methods [16]. In the absence of proper CAs, an imposter might be able to advertise his or her public key as the AS's public key because there is no CA to verify that the key belongs to the AS. We discuss three public-key authentication protocols: EAP-TLS, Identity-Based Authentication, and Greenpass, which is based on the Simple Public Key Infrastructure (SPKI).

EAP-Transport Layer Security (EAP-TLS): The IETF RFC [25] defines the EAP-TLS. It is based on a certificate approach, and requires trusted CAs. The TLS is a standardized version of the Secure Socket Layer (SSL) Protocol, which was developed by Netscape. The EAP-TLS extends the EAP to provide certificate-based authentication for WLANs. The client sends a random number c to the AS. Then AS responds by sending its certificate, cert AS, and another random number s. If the AS wishes to authenticate the client, it also sends a certificate request message at this stage, notifying the client that it should send the client's certificate and digital signature in response. Receiving the certificate from the AS, the client verifies the certificate using the CA's public key. If valid, the client selects another random value, p, encrypts it with the AS's public key, and sends it back to the server. This third random value is called "pre-master secret", to reflect that the value is secret and that it will be used to create the session keys. If the network requires mutual authentication, the client also sends its certificate, certClient, along with the certificate verify message. The former contains the client's public key and the latter is the digital signature of the handshake messages signed by the client's private key, so that the AS can authenticate the client by verifying that the client knows the private key that corresponds to the public key in the certificate. The AS and the client derive the same session key using the random numbers they exchanged and the pre-master secret. At the end of the handshake message, the AS sends a TLS-Finished message which contains the message digest of the handshake messages, including the pre-master secret. The client authenticates the AS by checking to see if the message digest that the AS sent matches the one the client computed. If the AS does not know the private key that corresponds to the server's certificate, then it would not have been able to obtain the pre-master secret and compute the same message digest as the client.

EAP-TLS supports mutual authentication between the client and the AS if the client also has a certificate signed by a CA that the AS trusts. EAP-TLS resists most attacks, including replay and MITM attacks but does not provide a way to authenticate clients who do not have a certificate that are signed by the CAs that the AS trusts.

ID-based cryptography: ID-based cryptography takes advantage of public-key authentication without the complication of certificates. ID-based cryptography is a form of public-key encryption for which the public key can be one's e-mail address or any arbitrary string that identifies the one who holds the associated private key. Gagne in Ref. [26] surveys ID-based cryptography and discusses possible applications of ID-based cryptography. An ID-based encryption scheme consists of four algorithms. The Setup generates the system parameters and a master key. The Extract uses the master key to generate the private key corresponding to an arbitrary string ID, which is the public key. The Encrypt encodes the plaintext using the public-key ID. The Decrypt decodes the cipher texts using the corresponding private key. A trusted authority, namely the private key generator (PKG), can run the algorithm Setup to get the master key. The PKG then runs the Extract at the request of a user who wishes to obtain the private key. After obtaining the private key from the PKG, users can then run the Decrypt to decode the encrypted messages. The authentication protocol is vulnerable to replay attacks because an eavesdropper can simply sniff the random number and the corresponding digital certificate and replay it to gain access to the network.

11.3.3.3 Tunneled Approaches

Finally, we discuss two tunneled methods: PEAP [27] and EAP-TTLS [28]. These authentication protocols have two phases (Figure 11.8). In the first phase, the client authenticates the AS using EAP-TLS, and uses the resulting session key to establish an encrypted tunnel to encrypt their communication. In the second phase, the AS authenticates the client through the encrypted tunnel.

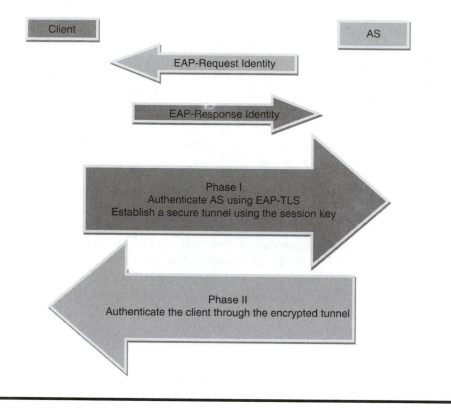

Figure 11.8   Message flow for tunneled approaches.

The choices of the client-authentication methods separate EAP-TTLS from PEAP. While PEAP supports any EAP method, EAP-TTLS supports not only EAP methods but also legacy password protocols such as MSCHAP. The tunnel allows the use of a less secure legacy protocol for client authentication in the second phase. Also, using the tunnel hides the client's identity from an eavesdropper by hiding the EAP Response-Identity message in the encrypted tunnel. To do so, in the first phase of the authentication process, the client's EAP Response-Identity message contains a generic domain name instead of the username. Because the AS does not authenticate the client in the first phase, the AS ignores the client's identity in the EAP Response-Identity message. The client authenticates the AS by the standard EAP-TLS method. When the TLS handshake is finished with the TLS Finished message, the client initiates the second phase by sending his or her username through the encrypted tunnel. PEAP and EAP-TTLS have many advantages. Not only does the tunnel provide identity privacy but also can provide delegation if the client-authentication method that is used in the second phase provides delegation. Moreover, even when the client-authentication protocol is vulnerable to dictionary attacks or replay attack, in the tunneled second phase it becomes no longer vulnerable to these attacks because the eavesdropper sniffing the tunneled session must break the secure EAP-TLS tunnel to mount these attacks on the client authentication.

To conclude, the tunneled protocols satisfy additional desired properties such as identity privacy and delegation. ID-based crypto has the potential to simplify revocation and delegation but missing features, such as lack of implementation and lack of session key derivation, make it an inappropriate choice for securing WLANs.

11.4   Authentication in the Cellular World

New paradigms are now appearing and the computer infrastructure will become important for the mobile phone industry. New phones face all the same issues of security found in a conventional computer network. Products are now being deployed that have both cellular phone data capability and an IEEE 802.11 wireless LAN capability built in. When you are within an AP, you can connect to the Internet using the wireless LAN and at other times you can use the cellular data network at a lower data rate. In Kerberos, for example, it is assumed that users remember passwords. For a large proportion of the world's cellular phones, the secret information is held in a smart card, often referred to as a SIM card. A SIM card or subscriber identity module is a portable memory chip used in some models of cellular telephones. The SIM card makes it easy to switch to a new phone by simply sliding the SIM out of the old phone and into the new one. The SIM holds personal identity information, cell phone number, phone book, text messages, and other data. It can be thought of as a mini hard disk that automatically activates the phone into which it is inserted. The secret information in the SIM card is not known by the subscriber. It is known only by the cellular phone company. When you subscribe to the phone service, the phone company programs a unique SIM card for you and installs the secret information onto it. It can then authenticate you as a subscriber and also encrypt the data going between your phone and the network. When a mobile phone with Wi-Fi LAN capability wants to connect to an AP and authenticate to the network, the phone company can be given the ability to charge you for Wi-Fi LAN network access using the secret information stored in the SIM card.

The GSM standard was designed to be a secure mobile phone system with strong subscriber authentication and over-the-air transmission encryption. The GSM Security Model, as depicted in Figure 11.9, is based on a shared secret between the subscriber network's home location register

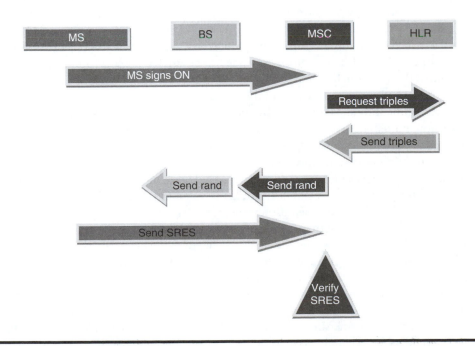

Figure 11.9   GSM security model.

(HLR) and the subscriber's SIM. The shared secret, called Ki, is a 128-bit key used to generate a 32-bit signed response, called SRES, to a random challenge, called RAND, made by the mobile switching center (MSC), and a 64-bit session key, called cipher key (Kc), used for the encryption of the over-the-air channel. When an MS first signs on to a network, the HLR provides the MSC with five triples containing a RAND, a SRES to that particular RAND based on the Ki, and a Kc based again on the same Ki. Each of the triples is used for one authentication of the specific MS. When all triples have been used, the HLR provides a new set of five triples for the MSC [29]. When the MS first comes to the area of a particular MSC, the MSC sends the Challenge of the first triple to the MS. The MS calculates a SRES with the A3 algorithm using the given Challenge and the Ki residing in the SIM. The MS then sends the SRES to the MSC, which can confirm that the SRES really corresponds to the Challenge sent by comparing the SRES from the MS and the SRES in the triple from the HLR. Thus, the MS has authenticated itself to the MSC.

The MS then generates a session key, Kc, with the A8 algorithm using, again, the Challenge from the MSC and the Ki from the SIM. The base transceiver station (BTS), which is used to communicate with the MS, receives the same Kc from the MSC, which has received it in the triple from the HLR. Each frame in the over-the-air traffic is encrypted with a different keystream. This keystream is generated with the A5 algorithm. The A5 algorithm is initialized with the Kc and the number of the frame to be encrypted, thus generating a different keystream for every frame. This means that one call can be decrypted when the attacker knows the Kc and the frame numbers. The frame numbers are generated implicitly, which means that anybody can find out the frame number at hand. The same Kc is used as long as the MSC does not authenticate the MS again, in which case a new Kc is generated. Only the over-the-air traffic is encrypted in a GSM network. Once the frames have been received by the BTS, it decrypts them and send them in plaintext to the operator's backbone network. [29]

The GSM authentication is based on a challenge–response mechanism. The A3/A8 authentication and key derivation algorithms that run on the SIM can be given a 128-bit random number (RAND) as a challenge. The SIM runs operator-specific algorithms, which take the RAND and a secret-key Ki (stored on the SIM) as input, and produce a 32-bit response (SRES) and a 64-bit long key Kc as output. The Kc key is originally intended to be used as an encryption key over-the-air interface, but in this protocol, it is used for deriving keying material and is not directly used. Hence, the secrecy of Kc is critical to the security of this protocol.

Extensible Authentication Protocol-Subscriber Identity Module (EAP-SIM) [30] specifies a mechanism for mutual authentication and session key agreement using the GSM-SIM and by proposing enhancement to the GSM authentication procedures. The lack of mutual authentication is a weakness in GSM authentication. The derived 64-bit Kc is not strong enough for data networks in which stronger and longer keys are required. Hence, in EAP-SIM, several RAND challenges are used for generating several 64-bit Kc keys, which are combined to constitute stronger keying material. In EAP-SIM, the client issues a random number NONCE-MT to the network to contribute to key derivation, and to prevent replays of EAP-SIM requests from previous exchanges. The NONCE-MT can be conceived as the client's challenge to the network. EAP-SIM also extends the combined RAND challenges and other messages with a message authentication code to provide message integrity protection along with mutual authentication. EAP-SIM specifies optional support for protecting the privacy of subscriber identity using the same concept as the GSM, which uses pseudonyms/temporary identifiers. It also specifies an optional fast re-authentication procedure.

The 3rd generation partnership project (3GPP) has specified an enhanced Authentication and Key Agreement (AKA) Architecture for the Universal Mobile Telecommunications System

(UMTS). The third generation AKA mechanism includes mutual authentication, replay protection, and derivation of longer session keys. EAP-AKA [31] specifies an EAP method that is based on the third generation AKA. EAP-AKA, which is a more secure protocol, may be used instead of EAP-SIM, if third generation identity modules and 3G network infrastructures are available. AKA works in the following manner:

1. Identity module and the home environment have agreed on a secret key beforehand.
2. Actual authentication process starts by having the home environment produce an authentication vector, based on the secret key and a sequence number. The authentication vector contains a random part RAND, an authenticator part AUTN used for authenticating the network to the identity module, an expected result part XRES, a 128-bit session key for integrity check IK, and a 128-bit session key for encryption CK.
3. RAND and the AUTN are delivered to the identity module. The identity module verifies the AUTN, again based on the secret key and the sequence number. If this process is successful (the AUTN is valid and the sequence number used to generate AUTN is within the correct range), the identity module produces an authentication result RES and sends it to the home environment.
4. Home environment verifies the correct result from the identity module. If the result is correct, IK and CK can be used to protect further communications between the identity module and the home environment.

When verifying AUTN, the identity module may detect that the sequence number the network uses is not within the correct range. In this case, the identity module calculates a sequence number synchronization parameter AUTS and sends it to the network. AKA authentication may then be retried with a new authentication vector generated using the synchronized sequence number.

11.5   Summary and Open Issues

Wireless LANs growth has made it the fastest growing sector of the communications industry. Unlike a wired end user, a mobile end user can attend meetings, collaborate with other end users, or move to other locations and still be part of the network and access the network resources. Moreover, the ability of wireless LANs to serve as the last mile in broadband access technology will not only increase sales but will enlarge its footprint on the business industry. No single security solution is likely to address all security risks. Organizations should implement multiple approaches to completely secure wireless application access. The most recent challenge for the wireless industry has been to make the real-world deployment of EAP-based wireless networks easier to manage and more secure. To conclude, wireless LAN security has a long way to go. Current implementation of WEP has proved to be flawed. Further initiatives to come up with a standard that is robust and provides adequate security are urgently needed. The 802.1x and EAP are just midpoints in a long journey. Because products are now being deployed that have built in cellular phone data capability and an IEEE 802.11 wireless LAN capability, the GSM security model directly reflects on the security of the wireless connection. The current GSM standard and implementation enables both subscriber identity cloning and call interception. Research in the field of wireless security is still in its infancy and a lot of work still has to be done.

REFERENCES

1. http://www.ieee.org.
2. E. D. Zwicky, S. Cooper, and D. B. Chapman, *Building Internet Firewalls*, Second Edition, O'Reilly and Associates, Inc., 2000.
3. J. Edney and W. Arbaugh, Real 802.11 Security: Wi-Fi Protected Access and 802.11i. ISBN: 0-321-13620-9, 2003.
4. LAN MAN Standards of the IEEE Computer Society. Wireless LAN medium access control (MAC) and physical layer(PHY) specification. IEEE Standard 802.11, 1997 Edition, 1997.
5. J. R. Walker, Unsafe at any Key Size: An Analysis of the WEP Encapsulation, IETF, October 2000.
6. W. A. Arbaugh, N. Shankar, and Y. J. Wan, Your 802.11 Wireless Network Has No Clothes, available at www.issac.cs.berkely.edu/issac/mobicom.pdf, March 2001.
7. J. Walker, Unsafe at any key size: An analysis of the WEP encapsulation, Tech. Rep. 03628E, IEEE 802.11 Committee, available at http://grouper.ieee.org/groups/802/11/Documents/ DocumentHolder/0-362.zi, March 2000.
8. http://standards.ieee.org/getieee802/802.11.html.
9. J. Edney and W. Arbaugh, Real 802.11 Security: Wi-Fi Protected Access and 802.11i. ISBN: 0-321-13620-9, Addison-Wesley, 2003.
10. B. Bowman, WPA Wireless Security for Home Networks, available at http://www.microsoft.com/WindowsXP/expertzone/columns/bowman/03july28.as, 2003.
11. Microsoft, Overview of the WPA Wireless Security Update in Windows XP, available at http://support.microsoft.com/?kbid=815485-8, 2004.
12. L. Blunk and J. Vollbrecht, PPP Extensible Authentication Protocol (EAP), IETF RFC 2284, March 1998.
13. Func Software Inc, EAP Tunneled TLS Authentication Protocol, available at http://www.ee.oulu.fi/research/ouspg/frontier/sota/whitepaper-wots/specs/draft-ietf-pppext-eap-ttls-03.txt, 2003.
14. Port-Based Network Access Control, IEEE Std 802.1X, 2001 Edition.
15. N. Ferguson, R. Schroeppel, and D. Whiting, A simple algebraic representation of Rijndael, Proceedings of Selected Areas in Cryptography, *Lecture Notes in Computer Science*, pp. 103–111, 2001 Springer-Verlag, Ontario, Canada, 2001.
16. IEEE. IEEE Standards for Local and Metropolitan Area Networks: Standard for Port Based Network Access Control. IEEE Std 802.1x-2001, available at http://standards.ieee.org/ getieee802/download/802.1X-2001.pdf, 2001.
17. K.-H. Baek, S. W. Smith, and D. Kotz, A survey of WPA and 802.11i RSN authentication Protocols, Dartmouth Computer Science Technical Report TR2004-524, 2004.
18. T. Wu, A real-world analysis of Kerberos password security, Proceedings of the 1999 Internet Society Network and Distributed System Security Symposium, San Diego, California, February 1999.
19. T. Wu, The SRP Authentication and Key Exchange System. IETF RFC 2945, September 2000.
20. C. Macnally, Cisco LEAP Protocol Description, available at http://www.missl.cs.umd.edu/wireless/ethereal/leap.txt, 2006.
21. Cisco, Dictionary Attack on Cisco LEAP. Tech Note, available at http://www.cisco.com/warp/public/707/cisco-sn-20030802-leap.shtml, August 2003.
22. J. Kohl and B. C. Neuman, The Kerberos Network Authentication Service (Version 5). IETF RFC 1510, September 1993.
23. J. Trostle, M. Swift, B. Aboba, and G. Zorn, Initial and Pass through Authentication Using Kerberos V5 and the GSS-API (IAKERB). IETF Internet Draft, draft-ietf-cat-iakerb-09.txt, October 2002.
24. W. Diffe and M. E. Hellman, New Directions in Cryptography. *IEEE Transactions on Information Theory*, 19(3), 644–654, 1976.
25. B. Aboba and D. Simon, PPP EAP TLS Authentication Protocol. IETF RFC 2716, October 1999.
26. M. Gagne, Identity-based encryption: A survey. *RSA Laboratories Cryptobytes*, 6(1), 10–19, 2003.

27. A. Palekar, D. Simon, G. Zorn, J. Salowey, H. Zhou, and S. Josefsson, Protected EAP Protocol (PEAP) Version 2. IETF Internet Draft, draft josefsson-pppext-eap-tls-eap-07.txt, 2003.

28. P. Funk and S. Blake-Wilson, EAP Tunneled TLS Authentication Protocol (EAP-TTLS). IETF Internet Draft, draft-ietf-pppext-eap-ttls-03.txt, August 2003.

29. M. David, GSM Security and Encryption, available at http://www.net-security.sk/telekom/phreak/radiophone/gsm/gsm-secur/gsm-secur.html, 2004.

30. Nokia and Cisco, EAP SIM Authentication, available at http://www.ee.oulu.fi/research/ouspg/frontier/sota/whitepaper-wots/specs/draft-haverinen-pppext-eap-sim-12.txt, 2003.

31. Ericsson and Nokia, EAP AKA Authentication, available at http://www.ee.oulu.fi/research/ouspg/frontier/sota/whitepaper-wots/specs/draft-arkko-pppext-eap-aka-11.txt, 2003.

Chapter 12

Interference Mitigation in License-Exempt 802.16 Systems: A Distributed Approach

Omar Ashagi, Seán Murphy, and Liam Murphy

CONTENTS

Operating in the license-exempt IEEE 802.16 wireless spectrum is a challenging research issue. The research focus is on deriving intelligent algorithms to mitigate the interference that occurs between different users. In this chapter, we propose an enhancement to our previously published distributed approach to mitigate interference between 802.16 systems operating in close proximity, by introducing a re-listening mechanism to determine whether there are more subcarriers available in the channel than what the base stations (BSs) are currently using.

Simulation results show that the re-listening mechanism offers a 100 percent throughput increase for some BSs. Our results also show the general trend of throughput variations between the downlink (DL) and the uplink (UL), due to the differences in transmission power.

12.1 Introduction

The available wireless spectrum is mainly divided into two types: licensed spectrum and License-exempt spectrum. The licensed spectrum is tightly controlled by a regulator, where licenses are issued to the different operators to use different frequency bands. After issuing a license, an operator will have an exclusive access to a specific frequency band, guarantees that no interference from other operators will occur. In license-exempt spectrum the regulator stipulates some rules regarding the transmission power and coexistence policies, but users do not have to obtain licenses to start using it, hence interference in this type of spectrum is anticipated.

It is apparent that the licensed spectrum has advantages over the license-exempt spectrum. However, measurements in [1] shows that the current licensed spectrum is under utilised. Therefore, due to the lack of available spectrum, new approaches to improve and manage the currently licensed spectrum are needed. As a result, a cognitive radio concept [2–4] was introduced to allow other users (secondary users) to share the spectrum with the licensed user (primary user) provided that the secondary users do not interfere with the primary.

Cognitive radio concept is still evolving right now. At the same time, the regulators are becoming more interested in license-exempt spectrum, due to the proliferation of Wireless LAN (WLAN) and Bluetooth and other technologies that use the same 2.4 industrial, scientific, and medical (ISM) band. For example, very recently OFOM the UK regulator has announced the release of two bands in the higher frequency between 59–69 and 102–105 GHz for unlicensed use.

The 802.16 standard, which is the focus of this chapter, supports the use of license-exempt mode of operation. However, the standard does not specify mechanisms either to share the spectrum or to mitigate interference between different users operating in the same channel, apart of a rudimentary dynamic frequency selection mechanism, which is used to scan the channels at the system startup. For successful operation, this mechanism requires a number of free channels to be available, which contradicts the essential nature of license-exempt operation.

The 802.16 standard [5] stipulates that the orthogonal frequency division multiplexing (OFDM) [6] mode of operation is used in unlicensed-spectrum deployments. In previous work [7], we have developed architectures and algorithms based on controlling the OFDM subcarriers to support the operation of a number of license-exempt 802.16 systems. The algorithms include listening, broadcasting, adapting, synchronization, and backoff. In this chapter, we propose a re-listening algorithm to allow some BSs to increase their resources after adapting their subscribers. Also, we added a functionality to the simulator to determine the modulation and coding scheme for the BSs and subscriber stations (SSs) transmissions.

The rest of the chapter is organized as follows. In Section 12.2 we describe some of the related works in the area. In Section 12.3 we give an overview of the 802.16 system. In Section 12.4 we discuss our proposed distributed approach to mitigate the interference between 802.16 systems. We describe the simulator and discuss the results in Section 12.5. Finally, we conclude the chapter in Section 12.6.

12.2   Related Work

A large amount of research has been conducted to address the problem of interference occurring between license-exempt wireless systems. Some of these research findings are discussed in the following subsection. Also, the cognitive radio concept and some of the proposed mechanisms to share the licensed spectrum are described later in this section.

12.2.1   *Interference between Wireless Systems Operating in License-Exempt Bands*

In previous work, we proposed a MAC layer-based approach to mitigate the interference between different IEEE 802.16 systems operating in license-exempt spectrum [8,9]. This approach was influenced by the thinking in 802.16h standard. It allows a BS and its associated SSs to transmit at random times while others remain silent. After further investigation, it became apparent that this scheme is too limited and requires too many assumptions regarding the PHY layer. Therefore, we proposed a PHY layer-distributed approach based on OFDM to address this problem [7]. It operates by dividing the OFDM subcarriers between the interfering 802.16 systems. To evaluate the performance of this approach, we proposed three centralized approaches [10]. Unlike the distributed approach, they are based on full knowledge of the system, especially the nodes locations. These three approaches have emphasis on throughput maximization, fairness, and a combination of fairness and throughput maximization.

Interference between different wireless technologies operating in the 2.4 GHz ISM band has been widely addressed in the literature, particularly in the 802.11 WLAN and Bluetooth contexts. These efforts vary between experimental, simulation, and modeling approaches. Experimental results that demonstrated the mutual interference between WLAN and Bluetooth are presented in Ref. [11]. The results showed that 802.11b signals have a significant impact on the performance of the Bluetooth. This is because the latter does not implement a carrier-sensing protocol. Results of simulation and modeling approach of the same type of interference has been published in Ref. [12]. The authors showed the impact of both technologies on each other's performances. They also indicated that the WLAN device suffers most from interference when the Bluetooth device is transmitting voice traffic. Solutions to enable coexistence between these technologies have been demonstrated in many papers, such as in Refs. [13–16].

Interference between other technologies operating in license-exempt bands has also been studied. In Ref. [17] the authors addressed the coexistence issues between 802.11a and Hiper Lan/2 in 5 GHz UNII band. They proposed a solution based on interworking or communications between the two technologies, which requires changes in these two existing standards. Coexistence issues in the same 5 GHz band, but this time between 802.16 and 802.11a, have been addressed in Ref. [18], in which an approach to solve this type of interference problem is proposed. The proposal requires some modifications to the 802.16 MAC to not to interfere with 802.11a system.

The above solutions are interesting. However, they are not suitable to mitigate the interference in 802.16 systems discussed in this chapter. This is because the 802.16 system architecture is different from 802.11 and Bluetooth. Moreover, 802.11 employs carrier sensing and collision-avoidance mechanisms and Bluetooth uses a spreading technique to avoid interference, whereas 802.16 does not implement any of these techniques.

12.2.2 Cognitive Radio and Spectrum Sharing

Cognitive radio is built on software defined radio (SDR). It aims to achieve an efficient spectrum utilization, by which cognitive radio users or secondary users will be able to operate in licensed spectrum without interfering with the licensed holders or the primary users. It operates by allowing the secondary users to intelligently avoid primary users' activities, and adapt their spectrum accordingly.

The standard community is currently looking into developing the first cognitive radio-based standard IEEE 802.22 [19,20] for wireless regional area network (WRAN). The goal is to use the television spectrum, which is mostly idle. After taking this goal into account, the authors of Ref. [21] have taken a step ahead and proposed cognitive PHY and MAC layers to achieve dynamic spectrum access to share the licensed TV bands. The authors in Ref. [22] described a mechanism for ultra-wideband (UWB) systems to share the licensed 802.16 spectrum.

Although the cognitive radio-based solutions discussed also above solve interference issues, but they differ from the solution proposed in this chapter. This is because they focus on avoiding the interference, which may be caused by the secondary user to the primary user. Moreover, they focus on utilizing the underutilized licensed spectrum and not on license-exempt spectrum operation. However, in Ref. [23] the authors proposed a solution based on cognitive radio to achieve spectrum sharing between 802.16a and 802.11b, which operates in license-exempt spectrum. Their approach is based on assuming a control channel to signal information related to spectrum usage by the different systems close to each other: they can then use this information to adapt their spectrum usage, and power control, to reduce the amount of interference. Again, this solution is different from our solution as it assumes coordination between the systems that is not assumed in our chapter.

12.3 IEEE 802.16 System Overview

The 802.16 system consists of a BS and number of SSs associated with it. The BS controls the uplink and the downlink transmissions. The system is designed to provide high-bandwidth broadband wireless connectivity for residential and enterprise users. The standard supports a wide range of frequency bands, which include licensed and license-exempt bands. As a result, the standard proposes different physical layers. More details about the Physical (PHY) and the MAC layers of the IEEE 802.16 system are given in the following subsections.

12.3.1 MAC Layer

The MAC layer is designed to support point-to-multipoint (PMP) connection. In addition, the standard also considers mesh connection. Mesh MAC protocol is outside the scope of this chapter [24]. In the PMP MAC, the data is transmitted using time division duplexing (TDD) or frequency division duplexing (FDD) frames. The standard stipulates the use of a TDD scheme in the license-exempt mode of operation. Different frame sizes are supported that vary between 2.5 and 20 ms.

The TDD frame is divided into two subframes, an uplink and a downlink subframe. The downlink subframe is broadcast by the BS, which include data and control information to the subscriber: access these data in a TDM fashion. The SSs receive the control information from the BS which include their data slots location in the downlink subframe and uplink grants location. TDMA is used in the uplink for the subscribers to access the channel.

Bandwidth request slots are allocated by the BS in the uplink subframe for the subscribers to request bandwidth. The BS receives these requests and grants bandwidth to the SSs according to their quality-of-service (QoS) class. The standard supports four different QoSs classes: unsolicited grant service (UGS), real-time polling service (rtPS), non-real-time polling service (nrtPS), and best-effort (BE) service.

12.3.2 PHY Layer

The IEEE 802.16 standard defines four different PHY layers that cover a wide frequency range, and support line-of-sight and non-line-of-sight connections. These PHY layer specifications are

- WirelessMAN-SC: 10–66 GHz uses a single carrier modulation and requires line-of-sight connection
- WirelessMAN-SCa: 2–11 GHz uses a single carrier modulation and supports non-line-of-sight connections
- WirelessMAN-OFDM: 2–11 GHz based on 256-carrier OFDM, robust against multipath and supports non-line-of-sight connection
- WirelessMAN-OFDMA: 2–11 GHz a 2048-carrier orthogonal frequency division multiple access (OFDMA) based on OFDM scheme, which provides multiple access by assigning a different set of subcarriers to each user

Of these four PHY layers, the OFDM PHY layer is the most common, also it is mandatory in the license-exempt mode of operation because of its immunity to multipath and channel fading, which causes inter-symbol interference. The OFDM PHY provides the setting of this chapter, hence the rest of this chapter will focus primarily on the OFDM PHY.

In OFDM, the channel is divided into a number of independent subcarriers. The OFDM system can be designed to be adaptive to allow the use of different modulation and coding schemes on different subcarriers according to their interference conditions. Also, in extreme interference conditions (or optionally), some of these subcarriers can be deactivated. The adaptive modulation and coding scheme features can be used to achieve various bit rates for different users, to enhance the overall system performance.

The OFDM PHY supports various channel sizes 1.25–20 MHz with 256 OFDM subcarriers. Of these 256 subcarriers, 192 are used for user data subcarriers and 8 pilot subcarriers are used for various estimation purposes. 28 lower- and 27 upper-guard subcarriers used to protect the OFDM signal from interference caused by adjacent channel transmissions. Owing to the fact that the OFDM PHY of interest operates in license-exempt spectrum, interference between different 802.16 systems may arise. The current version of the standard does not propose a mechanism to mitigate this interference. Therefore, Section 12.4 discusses our proposed distributed OFDM-based mechanism to mitigate this interference.

12.4 Distributed Algorithms for Interference Mitigation in 802.16 Systems

The type of interference considered here is the interference that occurs when two or more 802.16 systems operate in close proximity on the same channel. The OFDM modulation used in these systems divides the channel into a number of independent subcarriers; therefore, if these subcarriers could be allocated appropriately between the interfering systems, then interference could be eliminated. In Ref. [7] we proposed a distributed approach to apportion these subcarriers between the interfering systems. To determine the performance of this approach we proposed a centralized approach [10], which requires full knowledge of the system parameters, especially the stations' locations. The distributed approach differ from the centralized approach where it does not require any knowledge of the network topology. It also assumes that there is no synchronization between the interfering systems. Moreover, the BSs operate independently by selectively activating and deactivating the OFDM subcarriers to avoid interference, and to achieve spectrum sharing between the interfering systems.

In the distributed approach, M guard subcarriers are introduced between the data subcarriers of the interfering systems. The function of these subcarriers is the same as for the lower- and the upper-guard subcarriers in OFDM systems: to avoid the leakage of the station's signal to the adjacent station subcarriers that can cause harmful interference. In the proposed system, the total number of OFDM subcarriers K are divided into the standard lower- and upper-guard subcarriers at either end of the band, data subcarriers, and M guard subcarriers between each group of subcarriers: the number of M guard subcarriers and the number of data subcarriers are related to the number of interfering systems in the channel. In this chapter, the set of subcarriers a BS operates on are referred to as a subchannel. The number of subcarriers in a subchannel should not be less than Y. For example, if there are two BSs operating in close proximity and on the same channel as depicted in Figure 12.1, the K OFDM subcarriers of these BSs stations will be divided as follows:

- L lower-guard subcarriers
- $[(K-L-U)-M]/2$ data subcarriers for the first BSs
- M guard subcarriers
- $[(K-L-U)-M]/2$ for the second BS
- U Upper-guard subcarriers

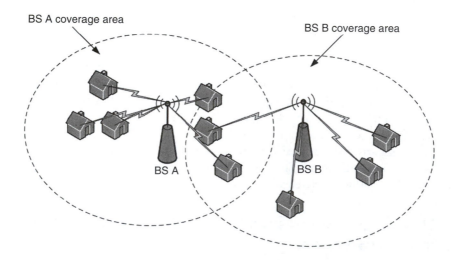

Figure 12.1   Interference scenario.

To generalize this, the following equation can be used to determine the number of data subcarriers D a BS can have in the channel:

$$D \quad \frac{(K-L-U)-(N-1) \times M}{N} \tag{12.1}$$

where N is the number of BSs in the channel

A number of algorithms are implemented in this approach. These algorithms are used in the BSs and in the SSs to perform or help the subcarriers allocation and hence to mitigate the interference. These algorithms are described in the following subsections where the BS algorithms follow first and then the SS algorithms are described. In these algorithms we assume that Y is the least number of subcarriers a BS can operate on.

12.4.1 BS Algorithms

In a high level description of the system operation, when a BS is activated, it first starts listening on the channel to determine its conditions. If the BS finds the channel or subchannel empty, it obtains it and starts to operate; otherwise, it starts to broadcast to make the other stations aware of it. The other stations will backoff and adapt their subcarriers to allows the broadcasting BS to join in. After that, the BS starts listening again to determine the empty subchannel vacated by the stations and obtains it, then it starts to operate. The following gives more details about the listening, the broadcasting, the backoff, and the adapting algorithms.

Listening: The BS initiates the listening process immediately after its activation or after a broadcasting process. If this is so, the BS will be able to determine the channel conditions that are one of the following: the channel is fully occupied, the channel is empty, or the channel is occupied but there is an empty subchannel or subchannels. The BS carries out this process by sensing the channel using an energy detector for a time t_l and determines the status of each subcarriers whether they are used or not. In the case where all the subcarriers are not used, the BS obtains them and starts to operate. If a number of consecutive subcarriers greater than or equal to E (where E is greater than or equal $2M$ plus Y) were not used then a subchannel is available, in that case the BS obtains the subchannel and starts to operate. In the case where the BS finds all the subcarriers busy and no available subchannel, it starts a broadcasting process.

Broadcasting: In the case when the channel is fully occupied, where the BS cannot obtain a subchannel after the listening process, it initiates a broadcasting process. The objective of this process is to make the other BSs in the area aware that a new BS has been activated, and it is looking for subcarriers to operate on. This broadcasting is done by activating a specific set of subcarriers. This set is specified for the BSs. Moreover, the BS starts transmitting signals on those subcarriers with full power for a time t_b. During this, when BSs and SSs in the area of the broadcasting BS hear the broadcasting signal, they too start broadcasting. This process repeats until the broadcasting signals propagate throughout the network. After the broadcasting process, the SSs that have been involved in the broadcasting backoff. At the same time, the BSs start adapting their subcarriers to allow the new BS to join in. In this approach, the new BS is allowed to broadcast only once, if it does not obtain a subchannel, then it is permanently deactivated. This is done to avoid introducing instability to the system. This situation occurs when the number of users in the channel has reached its maximum.

Backoff: The BSs initiate the backoff process after hearing an SS broadcasting signal. The BSs that hear this signal recognize that there is a new SS looking for its BS. Therefore, the BSs stop their activity and randomly backoff for a time t_{bf}. When time t_{bf} expires for a BS, it becomes active to allow the SS to connect to it. If the SS successfully connects to the BS, it sends its channel information that is the neighbors subchannel locations in the channel to the BS. The BS starts analyzing this information together with its channel information. After this, the BS determines whether there is an empty subchannel available or not. If there is, the BS obtains it and starts to operate; otherwise it starts broadcasting when it fails to obtain a subchannel. In the case where the BS fails to connect to the SS, the broadcasting process will be repeated.

Reconfiguration: After the broadcasting process, the BSs start to reconfigure their subcarriers to allocate a subchannel for the new BS. To do this, the BSs first determine the number of BSs in the channel. This is done by analyzing the OFDM subcarriers in the channel. After determining the number of BSs in the channel, the new BS calculates the number of subcarriers in its subchannel by applying Equation 12.1. It then uses Equation 12.2 to determine the subchannel location in the channel. According to these calculations, all the BSs will decrease the number of subcarriers in their subchannels and shift their subchannels' location to the left. As a result, a subchannel will be located at the rightmost subcarriers.

The following equation is used to determine the index of the data subcarriers in the subchannel:

$$\xi(\phi) \quad [(\phi - 1) \times D] \quad [(\phi - 1) \times U] \tag{12.2}$$

where
ξ is the first data subcarrier index
ϕ is the BS subchannel ID

In a case where the number of subcarriers allocated to a particular BS is Y, this BS will not perform any reconfiguration. If this was the case for the BS occupying the rightmost subcarriers but not for some of the others, therefore, the newly allocated subchannel will not be at the rightmost subcarriers; rather it could be any where else in the channel. For this, the new BS always starts a listening process to determine its subchannel location. The listening process is performed after a backoff period to allow all the BSs and their backedoff SSs to become active again, so it will be easy for the new BS to determine the subchannel location and avoid using other BSs subchannels. However, there is a possibility that no subchannel can be allocated for the new BS. This happens when the number of interfering BSs of all the neighboring BSs has reached a maximum.

Re-listening: The BSs initiate this process to determine whether there are more subcarriers available in the channel than what they are currently using. This process is started after a new SS connects to the BS. Each BS performs this process and instructs its SSs to do the same after the expiry of its re-listening timer (which was set beforehand). This timer is a random time generated by the BS multiplied by the number of subcarriers in its subchannel. This is done so that the BS with the lower number of subcarriers can get a re-listening chance first. After a BS and its SSs perform this process, the BS checks if there are more subcarriers available or not. If there are, the BS inform its SSs about the change of its subchannel location or the increase of its subchannel subcarriers and starts operating on them.

12.4.2   SS Algorithms

When an SS is activated, it starts listening and scans the channel to determine the subchannel locations of the BSs operating in its area. After that, the SS starts to synchronize and attempts to connect to each of the BSs in the area. If the SS does not connect successfully to its BS, it starts a broadcasting process. The BSs that hear the broadcasting signals will backoff to allow the SS to find its BS. The listening, synchronization, broadcasting, and backoff algorithms are explained in more detail below.

Listening: The SS listens on the channel to determine the BSs, subchannels locations. The SS initiates this process for a time t_l after its activation or after a backoff. The SS stores the channel information and communicates it to its BS when it connects to it. In the listening process, the SS begins by sensing the channel and then monitoring the subcarriers to determine the locations of the subchannels and the number of subcarriers in each channel. The SS realizes that there is a subchannel when it finds a set of consecutive subcarriers bounded with M guard subcarriers. If two BSs' subchannels overlap (due to the overlap of their coverage areas) the SS does not realize this overlap until it attempts to connect to them. After the expiry of t_l, the SS starts a synchronization process and then attempts to connect to each of the BSs until it finds its own BS.

Synchronization: The synchronization process is for the SS to synchronize to its BS. The synchronization process is performed after the listening process. During Synchronization the SS attempts to synchronize to the neighboring BSs. When the SS successfully synchronizes and connects to its BS, they exchange channel information and higher layer information and start operating. If the SS fails to synchronize to a BS it realizes that this BS subchannel is overlapping with another BS, so it stores this subchannel location. In the case where the SS fails to connect to its BS after synchronizing with the rest of the BSs, the SS then starts to broadcast on this subchannel.

Broadcasting: The SS starts to broadcast when it does not find its BS after performing the synchronization process. The broadcasting process is done by activating a specific set of subcarriers in the subchannel that failed to synchronize to the BSs operating in it due to interference between them. These specific set of subcarriers are specified for the SSs. After determining the set of subcarriers in the subchannel, the SSs starts to transmit signals with full power on them. From this, the BSs operating in this subchannel will realize that there is an SS looking for its BS; so they will backoff to allow the SS to find its BS.

Backoff: An SS backs off for a time t_{bf} in two cases: when it hears another SS broadcasting signals, or after a BS broadcasting process finishes. This is done to avoid interfering with the subcarrier adaptation in the case of a broadcasting BS, or to allow the broadcasting SS to connect to its BS. Before the backoff, the SS stores the subchannels locations of BSs operating in its area. During the backoff, the SS listens on the channel to determine if these BSs become active, then it starts a synchronization process to connect to its BS again.

12.5   System Performance Evaluation

This section describes the simulator we have developed to evaluate the approach proposed in Section 12.4. It is followed by a description of the test scenarios and then presentation and discussion of the results obtained from the simulator.

Table 12.1   Receiver SNR Assumptions

Modulation	Coding Rate	Receiver SNR (dB)
QPSK	1/2	9.4
	3/4	11.2
16-QAM	1/2	16.4
	3/4	18.2
64-QAM	2/3	22.7
	3/4	24.4

12.5.1 Simulator

To determine the performance of our approach a C simulator was implemented. The simulator is designed to model topologies with different numbers of BSs and SSs. The BSs and the SSs are generated randomly in an area of 250 km². The maximum radius of a BS is 3 km and the SSs are generated within this limit. One license-exempt channel is assumed. Free-space channel model based on Friis propagation model is used. Interference arises at an SS or BS receiver when its SNR is less than the SNR threshold. In the simulator, the BS transmits with full power in the downlink. The SS uses a rudimentary power control algorithm based on the Friis to send data on the uplink.

In the simulator, the BSs and the SSs are activated at random times. When a BS obtains a sub-channel and starts to operate or when an SS station connects to its BS, the other BSs and SSs measure their SNRs and then adapt their modulation, if there are any changes between their current SNR measurements and their previous SNRs. The supported modulation and coding schemes, and their SNR thresholds, are given in Table 12.1. A new OFDM parameters are used to increase the system capacity and performance. The new OFDM parameters together with other simulation parameters are given in Table 12.2. The system is assumed to be operating under saturation conditions where the BSs and SSs stations are active all the time.

Table 12.2   Simulation Parameters

OFDM Symbol Time	$94.6\,\mu s$
K	2048
L	38
U	38
M	38
Y	62
Antenna gain	7 dB
Receiver sensitivity	−69 dBm
Maximum Tx power	1 W
Channel bandwidth	20 MHz

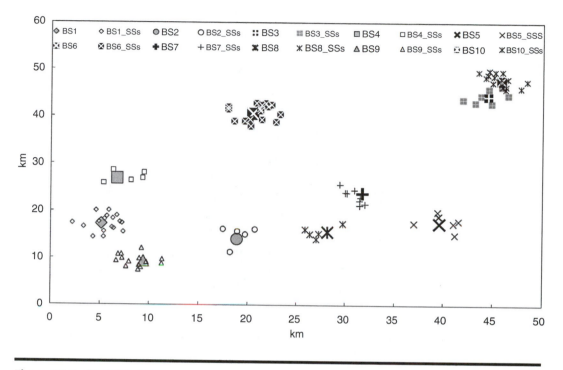

Figure 12.2   Test topology.

In this chapter, a topology of 10 BSs and 100 SSs depicted in Figure 12.2 is used. A total of ten simulations were performed on this topology. Each simulation was run for one minute. The results are presented and discussed in Section 12.5.2.

12.5.2   Results and Discussion

In this subsection the BS throughput, the BSs DL/UL throughput, the stations starting order, and the re-listening results are discussed.

BS throughtput: Figure 12.3 shows the average system throughput per BS. It can be seen that the average throughput of the BSs located at the edge of the topology is greater than the average throughput of the BSs located in the middle, e.g. BS1 and BS9 throughput is low compared to the other BSs. However, this is not always the case, as some BSs located at the edge have an average throughput, which is less than some BSs located in the middle. This is clearly demonstrated by comparing BS6 and BS2 average throughput. This is because BS6 has 15 SSs and BS2 has only 5 SSs.

DL/UL throughput: The results also show the general trend of the variation of the downlink and the uplink throughput due to the differences in their transmission power, which is related to the SSs locations in the uplink. In our approach, the downlink throughput is always greater than the uplink throughput, which depends on how close the SS is to its BS. Further, a maximum power is used in the downlink transmission and controlled power in the uplink transmissions and therefore, when an SS is located close to its BS, its downlink power will be much greater than its uplink power. As a result, a higher order modulation scheme is used in the downlink while the uplink modulation

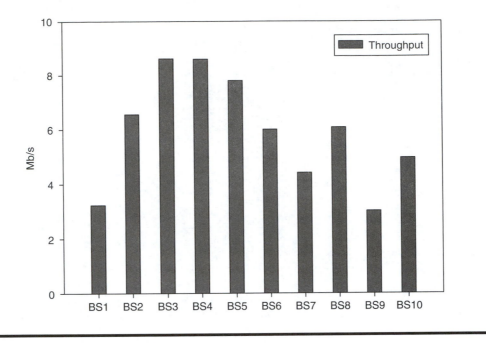

Figure 12.3 Average BSs throughput.

scheme will depend on how close the neighboring stations are. This trend is clearly illustrated in Figure 12.4 in which the average downlink throughput for all the BSs is greater than their uplink throughput.

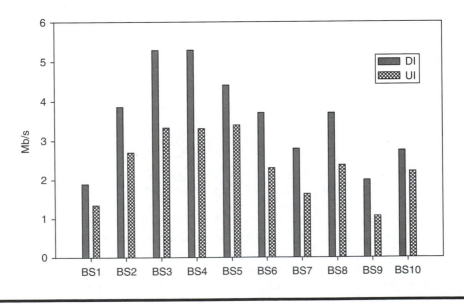

Figure 12.4 Average BSs DL and UL throughput.

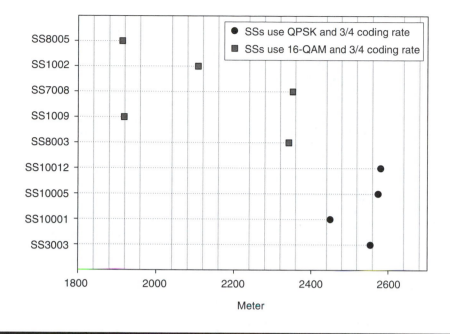

Figure 12.5 **Subscriber stations with different distance from their BSs use 16-QAM (Quadrature Amplitude Modulation) and QPSK (Quadrature Phase Shift Keying) modulation.**

In addition to the distance between the SSs and their BSs, the DL and the UL throughput is also affected by the transmissions of the far neighboring stations that use the same subcarriers. More specifically, these neighbors may have an impact on the SNR calculations, hence on determining the modulation and coding scheme. In some scenarios a lower order modulation is used for SSs closer to their BSs from other SSs who are farther away from their BSs. This is demonstrated in Figure 12.5 in which we can see comparison between two sets of SSs. In the first set, the SSs use QPSK modulation and 3/4 coding rate and they are close to their BSs compared to the second set who are farther away from their BSs and they use 16-QAM modulation and 3/4 coding rate.

Stations starting order: The stations' starting order is observed to be a significant issue, which impacts the subcarriers distributions and hence the overall system performance. Two experiments of the ten we carried out are considered as an example here. In these experiments, when the BSs with the fewest interfering neighboring are activated last, the system performs better than when they are activated first. These occurrences are demonstrated in Figures 12.6 and 12.7. This is because, when the BSs with the fewest neighboring are activated first, their subcarriers will be affected by all the broadcasting processes performed to add new BSs to the system, because the broadcasting signals propagate throughout the network as explained before. On the other hand, when they are activated last, less broadcasting will occur in the network, as free subchannels will be available for them to operate in. The results shown in Figure 12.8 are comparable with the previous results in which throughput improvement was achieved when the BSs with the fewest interfering neighbors are activated last.

Re-listening results: The re-listening enhancement introduced in this chapter considerably improves the system performance by increasing the number of subcarriers for some BSs. In the given scenario, some BSs have doubled their throughput after re-listening. For example, BS2 throughput

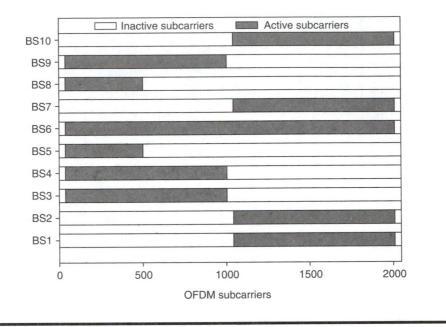

Figure 12.6 Stations with less interfering neighbors started first: worse subcarriers distribution.

illustrated in Figure 12.9 has improved by around 100 percent. Apparently, this corresponds to increase of its subcarriers as shown in Figure 12.10 in which they also increased by around 100 percent after the re-listening process. Although, the increase or the decrease of a BS throughput is not necessary related to the increase of its subcarriers. This can sometimes happen when an SS

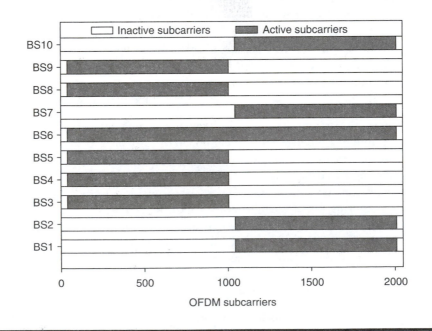

Figure 12.7 Stations with less interfering neighbors started last: better subcarrier distribution.

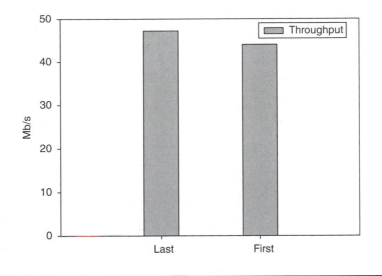

Figure 12.8 Average system throughput per BS.

connects to its BS after a broadcasting process. Furthermore, the BS subcarriers location might change after the SS connects to it. In some cases, when the BSs changes its subcarriers location, the SSs change their modulation and coding schemes to lower or higher order, which results in throughput variations for the BSs. For example, a throughput increase because of the use of higher order modulation is shown in Figures 12.11 and 12.12. In contrary, a throughput decrease because of the use of lower order modulation is shown in Figures 12.13 and 12.14.

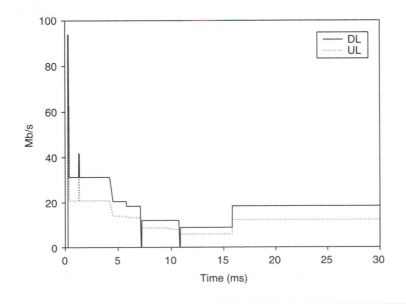

Figure 12.9 BS2 throughput.

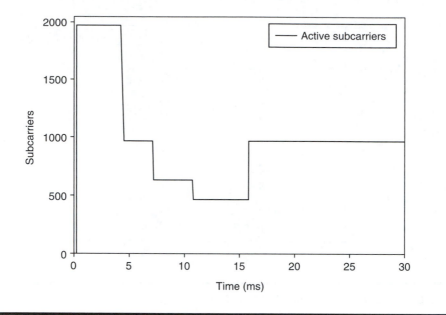

Figure 12.10   BS2 subcarriers.

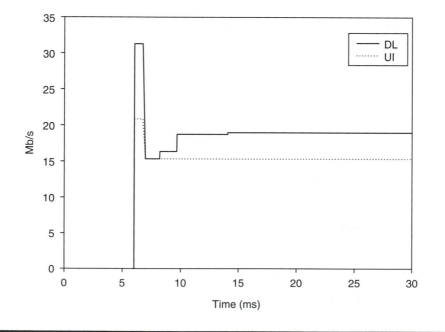

Figure 12.11   BS5 throughput.

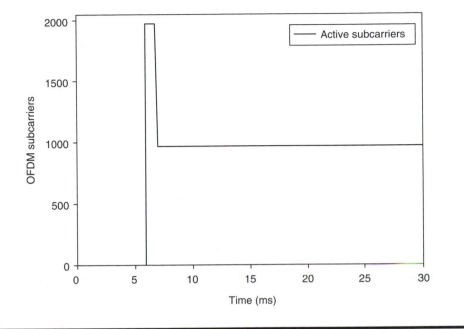

Figure 12.12   BS5 subcarriers.

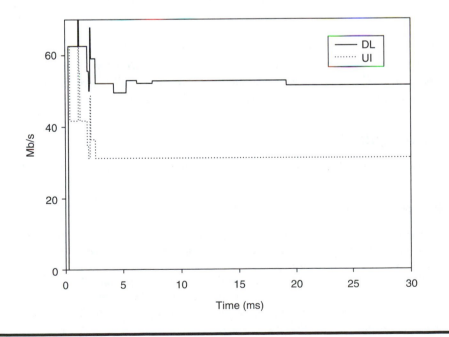

Figure 12.13   BS6 throughput.

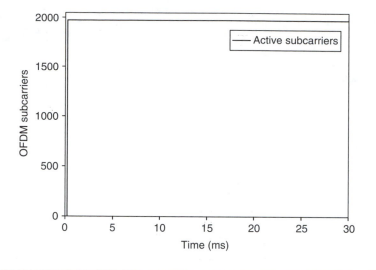

Figure 12.14 BS6 subcarriers.

12.6 Conclusion

In this chapter, a distributed approach to mitigate the interference between 802.16 systems operating in license-exempt spectrum is discussed. The approach is based on a number of algorithms implemented in the BS and the SS stations to selectively activate and deactivate the OFDM subcarrier to mitigate the interference impact and achieve spectrum sharing between the interfering systems.

The results showed a general trend of variation of the downlink and the uplink throughput. It also showed that the system is sensitive to the stations' starting order where better throughput is achieved when the BSs with the fewest neighbors are activated last. By introducing the re-listening mechanism, the throughput of some BSs has increased by around 100 percent.

Acknowledgment

The support of the Informatics Research Initiative of Enterprise Ireland is gratefully acknowledged.

REFERENCES

1. Shared Spectrum Company, *New York City Spectrum Occupancy Measurements*, September 2004.
2. J. Mitola and G. Q. Maguire, Cognitive radio: Making software radios more personal, *IEEE Personal Communications*, 6, 13–18, August 1999.
3. J. Mitola, Cognitive radio for flexible mobile multimedia communications, IEEE International Mobile and Communications (MoMuc 1999), San Diego, California, USA, November 1999.
4. S. Haykin, Cognitive radio: Brain-empowered wireless communications, *IEEE Journal on Selected Areas in Communications*, 23, 201–220, February 2005.
5. IEEE 802.16-2004, IEEE standard for metropolitan area network, Air Interface for Fixed Wireless Access, October 2004.
6. I. Koffman and V. Roman, Broadband wireless access solutions based on OFDM Access in IEEE 802.16, *IEEE Communications Magazine*, 40, 96–103, April 2002.

7. O. Ashagi, S. Murphy, and L. Murphy, A distributed approach to interference mitigation between OFDM-based 802.16 systems operating in licence-exempt spectrum, IEEE ICC, Glasgow, June 2007.

8. O. Ashagi, A. G. Ruzzelli, S. Murphy, L. Murphy, and J. Murphy, Performance modeling of a distributed approach to interference mitigation in licensed exempt IEEE 802.16 Systems, IEEE 1st International Symposium on New Frontiers in Dynamic Spectrum Access Networks (IEEE DySPAN 2005), Baltimore, Maryland, USA, November 2005.

9. O. Ashagi, S. Murphy, and L. Murphy, Mitigating interference between IEEE 802.16 systems operating in license-exempt mode, 3rd International Conference, Wired/Wireless Internet Communication (WWIC 2005), Xanthi, Greece, May 2005.

10. O. Ashagi, S. Murphy, and L. Murphy, Centralised approaches to subcarrier allocation for OFDM-based 802.16 systems operating in license-exempt mode, 2nd International Conference on Cognitive Radio Oriented Wireless Networks and Communications (CROWNCom 2007), Orlando, Florida, USA, August 2007.

11. R. Punnoose, R. Tseng, and D. Stancil, Experimental results for inteference between Bluetooth and IEEE 802.11b DSSS systems, IEEE 54th Vehicular Technology Conference (IEEE VTC 2001/Fall), Atlantic City, New Jersey, USA, October 2001.

12. N. Golmie, R. E. Van Dyck, and A. Soltanian, Interference of Bluetooth and IEEE 802.11: Simulation modeling and performance evaluation, ACM 4th International Symposium on Modelling Analysis and Simulation of Wireless and Mobile Systems (MSWiM 2001), Rome, Italy, July 2001.

13. J. Lansford, A. Stephens, and R. Nevo, Wi-Fi (802.11b) and Bluetooth: Enabling Coexistence, *IEEE Network Magazine*, 15, 20–27, September/October 2001.

14. A. Conti, D. Dardari, and G. Pasolini, Bluetooth and IEEE 802.11b coexistence: Analytical performance evaluation in fading channels, *IEEE Journal on Selected Areas in Communications* (IEEE JSAC) 21, 259–269, February 2003.

15. C. F. Chiasserini and R. R. Rao, Coexistence mechanisms for interference mitigation between IEEE 802.11 WLANs and Bluetooth, 21st Conference of the IEEE Communications Society (IEEE INFOCOM 2002), New York, USA, June 2002.

16. A. K. Arumugam, A. Doufexi, and P. N. Fletcher, An investigation of the coexistence of 802.11g WLAN and high data rate Bluetooth enabled consumer electronic devices in indoor home and office environments, *IEEE Transactions on Consumer Electronics*, 49, 587–596, August 2003.

17. S. Mangold, J. Habetha, S. Choi, and C. Ngo, Co-existence and interworking of IEEE 802.11a and ETSI BRAN HiperLAN/2 in multiHop scenarios, IEEE Workshop on Wireless Local Area Networks, Boston, USA, September 2001.

18. L. Berlemann, C. Hoymann, G. Hiertz, and B. Walke, Unlicensed operation of IEEE 802.16: Coexistence with 802.11a in shared frequency bands, 17th Annual IEEE International Symposium on Personal, Indoor and Mobile Radio Communications (IEEE PIMRC 2006), Helsinki, Finland, September 2006.

19. C. Cordeiro, K. Challapali, and D. Birru, IEEE 802.22: An Introduction to the first wireless standard based on cognitive radios, *Journal of Communications*, April 2006.

20. IEEE 802.22 Working Group on Wireless Regional Area Networks, http://www.ieee802.22.org/22/.

21. C. Cordeiro, K. Challapali, and M. Ghosh, Cognitive PHY and MAC layers for dynamic spectrum access and sharing of TV bands, IEEE International Workshop on Technology and Policy for Accessing Specturm, (IEEE TAPSA 2006), Boston, USA, August 2006.

22. S. M. Mishara, R. Mahadevappa, and R. W. Brodersen, Detect and avoid: An ultra-Wideband/WiMax coexistence mechanism, *IEEE Communications Magazine*, 45, 68–75, June 2007.

23. X. Jing and D. Raychaudhuri, Spectrum co-existence of IEEE 802.11b and 802.16a networks using the CSCC etiquette protocol, IEEE 1st International Symposium on New Frontiers in Dynamic Spectrum Access Networks (IEEE DySPAN 2005), Baltimore, Maryland, USA.

24. B. Han, W. Jia, and L. Lin, Performance evaluation of scheduling in IEEE 802.16 based wireless mesh networks, *Computer Communications*, 30, 782–792, November 2006.

Chapter 13

QoS Capabilities in MANETs

Bego Blanco, Fidel Liberal, Jose Luis Jodra, and Armando Ferro

CONTENTS

13.1   Introduction

Unlicensed mobile access (UMA) technology addresses the problem of seamless transitions (roaming and handover) between public GSM/GPRS networks and unlicensed spectrum networks using dual-mode mobile phones, thus providing an important step toward the convergence of different wireless technologies. Traditionally, UMA has been associated to cellular/WiFi handsets to provide subscribers high-performance/low-cost voice, data, and IMS service delivery where they spend most of the time, i.e., at home and office [1]. However, unlicensed spectrum technologies are not only circumscribed to infrastructured WLANs but also include a wide variety of emerging ad hoc approaches, which completely reverse the cellular model.

Among these, research and enterprise interest on mobile ad hoc networks (MANETs) has increased exponentially in recent years. These kinds of networks extend mobility to any scenario, as all components of the network play the roles of both node and router. In this way, the nodes can dynamically set up a new network without previously installing any fixed network infrastructure. As a matter of fact, the main goal of MANETs is to provide real movement capability due to its lack of infrastructure.

These features have led both industry and research communities to focus their attention on MANETs, because they are very useful for many applications, such as collaborative environments and communications inside limited areas (i.e., conferences) or communications along battlefields or catastrophic areas.

In the near future, MANET nodes will be installed in cars, ships, trains, and even in every portable device, so that users who share common interest will be able to configure self-organized networks everywhere. In addition, although the current UMA situation depicts the need for handover between "everywhere available" GSM service and "flat rate but discontinuous" WiFi coverage, this will likely change in the near future (due to MANETs and enabling technologies such as WiMax). With the widespread WiFi access, most UMA scenarios will include handover from an almost permanent WLAN access and spurious need for traditional cell coverage. As a result of this, demanded requirements for providing GSM equivalent service in this kind of networks must be considered.

MANET research is gaining ground due to the ubiquity of small, inexpensive wireless communicating devices. Traditional research efforts on MANETs have been mainly focused on finding optimal solutions for MANET routing. The challenge in this field is to design an effective routing protocol that responds to the typical limitations of MANETs, which include high power consumption, low bandwidth, and high error rates, all of them due to the dynamic nature of this kind of network. To overcome these limitations, three types of routing protocols have been proposed: proactive, reactive, and hybrid protocols. Each group applies different routing strategies, which may employ a flat or a hierarchical routing structure.

Nevertheless, there are aspects other than routing that must be addressed to design highly functional and dependable UMA networks capable of replacing GSM functionalities. This becomes especially relevant because, provided the transitions between unlicensed- and licensed-access networks must be transparent to the subscriber, service quality must be the same at both sides. Besides, MANETs will have to work over rather diverse and sometimes adverse situations, which result in additional complexity. Lately, new scenarios like vehicular or sensor networks and portable videogame platforms require reliable MANETs providing a minimum quality of service (QoS). Moreover, the evolution of PDAs, mobile phones, and multimedia devices for vehicles demand more and more stable QoS requirements to the network and this is the aspect that cannot be ignored. At the same time, most dual UMA phones are indeed smartphones, closer to PDAs than to traditional handsets, that makes it possible to include MANET-related technology into them while demanding additional QoS guarantees.

On the other hand, in this kind of environment, there appear additional constraints apart from traditional QoS requirements. For example, battery life, stability, and location must also be taken into account. According to this, MANET routing protocols should work together with a signaling mechanism to achieve end-to-end service quality. Hence, as traditional networks and QoS models do not seem to be suitable to handle these new scenarios, the development of new specific QoS network models become essential and must be carefully analyzed if we want to ensure equivalent service for dual devices in WiFi.

13.2   MANET Environments and QoS

A MANET consists of a group of potentially mobile nodes that coincide occasionally to share information. As the interchange goes by, nodes can be in constant and random movement, so the network must be constantly prepared to adapt. Because of the lack of infrastructure, nodes must organize themselves in the network and set the routes among them without any external help. In general, ad hoc networks must allow the communication between nodes indirectly connected by a sequence of hops across other nodes in a peer-to-peer manner (Figure 13.1). Intermediate nodes act as routers, so the nodes can play both roles, router and host. According to this description, MANET, mesh network, mobile radio packet network, and mobile multi-hop wireless network terms can be considered as synonymous to each other.

MANETs can operate in two ways. The first one is as an isolated network composed only of mobile nodes. The second and most usual one is as a hybrid network composed of mobile and fixed wireless nodes. In the latter case, the function of the fixed nodes is to forward traffic toward the mobile nodes.

MANETs were first used in battlefields and disaster area recovering. Nowadays, however, MANETs are ideal in scenarios that need fast and low-cost deployment without the installation of any infrastructure due to their dynamic and flexible features. That is why different industrial sectors have focused their attention on its market potential and advantages that can be provided to the general public. All in all, MANETs are used where there is a need for a temporary network, without time or chance to build any infrastructure. Table 13.1 shows different application areas for MANETs.

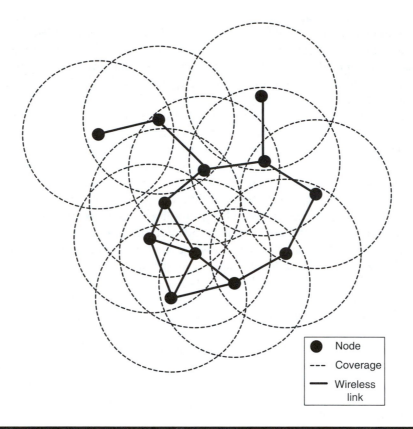

Figure 13.1   MANET.

Table 13.1   Application Areas of MANETs

Military and Emergencies	Industry and Business	Technology Research and Education	Home and Entertainment
Military communications and operation in battlefields	Electronic payment from anywhere	Mobile sensor networks	Wireless access to the Internet
Catastrophic areas	Mobile office	VANETs (vehicular ad hoc networks)	PANs (personal area networks)
Rescue operations	Wireless access to corporate network	Wireless network access in conferences, schools, and universities	Community networks to share Internet access
	Low-cost extension of corporate fixed network		Multiuser online games
			User location services

The primary aim of MANETs is to provide robust and efficient communication by incorporating routing functionality into mobile devices [2]. This mobility is achieved by means of the lack of infrastructure, which allows MANETs to be more flexible and to adapt to those scenarios where fixed technologies cannot arrive, such as catastrophic areas.

The dynamic features of MANETs offer many advantages (see Refs. [3–6] for a detailed compilation and description):

- Ad hoc technology allows users to form new networks or to join existing networks easily.
- Minimum infrastructure support is needed, which implies faster and simpler network deployment so that communication costs come down.
- Creation of user communities to promote cooperation and resource sharing is allowed.
- Ad hoc networks may be used to extend fixed networks to areas not reachable earlier or where it was economically not feasible.
- Lower deployment costs allow new operators to enter the market to offer wireless services.

Unfortunately, this flexibility brings some trade-offs:

- Mobility implies a highly dynamic topology, making route maintenance difficult.
- Heterogeneous devices may have capacity differences.
- Bandwidth of wireless links is much narrower than those in fixed networks.
- Multi-hop communication implies using more than one wireless link, which can worsen the QoS even more, because of the unpredictable links or hidden terminal problems.
- Limited battery life forces to design power-aware and power-efficient protocols.
- There may be selfish nodes that do not cooperate or allow its resources to be used in communications between other nodes.

Main research efforts, led by IETF MANET Working Group, have been focused principally on routing issues. However, it must be borne in mind that MANETs operate in the most diverse situations, which adds even more complexity to the communication. As showed before, new application scenarios have emerged demanding a minimum QoS: vehicular and sensor networks, portable game consoles, or general multimedia (video, voice, etc.) applications (i.e., UMA scenarios). Consequently, it is no longer enough to bring data packets from a source to a destination, but it is also necessary that they fulfill some minimum QoS constraints.

Together with the QoS problems of any network Figure 13.2, MANETs involve dealing with a new set of challenges to guarantee a minimum level of QoS, due to the dynamic behavior and limited resources of such networks. On one hand, these limitations reinforce the need for some kind of QoS support, almost at any level. On the other hand, this sort of environment demands new quality requirements, such as battery supply [7], link stability, [8] or location [9].

13.3 Applicability of Traditional QoS Models to MANETs

The Internet was designed to provide just a best-effort service, to process the traffic as fast as possible, but with neither delay bounds nor losses guarantees. The fast transformation of the Internet into a commercial infrastructure has generated the appearance of some QoS demand. Users making business through the net demand a robust and predictable service. Other users, for example, demand

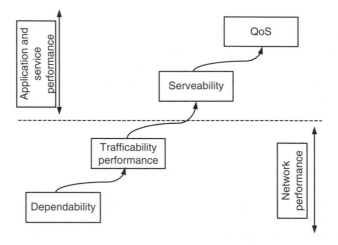

Figure 13.2 Main QoS components according to International Telecommunication Union (ITU).

low-delay-and-jitter service for IP telephony or videoconference (those UMA has to deal with). It is obvious that a QoS model will have to include different classes of service to satisfy the diverse requirements of each application type.

The objective of any QoS model is to define a set of configurable QoS interfaces, which formalize the QoS in the end system and the network, providing an integration frame for the QoS control and management mechanisms [10].

Next, we will expose a brief analysis of the situation of QoS management models for traditional networks, and will evaluate their behavior in wireless environments, and more precisely in ad hoc multi-hop networks. In this way, we will show the adaptability of the models proposed by IETF in the last years to the environments discussed in this chapter.

13.3.1 Integrated Services

The main feature of Integrated Services (IntServ) model is the reservation of resources before data is transmitted. With that objective, a signaling protocol is used to check whether the requirements of each flow can be served all through the path to destination. In that case, a path is assigned to each specific flow, and a new entry with information about the requested QoS requirements (bandwidth, delay, etc.) is added to the tables of the intermediate nodes.

IntServ architecture meant an innovative revolution of the structure of the Internet, whose fundamentals rely on handling all the information about flow state in the end systems. However, IntServ does not adapt to MANETs properly due to two main reasons [11,12]:

■ Signaling: for MANETs with dynamic topology and link capacity, connection maintenance overload often is not worth the initial cost of connection establishment. Thus, Resource Reservation Protocol (RSVP) is not a suitable signaling protocol for MANET.
■ Routing protocols: IntServ imposes the implementation of its four components, that must exist in all the routers, which is not advisable in battery-dependent devices.

13.3.2   Differentiated Services

The operation of Differentiated Services (DiffServ) is based on traffic classification and aggregation of different flows into a limited amount of traffic classes that meet similar service requirements. Network routers handle each packet regarding the traffic class it has been assigned to. This way, each traffic class is managed in a differentiated way, making it possible to assign different priority levels, but without needing to cope with an undefined amount of individual flows, each one with a specific treatment. Given that the granularity of service depends on the classes, the quantity of state information will be proportional to the amount of classes and not to the amount of flows. Hence, DiffServ complies better with the characteristics of MANETs, because of its relative priority scheme to smooth out the severe requirements of other QoS models, providing a qualitative QoS for aggregated flows [11].

Still, there is a point that makes the application of DiffServ in MANETs difficult: classifying, marking, policing, and shaping mechanisms must only be implemented in boundary nodes but the dynamic topology of MANET makes it difficult to establish which nodes are at the boundary to perform these tasks.

13.3.3   G/MPLS

Generalized Multi-Protocol Label Switching (G/MPLS) [13,14] also arises a lot of interest for QoS management in data networks. MPLS appeared as an evolution of Cisco's label switching and operates between link and network layers, by defining a label-based retransmission scheme to establish a connection-oriented service emulation over a datagram network. The labels inserted between link and network layer at every packet entering an MPLS domain are used to access the switching table, which is a faster process than searching a route in an IP table.

However, even when MPLS allows the specification of mechanisms for the management of different kinds of traffic flows and makes link layer and network layer protocols independent, there can be problems in applying a connection-oriented mechanism in a network where the intermediate nodes operate with a high-mobility pattern, which causes rapid connectivity changes, or where the shared medium favors the appearance of interferences or collisions.

In that sense, WMPLS (wireless MPLS) has been defined [15] to adapt the functionalities of MPLS protocols not only to wireless networks but also to ad hoc multi-hop networks. This protocol will be commented on later in this chapter.

13.3.4   Traffic Engineering and Constraint-Based Routing

Flow-control mechanisms of TCP/IP model cannot ensure an efficient operation of the network. The main objective of traffic engineering (TE) is to facilitate reliable and efficient network operation, while optimizing the available resource usage and traffic throughput [16]. In this way, network load is balanced and congestion problems are avoided.

A TE mechanism stipulates the convenient criteria for the network to accomplish traffic performance issues. But these performance multi-dimensional requirements are complex and sometimes antagonistic, making the TE task very difficult.

There are three main components in a TE mechanism [17]:

■ Measurement subsystem: No network can be optimized without having an accurate vision of its state, so that measuring and monitoring prove to be a crucial issue for TE. The data measured and collected is later used to choose between different alternatives of action and also to evaluate the effectiveness of the TE policies. The monitoring can be done at different levels of abstraction, such as, packet level, flow level, or user level features.

◾ Modeling subsystem: The measures taken by the previous subsystem are used to build a representation of the network in terms of traffic characteristics and network attributes. There are two basic TE model approaches: Structural models focus on the organization of the network and its components, whereas behavioral models focus on the dynamics of the network and the traffic loads.

◾ Optimization subsystem: Network performance optimization involves choosing among the existing action alternatives to improve network operation based on the previous analysis of the measured data. The optimization can be corrective/reactive or perfective/proactive. The goal of reactive optimization is to correct a detected inconvenient behavior through the application of the appropriate policies. In proactive optimization, potential operation problems are anticipated and deflected before they happen.

As commented above, TE provides a way to avoid operation complications caused by unbalanced network utilization by means of arranging traffic flows through the network. Constraint-based routing (CBR) is an important tool for automating the TE process. Avoiding congestion and providing smooth performance degradation in the case of congestion are complementary.

CBR [18] minimizes manual intervention needed to reach TE objectives. The final aim of CBR is to make possible on-demand routing with resource reservation combined with conventional routing protocols. The resource availability information to take decisions in CBR is interchanged through routing protocol extensions. Signaling protocols (like RSVP) perform resource reservation and finally, once the route is calculated by means of a routing protocol, it can be commuted via any switching protocol.

When applying CBR protocols, there can be several feasible routes to a destination. But, if some restrictions, such as bandwidth or delay, are applied, the amount of available routes will be reduced. Recent research highlights the convenience of selecting routes that not only meet some QoS requirements but also contribute to optimize the resource utilization. In fact, this optimization eventually benefits long-term QoS provision maintenance, especially in MANETs, where resources all over the network are scarce and cooperation between nodes becomes crucial.

The use of CBR techniques allows a better fulfillment of QoS requirements and achieves the optimization of network resources. Nevertheless, it also has disadvantages such as the increase of control information load, the increase of the size of routing tables, and the establishment of longer routes that take up more resources and the probable link instability.

Because the bandwidth provision does not seem to be a problem for UMA-oriented MANETs, CBR protocols should focus on providing delay-bounded paths, while also trying to optimize bandwidth and power consumption.

13.4   General QoS Support in MANETs

Although MANETs offer a significant solution in many applications, several difficulties come up when these tools require some QoS support. The amount of this kind of applications is growing, mostly because of the arrival of multimedia traffic to MANETs.

Generally, QoS requirements demanded by ad hoc networks are those of classical IP networks plus a set of specific attributes associated to the singular nature of ad hoc networks. Within the first group of parameters, the most significant ones are delay, bandwidth, jitter, and loss packet rate. The main MANET-specific QoS parameters are power consumption and service range area.

Special features of MANETs hinder the QoS support, and therefore, the guarantee of the previous parameters. Reference [5] details a list of the reasons why QoS support in MANETs is such a complex task:

- Shared radio-electric medium, where collisions and interference are common, is very unpredictable, so parameters such as bandwidth or delay cannot be ensured.
- Node mobility causes constant changes in the links, and therefore, in the routing scheme. New links can have completely different features affecting the QoS of the routes.
- Routing protocols should not require excessive processing to save batteries.
- Hidden and exposed terminal problems of wireless networks may also cause interferences.
- Route maintenance and reconstruction should involve a minimum overload in the network.
- Finally, lower security of wireless links must be considered as well.

Next, we describe several proposals to provide QoS in MANETs. The first efforts directed to provide QoS in ad hoc networks were individual actions to improve QoS performance in a single layer. As this approach proved to be insufficient, other MANET-specific models were proposed to cover QoS requirements of the arising mobile applications. Thus, in this section we describe first a QoS model considering each layer individually and, after that, we explain the MANET-specific models.

13.4.1 QoS Models Considering Each Layer Individually

MANETs make use of traditional TCP/IP stack to establish end-to-end communications between nodes. Yet, due to the mobility features and resource shortage of wireless networks, each layer of the TCP/IP architecture requires a series of modifications to adapt its operation efficiently to this kind of network [5].

13.4.1.1 QoS Support in Physical Layer

The time-variant wireless channels of MANETs make channel estimation a difficult task. To accurately synchronize sender and receiver, the receiver must carry out precisely this estimation and make a reliable feedback to the transmitter. Nevertheless, a perfect synchronization in MANETs is almost impossible.

Communications over wireless channels experience noise and collision problems and increase of wireless applications demanding real-time audio and video transmissions will worsen this situation in the future. Thus, the communication improvement techniques over wireless media will not only be just a physical layer task but also a job that must be carried out by the upper layers, for example, by compressing data before being transmitted and so reduce the data amount to be sent. Another possibility would be to adapt transmission power to the dynamic conditions of the network to save battery by reducing transmission power or increase transmission rate by increasing it. These examples lead to the need of a cross-layer design that is discussed later.

13.4.1.2 QoS Support in Link Layer

The data link layer provides the functional and procedural means to transfer data between network nodes across the physical link. To simplify its analysis and implementation, this layer is further split into two sublayers: LLC (Logical Link Control) layer, which provides addressing and link control; and MAC (Medium Access Control) layer, which regulates the access to the physical media and states the frame structure to be sent. Because main challenges of ad hoc networking happen at the

MAC layer as a consequence of the wireless medium, most of the research effort has been focused on developing enhancements at this level.

The design of a MAC protocol lays on four strategic characteristics: channel access, transmission initiation, network topology, and power consumption [19,20].

■ Channel access: a key feature of a MAC protocol is how the channel bandwidth is used. The two main alternatives are single-channel protocols and multi-channel protocols. In single-channel protocols, all the nodes in the network share all the bandwidth for both control and data transmissions (e.g., Carrier Sense Multiple Access [CSMA]). Implicit particularities of these access methods are collisions, making this protocol unsuitable for voice and real-time applications, as transmission delay is damaged by the degradation of the performance with the increase of traffic load. Hence, a single-channel protocol does not fit UMA. However, because CSMA is the most usual MAC layer protocol in WLAN, some performance enhancements have been developed in Multiple Access Collision Avoidance (MACA) [21,22], MACA for Wireless (MACAW) [23], or Floor Acquisition Multiple Access (FAMA) [24]. Two typical problems of wireless networks are hidden and exposed terminal problems. A hidden terminal problem occurs when two nodes that are out of range of each other try to communicate with the same node resulting in a collision. An exposed terminal problem takes place when a node is prevented from sending packets to other nodes because of a neighboring transmission. To avoid hidden and exposed terminal problems in multi-hop wireless networks, there must exist a completely distributed scheme. MACA is a proposal to solve this problem through Request-to-Send/Clear-to-Send (RTS/CTS) dialogs, though this initiative does not entirely eliminate hidden window problem. As an extension of MACA, MACAW is proposed, with the objective of a faster recovery from the hidden terminal collisions. 802.11 specification [25–28] includes both MACA and MACAW proposals to avoid collision through distributed control function (DCF) [29]. This access method allows Carrier Sense Multiple Access with Collision Avoidance (CSMA/CA) to avoid collision and the handshake mechanism improves the performance of DCF in many cases, but also introduces a control overhead that involves suboptimal utilization of the channel and has no support for real-time traffic. This is the reason for the development of point coordination function (PCF) [30], which provides an optional centralized contention-free method for IEEE 802.11 that requires an access point (AP) to coordinate the access to the medium to avoid collisions. The AP polls the stations within its range to later assign a transmission slot without contention. The centralized nature of PCF makes this method unsuitable for the MANET environment, distributed by definition.

Multi-channel protocols separate one channel for control and one or multiple channels for data transmissions, by means of techniques such as time division multiple access (TDMA), frequency division multiple access (FDMA), or code division multiple access (CDMA). Real-time applications perform better with TDMA access, but has the disadvantage of the loss of efficiency in use of the channel, because unreserved slots are not reused by other communications. In Ref. [31] an algorithm is proposed to calculate the available bandwidth in an ad hoc network with terminal that employs TDMA. This algorithm addresses both the end-to-end available bandwidth calculation and the assignment of part of that bandwidth to different flows. Therefore, a possible application of this algorithm could be the establishment of an admission control. In traditional wired networks, the calculation of the available bandwidth along a route is reduced to the minimum bandwidth of a link through the whole path. But time slotted medium access of ad hoc networks makes the calculation of the bandwidth much more complex. In general, it is not only enough to know the available slot number but also the way to assign the free slots at every hop must be determined. The slot assignment can be performed during the control phase when every node learns the free slots with its neighbors. However, the problem of allocating the bandwidth together with the establishment of free slots is an Non-deterministic Polynomial-time (NP)-complete problem. This approach introduces a heuristic algorithm to solve it. Ref. [32] proposes another alternative for QoS providing QoS routing in MANETs. The new routing protocol is based on the previously commented Ad-hoc On-demand Distance Vector (AODV) protocol with QoS support and presupposes that the applications are session oriented and have constant bandwidth requirements during the session. Finally, symmetric links of the MANETs are estimated and the MAC Protocol must be able to assign TDMA slots according to the requirements.

Intermediate proposals are hybrid approaches that combine two or more of these techniques to improve channel utilization, such as hop reservation multiple access (HRMA, Hop Reservation Multiple Access) [33] that combines TDMA and FDMA, or Bluetooth [34], that combines CDMA and TDMA.

- Transmission initiation: Traditionally, sender-initiated protocols are more common (CSMA, MACA), because they are more intuitive and perform better for general traffic. In sender-initiated transmissions, when a node has data to transmit, it asks the receiver for permission to start communication. On the other hand, in receiver-initiated transmissions, when a receiver is ready to start a communication, it performs a poll to the other nodes asking for data to receive (MACA-BI [35]). Receiver-initiated protocols perform better in specialized high-loaded environments, because the control interchanges (only one control ready to run [RTR] message sent by the receiver before data transmission) are simpler than those of sender-initiated protocols [36].

- Topology: Flat topologies require less control overhead and are more appropriate for balancing load among similar nodes. On the other hand, hierarchical/clustered topologies scale better, and can take advantage of the most powerful nodes to take charge of the cluster/hierarchy header and perform control tasks. However, mobility of the nodes can affect the control of the clusters/hierarchy, as mobile nodes involve frequent topology changes that imply more control overhead.

- Power management: Another essential design criteria for any mobile network is power consumption, due to the limitation of the network devices. Several techniques have been proposed to improve power consumption behavior at MAC layer, such as adapting the transmission power to the minimum acceptable SNR, idle listening prevention through entering in sleep modes to save power when there is no data to interchange (the case of Bluetooth), or the power saving of Multiple Access with Reduced Handshake (MARCH) protocol [37] by maintaining the control overhead, unavoidable to minimize collisions, as low as possible.

13.4.1.3 QoS Support in Network Layer

Before the arrival of wireless networks, wired networks used two main types of routing algorithms: link state and distance vector [38]. The most relevant examples are RIP for link-state algorithms and OSPF for distance-vector algorithms.

These traditional protocol families are not suitable for MANETs because involved periodic or frequent route updates can consume much of the available bandwidth and resources from nodes. To get through the problems associated with link-state and distance-vector algorithms, a large number of protocols have been proposed for MANET routing, gathered in three main groups [39]:

- Proactive-routing protocols (PRP): route to destination is set at an initial stage, maintaining them later through a periodic update mechanism. When a packet is to be sent, the route is

already known and can be immediately used. However, as the network grows, the maintenance and update of the routing tables becomes more weighting.
- Reactive-routing protocols: routes are established on demand when a source needs to send a packet, by means of a route discovery process. This more efficient operation, though, adds the latency to calculate the routes the moment they are requested.
- Hybrid-routing protocols: they combine features of the previous protocols, to take advantage of their benefits and avoid inconveniences.

As routing aspects of MANETs have focused on main research efforts in this field, they are thoroughly discussed in the next section, even extending this classification with more categories.

13.4.1.4 QoS Support in Transport Layer

Although, as just commented in the section above, network layer has been the main research field for providing QoS support, transport layer has come in to play an essential role in providing services with QoS, granting a transparent end-to-end data transference and saving upper layers from functions such as error control and flow control. TCP and User Datagram Protocol (UDP) are the two main TCP/IP protocols for the transport layer, but they were not designed to guarantee a minimum end-to-end QoS level.

UDP is the most usual carrier for real-time and voice traffic, thanks to its lack of error recovery and ordering mechanisms resulting in better performance for this kind of applications. On the other hand, as TCP was originally designed for almost error-free wired networks, any packet loss is presumed to be caused by congestion. Consequently, TCP reacts adjusting the congestion windows to reduce the transmission rate. However, in highly dynamic networks, such as MANETs, this operation renders a poor performance [40–43], because packet losses may not be induced by congestion but by transmission errors, due to the effect of noise, frequent route changes, route failures, route recomputation procedures, network partitioning, and multipath routing [44,45]. Therefore, TCP must be modified to adjust it to the dynamic features of wireless networks. For example, in Selective ACKnowledgment (SACK) TCP [46], traditional TCP is adapted to make it wireless friendly by including selective acknowledgments (to avoid unnecessary retransmissions). Another attempt to enhance TCP is the explicit loss notifications [47,48], which notifies the sender the reason of the packet loss and calling the congestion control procedures only when needed. Other proposals are TCP-F [49], Explicit Link Failure Notification (ELFN) [50], the fixed Retransmission TimeOut (RTO) technique [51], TCP-DOOR [45], Split TCP [52] and A-TCP [44] (see Ref. [45] for a detailed review).

Another emerging trend is to include a signaling protocol at network layer to perform an end-to-end reservation of resources fulfilling the demanded QoS requirements. This solution forces all the data packets of the same communication to follow the same route during that connection. In short, a signaling mechanism allows to emulate a circuit-oriented communication over a connection-oriented network, so as to satisfy the end-to-end QoS needs of the application. This signaling protocol must work together with a routing protocol that proposes the available routes, so this method is not a pure transport layer modification to improve QoS operation, but a sort of inter-layer cooperation. For that reason, signaling is dealt with later in this chapter.

13.4.1.5 QoS Support in Application Layer

Applications operating over MANETs should have versatile strategies to adapt to the probable QoS variations. Real-time applications, such as voice and video streaming are examples of this kind of applications, which may use a flexible and simple user interface, provide dynamic ranges of QoS, use adaptable compression algorithms, etc.

13.4.2 MANET-Specific QoS Models

In Section 13.4.1, we have presented the individual actions oriented to provide QoS support separately in each protocol stack layer. However, more efficient QoS models have been also proposed to provide QoS support in MANETs, which promote the cooperation between the different layers, sharing information to optimize the process of QoS provision. Next, we introduce a brief survey of the most relevant QoS models.

13.4.2.1 FQMM

FQMM (Flexible QoS Model for mobile networks) [11] was probably the first QoS model designed specifically for MANETs and defines a hybrid provision scheme, combining both IntServ and Diff-Serv features. A minor portion of the total traffic (this of higher priority) is provisioned by flow, following IntServ model. The majority lower-priority classes are aggregated and provisioned by class, following DiffServ model.

In this way, the scalability problem of IntServ and the lack of granularity of DiffServ are avoided, while their advantages are maintained. The architecture proposed in FQMM is similar to Diff-Serv architecture, identifying ingress routers, core routers, and egress routers. The main difference is that the node denomination is not related to its physical location. A node is an ingress node when transmitting data, a core node when forwarding data, and an egress node when receiving data (Figures 13.3 and 13.4).

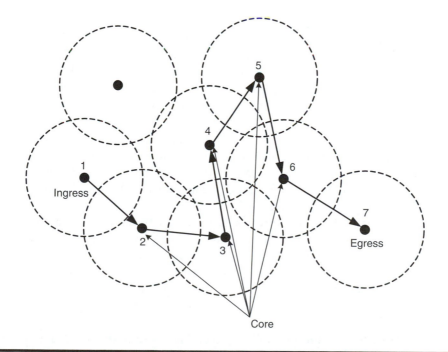

Figure 13.3   Node identification in FQMM.

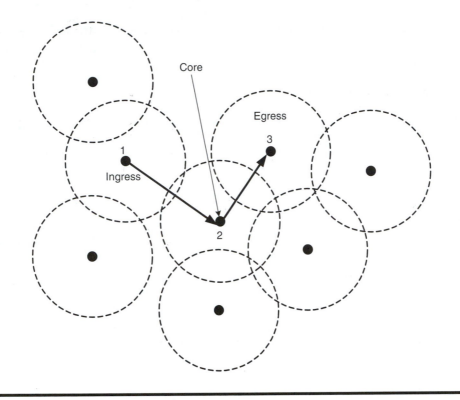

Figure 13.4 Node identification in FQMM.

A traffic conditioner in the ingress node where the traffic is originated polices it by profiles after a valid route is found. A conditioner consists of a traffic profile meter, a marker, and a dropper. FQMM suggests a relative and adaptive traffic classification because as the effective bandwidth of a wireless link is time variant, it is not feasible to try to determine absolute traffic profiles. If a traffic profile is defined as a relative percentage of the effective capacity of the link, class differentiation can be maintained predictably and consistently among sessions, and be adaptable to the dynamic conditions of the network.

Traditional MANET-routing protocols provide just best-effort routing, which is not enough to support QoS because it would be necessary to consider other requirements consistent with provision policy. Consequently, after the routing protocol finds all the routes, an additional QoS analysis should be performed. A routing protocol with QoS support would be more efficient, but FQMM uses previously existing routing protocols.

In FQMM, the main criteria in resource management are the capacity of the wireless link and buffer space. To achieve the best channel utilization according to the provision policy, the scheduler decides which flow accesses the channel every moment. When the network begins to become congested, the dropper discards some packets. These two criteria together make it possible to reach the desirable QoS requirements.

13.4.2.2 WMPLS

WMPLS [15] adapts the functionality of MPLS protocol both to infrastructured and ad hoc wireless networks. WMPLS offers DiffServ and TE support and recommendations for MANET interoperability [2] have been considered. The authors define WMPLS as a protocol applicable to all ad

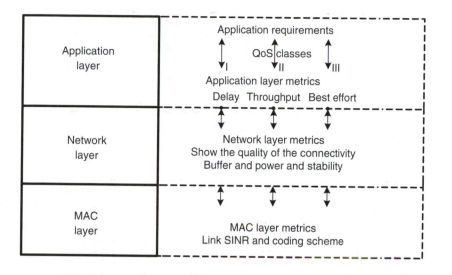

Figure 13.5 **QoS interfaces between protocol stack levels. (From Nikaein, N. and Bonnet, C., 2002.)**

hoc and MANETs, thanks to its soft-handover procedures, that are very reliable at high-speed communications and integrate the negotiation of QoS.

13.4.2.3 Cross-Layer Model

QoS provision in MANETs is usually tackled by modifying only one protocol layer, generally the routing protocol at the network layer. However, the variability of physical and link layer conditions of wireless media suggests the need for exploring other alternatives that combine adjustments in various layers of the protocol stack to provide QoS support. All this leads to cross-layer models.

In MANET environments, routing protocols should be feedbacked with metrics sampled in the lower layers, as an abstraction of physical and link layers [3]. At the same time, as the quality of a link can vary rapidly, routing algorithm should be able to damp the oscillating effects of the wireless medium.

The operation of cross-layer models [53] is shown in Figure 13.5. While MLN (MAC layer metrics) and NLM (network layer metrics) analyze link quality to build good quality paths, ALM (application layer metrics) selects the path that better satisfies the requirements of the application.

Ideally, a cross-layer QoS model would split up MLM, NLM, and ALM. To that effect, one of the proposals suggests the definition of a set of parameters per layer, that, depending on the priority class, gives emphasis to some parameters over the rest. Definition and mapping of these parameters are shown in Figure 13.6.

Next, as an example of a cross-layer network model, we describe INSIGNIA.

13.4.2.3.1 INSIGNIA

While other QoS approximations are based on a circuit model that needs explicit connection management and the establishment of a hard-state before the communication, INSIGNIA [54] brings about a new network model more adaptable to the changes in MANETs.

Generally, in MANETs the virtual circuits are established using an out-of-band signaling to perform resource reservation for the ongoing session. However, it may be more convenient to establish

	ALM	NLM	MLM
Class I	Delay	Buffer and hop count and power	SINR
Class II	Throughput	Buffer and hop count and power	SINR
Class III	Best effort	Stability and hop count	SINR

Figure 13.6   Class-parameter mapping in different layers. (From Nikaein, N. and Bonnet, C., 2002.)

and maintain the flows using a faster and more reactive mechanism based on soft-state and in-band signaling paradigms. Traditional virtual circuits lack flexibility to adapt to the dynamic nature of MANETs. Therefore, INSIGNIA is motivated by the need of developing new QoS architectures to provide a fast resource reservation, promptness in the restoration of flows after connection failures, and easy adjustment to the dynamics of the network, through the flexibility, robustness, and scalability implicit to IP.

The main objective of INSIGNIA is to provide adaptive services that support minimum quality guarantees to real-time flows, offering better service levels when the resources become available. Thus, INSIGNIA is a QoS model designed to adjust user sessions to the service level available in the network, while not explicitly signaling it between source–destination pairs. The network and the applications operate at different levels: the network makes the adjustments to the topology changes or variations of link states by restoration algorithms, while applications adapt themselves to the observed end-to-end QoS fluctuations within the preestablished maximum and minimum limits.

Another aspect considered in INSIGNIA is the time a MANET link maintains its characteristics stable, and consequently, route recalculation time is usually lower than the duration of a session. This is the reason to model independently signaling, resource management and routing functionalities. As a result, INSIGNIA can operate over a wide variety of routing protocols, which are handled as plug-ins.

In-band signaling is another pillar of INSIGNIA. As control information is sent together with data, it operates almost at packet transmission speed, so it is suitable for the necessity of high-response capacity of MANETs. Under ideal conditions, INSIGNIA can restore a data flow in response to a topology change within the interval of two consecutive IP packets.

Regarding resource management, and in contrast to centralized management of hard-state virtual circuits, soft-state model can join up the different time bases of the rapid changes of wireless environment, the moderately frequent ones due to the mobility and the longer session times. Data packets themselves refresh reservation information at each node, working in an absolutely decentralized manner (the architecture of INSIGNIA model is shown in Figure 13.7).

The main disadvantage of INSIGNIA is its need to store the information about the state machines of the mobile nodes, and it may become a problem as the network grows. On the other hand, INSIGNIA only provides two classes of service: real-time and best effort.

An example of the implementation of INSIGNIA with a MANET routing protocol is INORA [55], which combines INSIGNIA signaling with TORA routing protocol.

13.5   QoS Routing in MANETs

Although QoS provision could be carried out at different layers, because most MANETs particularities are derived from the fact that all the nodes behave as routers, network layer's QoS provision

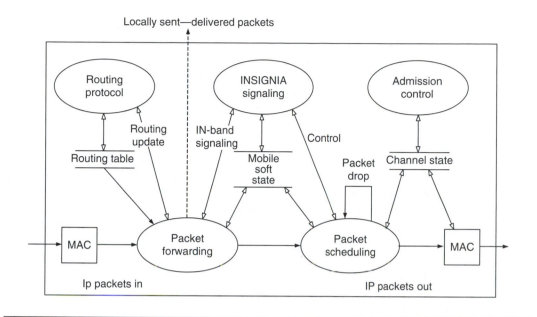

Figure 13.7   INSIGNIA architecture. (From Ahn, G.-S., Campbell, A.T., Lee, S.-B., and Zhang, X., 1999.)

capabilities are quite important in MANETs. Therefore, MANET routing protocols will be analyzed in depth in this section.

The now concluded QoS Routing Working Group of IETF was created in June 1996, to discuss some aspects about QoS supporting routing. This working group was paralyzed in 1999 basically because a lack of a global vision of the problem was detected. It was considered that the understanding of the theory and essential fundamentals of the problem to solve should precede the standards, and not the contrary.

Nevertheless, QoS routing is a functionality logically demanded by the architectures, because most standards are based on traditional routing without QoS support. From this point of view, QoS routing is a major piece of a complete QoS architecture. And besides, to the general QoS routing complexity, the ones derived from the application to MANET environments must be added. For this reason, we first analyze the problematic of MANET routing without QoS support and later we study the existing solutions to incorporate QoS into the routing protocols to provide reliable VoWLAN capabilities in MANETs equivalent to traditional GSM voice service.

13.5.1   Routing Protocols for MANETs without QoS Support

Since the late 1990s, a lot of routing protocols for ad hoc multi-hop networks have been developed. An ad hoc-routing protocol is a standard that controls how the nodes organize themselves to route packets through MANET devices.

To use the limited resources of a wireless network efficiently, we need a routing strategy that suits and self-adapts to the network-varying conditions, such as its size, traffic, and node density or the partition of the network into separate zones. At the same time, as we analyze more in depth in the next section, the routing protocol may have to provide different levels of QoS to the different types of traffic and users. Though the objective of this chapter is not to describe exhaustively this routing protocol, a brief comparison of the different proposed solutions may be useful.

As explained in Section 13.4.1, MANET-routing protocols are classified into three main categories: proactive (table driven), reactive (on demand), and hybrid protocols. But in addition to this general classification, each kind of protocol can implement either a flat routing strategy or an alternative strategy to improve its efficiency. Most of the protocols that follow an alternative routing strategy are based on zone divisions, but there are also other solutions using trees or clusters. The objective of these strategies is to reduce the amount of control load to discover a route or establish a connection, by designating some head-nodes, which organize the packet forwarding from and to the zone/cluster they are in charge of. Other routing strategies are based on positioning mechanisms, such as global positioning system (GPS) to perform a more efficient routing.

On the other hand, as technology progresses and the popularity of Internet grows, applications that need multicast support, such as videoconference, assume more importance. Multicast transmission emerges as one of the most prominent research areas. In a typical ad hoc environment, the nodes form working groups to carry out tasks, so multicast may play an important role in this kind of networks. Multicast protocols used in static networks (DVMRP, Distance Vector Multicast Routing Protocol; MOSPF, Multicast Open Shortest Path First; CBT, Core Based Tree; PIM, Protocol Independent Multicast, etc.) are not suitable for ad hoc networks, because multicast trees must be readjusted each time a connectivity change occurs, and this is a frequent event in MANETs. Therefore, specific multicast protocols for ad hoc mobile networks become necessary.

Table 13.2 shows a summary of the most relevant MANET routing protocols classified according to the concepts commented above.

13.5.2 Routing Protocols for MANETs with QoS Support

Most of the protocols listed above are based on using the number of hops to destination as the only metric. Similar to what happens with traditional routing protocols, this mechanism is inefficient and insufficient when route establishment must consider QoS aware aspects, such as bandwidth, delay, or jitter. Another aspect to be considered is that QoS routing will always add some overload to the network, because of

- Processing overload for more frequent and complex calculus
- Additional data storage for larger routing tables
- Communication overload for the increase of update interchanges

Furthermore, all these factors affect directly the network scalability, so it is very important to design lightweight protocols that minimize power requirements of mobile nodes and link usage for control messages. In that sense, we must recall that though a link can a priori provide an acceptable transmission rate with low delay, the high-mobility and interference probability may cause the collapse of the link. So, signal strength and error probability analysis have become leading research factors.

Regarding the works carried out around QoS routing mechanisms for MANETs, they can be classified into two groups. On the one hand, there are the proposals focused on gathering information about QoS parameters of the available routes, discovered with traditional techniques. Later, this information is used in the selection of the most suitable path for the requested QoS constraints. On the other hand, there is another research line focused on developing protocols capable of computing QoS-constrained routes. These studies are aimed at solving multi-constrained route computing problem, which is known to be NP-complete.

Table 13.2 MANET Routing Protocol Classification

Ad Hoc Routing Protocols							Multicast	
Unicast							*Multicast*	
Proactive			*Reactive*		*Hybrid*			
Plain	*Hierarch.*	*Position Based*	*Plain*	*Position Based*	*Plain*	*Hierarch.*		*Geocast*
DBF	CGSR	DFR	DSR	LAR	CBRP	CSR	ABAM	LBM
DSDV	DST	DREAM	AODV	LOTAR	DDR	IZR	ADMR	GeoGRID
IARP	DDR	GPSR	ABR		HARP	VBR	AMRIS	GeoTORA
MMRP	FSR	ZHLS	BSR		GLS	ZRP	AMRoute	MRGR
OLSR	GSR		CHAMP		(Grid)		CAMP	Mobicast
WRP	HSR		DNVR		GPSAL		CBM	
STAR	LANMAR		DYMO				DCMP	
TBRPF			FSDSR				DDM	
			IERP				DSR-MB	
			RDMAR				FGMP	
			ROAM				LAM	
			SSR				MAODV	
			TORA				MCEDAR	
							MZR	
							ODMRP	
							SOM	
							SPBM	
							SRMP	

The following sections describe some of the works proposed to provide a MANET routing protocol with QoS support.

13.5.2.1 *CEDAR*

CEDAR (core extraction distributed ad hoc routing) [38,56,57] was one of the first proposals addressing QoS management for small- or medium-sized MANETs. This protocol is based on the dynamic arrangement of a core network and on the later propagation of the state of the link with higher stability and capacity toward the core. Route establishment is on demand, for which core nodes only use local states.

CEDAR operation can be divided into three phases:

- Core extraction: a set of nodes is selected to maintain the local topology of all the domain nodes. These nodes will be in charge of route computation. Node selection is performed by and approach of the dominating nodes of the network, i.e., the subset of nodes that makes that any node not belonging to the core is adjacent, at least, to one of the core nodes.
- Link-state propagation: stable links inform the core nodes about their capacity. This way, information of good links is propagated to the furthest nodes, as dynamic or low-capacity links remain local.
- Route computation: this last phase establishes a path through the core from the domain of the source node to the domain of destination, satisfying the requested bandwidth.

The core creation in CEDAR provides an efficient low-overload infrastructure to perform the routing, while link-state propagation assures that the core nodes have the necessary information without overloading the network.

13.5.2.2   Ticket-Based Probing

The design objective of ticket-based probing (TBR) algorithm [58] is to maximize the probability of success while finding a feasible route in dynamic networks in the presence of inaccurate information, while restricting the amount of messages needed for route discovery with a flooding scheme.

This protocol is based on tickets, which constitute a permission to search one path. Route discovery is performed by means of a hop-by-hop probing mechanism. Each probe message must include at least one ticket. When this kind of message reaches a node, this can be split up into more probes and forwarded to the neighbors. Each child probe will contain a subset of the tickets of its parent. If the message only contains one ticket it will not be divided into more probes.

Once the messages reach the destination, the hop-by-hop path covered by each message is known, and the information about the delay or bandwidth can be used to perform the reservation of resources along the route that better fits the QoS requirements.

Because of the dynamic nature of MANETs, there is no accurate information about the bandwidth or delay of the links. That is the reason to propose an imprecise but simple model for the ticket-based probing algorithm. This algorithm estimates the delay using the information about the delay variation, represented as a (delay $-\delta$, delay δ) range.

To get adapted to the dynamic conditions of the network topology, the algorithm allows different route redundancy levels. Likewise, the algorithm defines route maintenance mechanisms through rerouting and path recovery. When a node detects a link failure, reports the source node, who will be in charge of starting the necessary mechanisms to find another suitable path and of notifying the nodes along the previous route that the reserved resources can be released. On the other hand, path recovery mechanism does not search for a new route that satisfies the traffic requirements, but tries to recover it with local reconstruction techniques.

13.5.2.3   Enhanced Ticket-Based Routing

DCLCR (delay-constrained least-cost routing) problem tries to find the lower cost path satisfying a delay constraint. The obstacles to solve it are that it is an NP-complete problem and that routing information may be inaccurate (even more in MANETs). TBR protocol, attempting a suboptimal solution, provides a heuristic approach, without optimizing ticket probing to identify the best routes.

An enhanced ticket-based routing (ETBR) algorithm [59] is proposed to work out the DCLCR problem, improving the effectiveness of the probe and the tolerance to imprecise end-to-end information by means of a color-based ticket distribution, and also ticket-based probe optimization techniques.

When a probe packet is sent, a number of green tickets and a number of yellow tickets are included. Green tickets will be sent through lower cost links, and the yellow ones through lower delay links. Tickets reaching the destination not exceeding a delay threshold, will state feasible routes and among all the feasible routes, the one with the lower cost will be selected.

Probe optimization technique consists of detecting ticket lops to discard the tickets that reached the node previously. Tickets can be also optimized by saving a probe historical with the results of previous tickets.

13.5.2.4 PANDA

PANDA (positional attribute-based next-hop determination approach) [60] is another algorithm that includes QoS mechanisms in the flooding-based route discovery process. This time, PANDA determines next hop based on the location or capabilities of neighboring nodes.

When a Route Request is received, instead of waiting for a random interval before the retransmission, reception node introduces a delay proportional to its capacity of satisfying the QoS requirements of the request. This way, the paths that better comply with the QoS requirements take precedence. The decision about the selected path is taken in the destination, in accordance to a set of predefined rules.

This algorithm fits end-to-end QoS requirements while the route does not suffer important damage. In that case, a mechanism for new route discovery with QoS support is defined.

13.5.2.5 QoS Multipath Routing

Ref. [61] proposes a QoS routing protocol that uses a multipath mechanism to meet the QoS requirements demanded from source to destination.

Contrary to the previously described QoS mechanisms, this protocol is based on detecting different feasible routes that altogether manage to satisfy the QoS requirements. The protocol defined is on demand and searches for different paths from source to destination that altogether guarantee a certain transference rate. The protocol fundamentals are similar to those of TBR.

13.5.2.6 AODV with QoS Extensions (QAODV)

An extended QoS-aware AODV protocol is proposed in Ref. [62]. The main novelty is the introduction of some extensions into Route REQuest (RREQ) and Route REPlay (RREP) messages during the route discovery phase. A node receiving an RREQ message with QoS extensions must verify QoS requirements before retransmitting the message or sending an RREP message to the source in case of having an available cached route that satisfies demanded prerequisites. If after the route establishment a node detects that the QoS requirements cannot be fulfilled any more, that node must send an ICMP QOS LOST message to the source.

To compute the routes according to QoS requirements, four new elements are included in the structure of the routing tables: maximum delay, minimum available bandwidth, list of sources requesting delay guarantees, and list of sources requesting bandwidth guarantees.

13.5.2.7 QOLSR

Quality Optimized Link State Routing (QOLSR) [63] is a research project addressed to the inclusion of QoS aspects in Optimized Link State Routing (OLSR) Protocol, supporting multi-constraint routing. Each network node performs measurements of the bandwidth and delay of the symmetric links to its neighbors and stores the information in a table using it later to establish the MPR (multi-point relay) through heuristics. Each MPR sends topology control messages, including its measurements of bandwidth and delay. Thus, this information is spread throughout the network, to be used by each node to build the routing tables according to the QoS parameters.

Figures 13.8 and 13.9 show the difference between classical flooding process and the one introduced in QOLSR. The reduction of the bandwidth consumption because of these control messages can be graphically checked.

QOLSR allows to make use of a QoS suitable route whenever it is demanded, so each node must know the QoS characteristics of each MRP before the data transmission.

Delay measurements are performed using only control messages. The limitations of wireless interfaces require that the amount of control messages remains low. Periodically, each node broadcasts a Hello message without acknowledgment. This message will be used to measure one-way delay. If the nodes are synchronized, a timestamp is used. Otherwise, another delay measurement will have to be carried out. Regarding bandwidth, nodes must hold the pertinent information about the link to each neighbor.

QOLSR suggests two algorithms for calculating routing tables. In environments where an only metric is going to be considered, a Dijkstra algorithm is run to optimize the parameter measured. If multiple metrics are going to be employed to build routes, NP-complete problem must be solved.

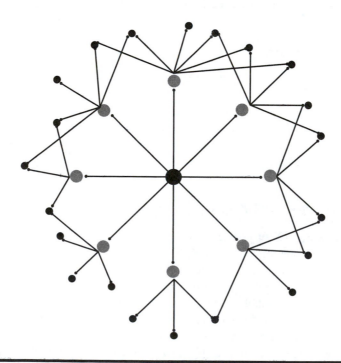

Figure 13.8 OLSR flooding procedure (classical flooding).

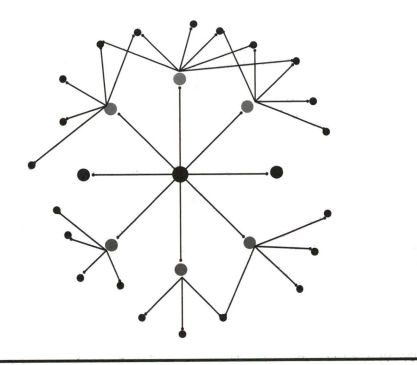

Figure 13.9   OLSR flooding procedure (MPR flooding).

If bandwidth and delay metrics are considered for calculating the routes, first those paths that maximize the bandwidth are identified and, afterward, the route among them that minimizes the delay is selected. For flows that need to comply with a number of QoS constraint (bandwidth, delay, cost, etc.) an efficient heuristic methods based on Lagrangian relaxation is proposed [64]. This method can handle two, three (delay and bandwidth constrained least hop path problem), and four (DBLCLH (delay, bandwidth, and loss probability constrained least hop path problem)) metrics providing a polynomial solution in finding feasible paths.

13.5.2.8   WARP

WARP (Wireless Ad-hoc Routing Protocol) is a hybrid protocol based on ZRP that was developed by the University of Cornell in New York. Contrary to ZP, WARP was designed to consider both link stability and power consumption of the nodes.

Neighbor Discovery Protocol (NDP) identifies the nodes at one hop distance (Figure 13.10). Each node defines a local zone that includes all the nodes located nearer than a certain number of hops. This number of hops is a configurable parameter and depends on the network node number. Within a local zone, the nodes use proactive protocols to build their routing tables. To send packets outside the local zone, reactive protocols are used. In Ref. [39] an initial implementation of WARP is described, providing best-effort and QoS routing. Routes are selected to optimize power consumption and ensure route stability.

The PRP in WARP, called PRP (Proactive Routing Protocol), is a timer-based link-state protocol, which provides a narrower control over introduced overload. PRP creates and maintains

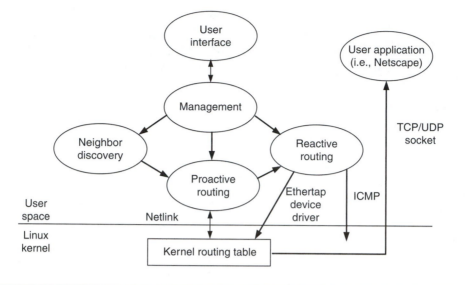

Figure 13.10 WARP architecture. (From Sholander, P., Yankopolus, A., Coccoli, P., and Tabrizi, S.S., 1, 513, 2002.)

an internal link-state table built from NDP information and LSA (link-state advertisements) announcements received from the other nodes belonging to the local zone of the source. A zone in WARP corresponds to the set of nodes at a certain number of hops.

WRP defines both best-effort routing, based on the number of hops, and QoS routing, based on specific metrics for the wireless environment, such as link stability and battery status of the node.

Battery status metric prioritizes the use of routes along nodes with more consistent power supply, penalizing those with nodes working on batteries. Sensor networks are an example of application where this metric becomes important.

On the other hand, link stability metric prioritizes the use of more stable links and is useful to support data flows that require some QoS for a period. It can be also an alternative for TCP transmissions, because packet losses reduce end-to-end throughput.

13.5.2.9 *Ad Hoc QoS Multicast*

In Ref. [65] a multicast ad hoc routing protocol is proposed. This protocol reaches multicasting efficiency by tracking the availability of networks resources in each node inside a neighborhood. Neighbor nodes send information about ongoing sessions and similar information to compute what the available bandwidth is. When a new session is initiated, current QoS status is announced and periodically updated, so nodes are prevented from applying for membership if there is no QoS path for the session. The protocol ensures that the QoS information is refreshed and used to select the most appropriate routes.

To assess the performance of ad hoc QoS multicast (AQM), authors include two new metrics: member and session satisfaction degrees. AQM is compared with a non-QoS scheme resulting in simulation that AQM significantly improves multicasting efficiency.

13.5.2.10   Predictive Routing

It is based on the same principles of page prediction in computer architecture to foresee route and link availability [3]. Extending this concept of page prediction, it will be necessary to maintain a historical record of the information about route occupation.

13.6   QoS Signaling in MANETs

To provide QoS in MANETs two main problems must be solved: QoS routing, or how to find a route through the network that satisfies the requested QoS level, and QoS maintenance, or how to guarantee that, once a route is found, QoS agreement is going to be granted against the network dynamics. The routing problem has been studied in Section 13.5, but route maintenance is particularly complex. Signaling protocol will perform the reservation and liberation of network resources, together with flow establishment to provide GSM equivalent service over MANETs in UMA environments.

Signaling mechanisms fall into two categories: those that include control information inside data packets (in-band signaling) and those that employ specific control messages (out-of-band signaling) [54].

13.6.1   RSVP

One of the most extended signaling protocols is Resource reSerVation Protocol (RSVP) defined by IETF. It is an out-of-band signaling protocol that allows end-to-end resource reservation for unicast and multicast traffic. It is based on the use of two messages Path and Resv that perform the reservations in the intermediate routes.

The source node sends a Path message to the destination specifying traffic characteristics (traffic characterizing). This message is forwarded through intermediate nodes following the route provided by the routing protocol. When receiving a Path message, the destination node answers with a Resv message to request the resource reservation for that flow. Each intermediate node can accept or deny the request. If accepted, bandwidth and buffer space are reserved and flow state information is installed.

However, this signaling protocol may not be suitable for a MANET due to the excessive protocol overload and the lack of adaptation capability to a dynamic topology.

13.6.2   dRSVP

dRSVP (dynamic Resource reSerVation setup Protocol) [66] is an evolution of RSVP aimed at providing a dynamic QoS over a distributed network protocol. This approach to QoS is based on the reservation of resources, but with a more extended meaning of the term reservation, representing an agreement with the network to provide a service level within a given range. The applications request QoS indicating the required minimum service level and the maximum level they can manage. After that, the applications complete the reservation in the range provided by the network, that may vary in time.

This way, the necessary flexibility to deal with the dynamics of MANETs is achieved. As available resources change, the network can adjust its resource distribution within the reservation range. If resource level decreases, but stays in that range, the network fits its operation, instead of switching to best effort. As the number of application flows competing for the resources grows, instead of rejecting new flows in the admission control, the network can try to distribute the resources to accommodate new requests, as long as service level is maintained within the specified range.

The use of dRSVP has two implications for the applications. On the one hand, they must know the QoS range they can manage. This scope can be programmed in the application or configured by the user. On the other hand, the applications will have to adapt their operation at runtime based on the feedback provided by the network.

13.6.3 Other Signaling Protocols

ASAP (Adaptive Reservation and Pre-Allocation Protocol) [67] is an adaptive reservation QoS protocol. Thanks to a simple signaling system and a two-phase reservation mechanism, ASAP provides adaptive QoS support, a fast route configuration, and local restoration, as well as processing optimization.

In Ref. [4] another signaling protocol for MANETs is proposed to satisfy bandwidth requirements. The available bandwidth in a coverage area is defined as the traffic generated and forwarded from neighbor nodes or other overlapping MANETs. This reservation scheme is applied to AODV and LSR, but could also be used with other routing protocols for ad hoc mobile networks. This method maintains the occupation of QoS traffic competing for the same shared medium in each node under Q bps. The Q parameter can be dimensioned to keep the delays at an acceptable value for QoS connections. The nodes only need to know the reservations agreed and the maximum available bandwidth of their neighbors. This information can be easily distributed by Hello messages. This scheme also includes an admission control mechanism.

13.7 Conclusions and Open Issues

As mobile devices have become more ubiquitous and less expensive, applications over wireless media have suffered a sudden increase. The proliferation of this kind of devices has resulted in a demand for seamless handover between public cellular networks and UMA networks. Traditionally, UMA is provided over infrastructured WLANs to support low-cost GSM services where the subscriber spends most of the time: home and office. However, this is not the only unlicensed wireless technology to be taken into account. One of the research fields that has experienced a greater growth within the wireless technologies area is the MANET environment. The possibility of setting up a network everywhere without the implementation of any previous infrastructure confers on MANETs a flexibility that has not been provided by any other technology before. This is the reason for the exponential growth in these kinds of networks over the last few years. There are a lot of possibilities for MANETs, from the first applications in battlefields and disaster areas to the current network access in conferences or fast network deployment where wired networks are difficult or not feasible. In the near future, the ubiquity of small, inexpensive wireless communicating devices will make it possible to configure a self-organized network anywhere users want to share information. By incorporating to UMA technology, these networks will also support GSM/GPRS/UMTS-equivalent services.

But precisely, the flexibility is also the source of the limitations of ad hoc mobile networks. The dynamic nature of the wireless media becomes more noticeable with the lack of infrastructure and the mobility of nodes. Applications have to deal with many difficulties such as connection failures, time-variable bandwidth, and delay or exposed and hidden terminal problems, that do not appear in traditional networks, or at least not as frequently as in MANETs. Therefore, the first necessity to be covered in this area was to design specific MANET-routing protocols to overcome the limitations of ad hoc mobile networks. This was the purpose of the IETF MANET Working Group, which has published many MANET-specific routing protocols and continues with the job of proposing and discussing new methods to improve MANET's general performance.

Most of the first MANET routing protocols were based just in number-of-hops metrics, which was enough for the basic first applications of ad hoc mobile networks. However, as previously commented, the great increase in this research area in the last few years has led to new applications, such as UMA, which require a minimal level of QoS. Traditional QoS models are not perfectly suitable by themselves for the MANET environment, due to the dynamic features of these networks. Hence, new models for ad hoc mobile networks have been proposed.

The initial attempts to face up to this new challenge included the design of new QoS-aware routing protocols including bandwidth or delay as metrics. In fact, one of the most promising research areas to provide QoS in MANETs is the multi-constraint routing, where several metrics are combined to optimize route discovery.

With QoS-aware routing protocols, nodes can find routes that satisfy certain QoS requirements, but cannot guarantee that this situation is not going to change during the communication. Therefore, there must be another mechanism that assures an end-to-end performance and this is where a signaling protocol comes into play. A signaling protocol allows the management and reservation of available network resources to balance their utilization to guarantee that end-to-end QoS is provided. The signaling protocol works together with the routing protocol, so that the routing protocol discovers available routes satisfying certain QoS requirements and the signaling protocol performs the end-to-end reservation of resources. So, an incipient relation between different layers can be guessed. This is the origin of cross-layer models, where different protocols at different levels share information to improve the overall QoS provision. An important research effort is being made around this idea and several proposals have been presented, most of them combining network and transport layer protocols, but there are also some suggestions considering link layer as well.

On the other hand, because not many MANETs are currently deployed, research in this area is mostly simulation based. Thus, another open issue is to develop good test beds to assure that results of simulations are as accurate and reliable as possible.

REFERENCES

1. Inc. Kineto Wireless. The dual-mode handset opportunity. Technical report, 2007.
2. M. S. Corson, S. Papademetriou, P. Papadopoulos, V. Park, and A. Qayyum. An Internet MANET encapsulation protocol (IMEP) specification, IETF Internet Draft, 1999.
3. X. Masip-Bruin, M. Yannuzzi, J. Domingo-Pascual, A. Fonte, M. Curado, E. Monteiro, F. Kuipers, P. Van Mieghem, S. Avallone, and G. Ventre. Research challenges in QoS routing. *Computer Communications*, 29(5):563–581, 2006.
4. C. Lloren, V. Michael, G. Rafael, B. Jos M., G. Jorge, and B. Chris. *A Reservation Scheme Satisfying Bandwidth QoS Constraints for Ad Hoc Networks*. Lecture Notes in Computer Science: Wireless Systems and Mobility in Next Generation Internet. University of Catalonia, Barcelona, Spain, 2005.
5. P. Mohapatra, J. Li, and C. Gui. Qos in mobile ad hoc networks. *IEEE Wireless Communications*, 10:44–52, 2003.
6. S. Chakrabarti and A. Mishra. Qos issues in ad hoc wireless networks. *IEEE Communications Magazine*, 39(2):142–148, 2001.

7. H. I. Tanzeena, A. Chadi, and J. W. Atwood. Randomized energy aware routing algorithms in mobile ad hoc networks, *Proceedings of the 8th ACM international symposium on Modeling, analysis and simulation of wireless and mobile systems*, ACM Press: Montreal, Quebec, Canada, 2005.

8. R. Dube, C. D. Rais, K. Y. Wang, and S. K. Tripathi. Signal stability based adaptive routing (SSA) for ad-hoc mobile networks. *IEEE Personal Communications*, 1997.

9. M. Mauve, A. Widmer, and H. Hartenstein. A survey on position-based routing in mobile ad hoc networks. *IEEE Network*, 15(6):30–39, 2001.

10. C. Aurrecoechea, A. T. Campbell, and L. Hauw. A survey of QOS architectures. *Multimedia Systems*, 6(3):138–151, 1998.

11. X. Hannan, W. K. G. Seah, A. Lo, and K. C. Chua. A flexible quality of service model for mobile ad-hoc networks. *IEEE 51st Vehicular Technology Conference Proceedings*, Tokyo, vol. 1, pp. 445–449, 2000.

12. X. Xipeng and L. M. Ni. Internet QoS: A big picture. *IEEE Network*, 13(2):8–18, 1999.

13. X. Hesselbach, M. Huerta, O. Calderon, and M. Bajo. Introduction a las tecnologias MPLS, mplambdas y gmpls. *Electro- Electrnica*, 2004.

14. MPLS Forum. Gmpls interoperability event, 2002.

15. J.-M. Chung. Wireless multiprotocol label switching (WMPLS). *Conference Record of the Thirty-Fifth Asilomar Conference on Signals, Systems and Computers*. Pacific Grove, California, USA, vol. 1, pp. 679–683, 2001.

16. B. Fortz, J. Rexford, and M. Thorup. Traffic engineering with traditional IP routing protocols. *IEEE Communications Magazine*, 40(10):118–124, 2002.

17. D. Awduche, A. Chiu, A. Elwalid, I. Widjaja, and X. Xiao. Overview and principles of internet traffic engineering, *IEEE*, Editor, 2002.

18. O. Younis and S. Fahmy. Constraint-based routing in the internet: Basic principles and recent research. *IEEE Communications Surveys and Tutorials*, 5(3):2–13, 2003.

19. Sunil Kumar, V. S. Raghavan, and J. Deng. Medium access control protocols for ad hoc wireless networks: A survey. *Ad Hoc Networks*, 4(3):326–358, 2006.

20. R. Jurdak. *Wireless Ad Hoc and Sensor Networks: A Cross-Layer Design Perspective* (Signals and Communication Technology). Springer, Tokyo, Japan, 40–43, 2007.

21. P. Karn. MACA-a new channel access method for packet radio, *ARRL/CRRL Amateur Radio 9th Computer Networking Conference*, ARRL: Dallas, Texas, USA, 1990.

22. C. R. Lin and M. Gerla. Asynchronous multimedia multihop wireless networks. *INFOCOM'97. Sixteenth Annual Joint Conference of the IEEE Computer and Communications Societies. Proceedings IEEE*. Kobe, Japan, vol. 1, pp. 118–125, 1997.

23. V. Bharghavan, A. Demers, S. Shenker, and L. Zhang. Macaw: A media access protocol for wireless LAN's, *Proceedings of the conference on Communications architectures, protocols and applications*. ACM Press: London, United Kingdom, 1994.

24. C. L. Fullmer and J. J. Garcia-Luna-Aceves. Floor acquisition multiple access (FAMA) for packet-radio networks. *SIGCOMM Computer Communication Review*, Massachusetts, USA 25(4):262–273, 1995.

25. IEEE. ANSI/IEEE std 802.11 part 11: Wireless LAN medium access control (MAC) and physical layer (PHY) specifications, 1999.

26. IEEE. ANSI/IEEE std 802.11a part 11: Wireless LAN medium access control (MAC) and physical layer (PHY) specifications high-speed physical layer in the 5 GHz band, 1999.

27. IEEE. ANSI/IEEE std 802.11b part 11: Wireless LAN medium access control (MAC) and physical layer (PHY) specifications higher-speed physical layer extension in the 2.4 GHz band, 1999.

28. IEEE. ANSI/IEEE std 802.11g part 11: Wireless LAN medium access control (MAC) and physical layer (PHY) specifications amendment 4: Further higher data rate extension in the 2.4 GHz band, 2003.

29. A. Veres, A. T. Campbell, M. Barry, and S. Li-Hsiang. Supporting service differentiation in wireless packet networks using distributed control. *IEEE Journal on Selected Areas in Communications* 19(10):2081–2093, 2001.

30. A. Lindgren, A. Almquist, and O. Schelen. Evaluation of quality of service schemes for IEEE 802.11 wireless LANs. *26th Annual IEEE Conference on Local Computer Networks, Proceedings.* Tampa, Florida, USA, pp. 348–351, 2001.

31. J.-H. Ju and J. Li. Tdma scheduling design of multihop packet radio networks based on latin squares. *IEEE Journal on Selected Areas in Communications, Wireless Ad Hoc Networks,* 17(8):187–193, 1999.

32. C. Zhu. Medium access control and quality-of-service routing for mobile ad hoc networks. PhD thesis, CSHCN, University of Maryland, USA, 2001.

33. Y. Zhenyu and J. J. Garcia-Luna-Aceves. Hop-reservation multiple access (HRMA) for ad-hoc networks. In *INFOCOM '99.* Proceedings of the 18th Annual Joint Conference of the IEEE Computer and Communications Societies. vol. 1, pp. 194–201 1999.

34. J. Haartsen. Bluetooth the universal radio interface for ad hoc, wireless connectivity. *Ericsson Review,* 1998.

35. F. Talucci, M. Gerla, and L. Fratta. Maca-bi (maca by invitation)—a receiver oriented access protocol for wireless multihop networks. In the 8th IEEE International Symposium on Personal, Indoor and Mobile Radio Communications, 1997. 'Waves of the Year 2000 *PIMRC '97,* Helsinki, Finland, vol. 2, pp. 435–439, 1997.

36. A. E. Tzamaloukas. Sender- and Receiver-Initiated Multiple Access Protocols for Ad Hoc Networks. PhD thesis, University of California, Santa Cruz, California, USA, 2000.

37. C. K. Toh, V. Vassiliou, G. Guichal, and C. H. Shih. March: A medium access control protocol for multihop wireless ad hoc networks. In 21st Century Military Communications Conference Proceedings, MILCOM 2000, Los Angeles, California, USA, vol. 1, pp. 512–516, 2000.

38. M. Abolhasan, T. Wysocki, and E. Dutkiewicz. A review of routing protocols for mobile ad hoc networks. *Ad Hoc Networks,* 2:1–22, 2004.

39. P. Sholander, A. Yankopolus, P. Coccoli, and S. S. Tabrizi. Experimental comparison of hybrid and proactive MANET routing protocols. *MILCOM 2002. Proceedings.* Anaheim, California, USA, vol. 1, pp. 513–518, 2002.

40. A. Al Hanbali, E. Altman, and P. Nain. A survey of TCP over ad hoc networks. *IEEE Communications Surveys and Tutorials,* 22–36, 2005.

41. H. Balakrishnan, V. N. Padmanabhan, S. Seshan, and R. H. Katz. A comparison of mechanisms for improving tcp performance over wireless links. *IEEE/ACM Transactions on Networking (TON),* 5(6):756–769, 1997.

42. Y. Tian, K. Xu, and N. Ansari. TCP in wireless environments: Problems and solutions. *Communications Magazine, IEEE,* 43(3):S27–S32, 2005.

43. G. Xylomenos, G. C. Polyzos, P. Mahonen, and M. Saaranen. TCP performance issues over wireless links. *IEEE Communications Magazine,* 39(4):52–58, 2001.

44. J. Liu and S. Singh. ATCP: TCP for mobile ad hoc networks. *IEEE Journal on Selected Areas in Communications,* 19(7):1300–1315, 2001.

45. F. Wang and Y. Zhang. Improving TCP performance over mobile ad-hoc networks with out-of-order detection and response. Proceedings of the 3rd ACM International Symposium on Mobile Ad Hoc Networking and Computing, Lausanne, Switzerland, pp. 217–225, 2002.

46. N. K. G. Samaraweera and G. Fairhurst. Reinforcement of TCP error recovery for wireless communication. *ACM SIGCOMM Computer Communication Review,* 28(2):30–38, 1998.

47. H. Balakrishnan and R. H. Katz. Explicit loss notification and wireless web performance. In IEEE Globecom Internet Mini-Conference, Sydney, Australia, 1998.

48. H. Balakrishnan, V. N. Padmanabhan, S. Seshan, and R. H. Katz. A comparison of mechanisms for improving TCP performance over wireless links. *IEEE/ACM Transactions on Networking,* 5(6):756–769, 1997.

49. K. Chandran, S. Raghunathan, S. Venkatesan, and R. Prakash. A feedback-based scheme for improving tcp performance in ad hocwireless networks. *IEEE Personal Communications,* 8(1):34–39, 2001.

50. G. Holland and N. Vaidya. Analysis of TCP performance over mobile ad hoc networks. *Wireless Networks,* 8(2):275–288, 2002.

51. T. D. Dyer and R. V. Boppana. A comparison of TCP performance over three routing protocols for mobile ad hoc networks. Proceedings of the 2nd ACM International Symposium on Mobile Ad Hoc Networking and Computing, Long Beach, California, USA, pp. 56–66, 2001.

52. S. Kopparty, S. V. Krishnamurthy, M. Faloutsos, and S. K. Tripathi. Split TCP for mobile ad hoc networks. In IEEE Global Telecommunications Conference, 2002. GLOBE-COM'02, Taipei, Taiwan, vol. 1, 2002.

53. N. Nikaein and C. Bonnet. A glance at quality of service models in mobile ad hoc networks, *16eme Congrès DNAC (De Nouvelles Architectures pour les Communications)*, Paris, 2002.

54. S.-B. Lee, G.-S. Ahn, X. Zhang, and A. T. Campbell. Insignia: An IP-based quality of service framework for mobile ad hoc networks. *Journal of Parallel and Distributed Computing*, 60(4):374–406, 2000.

55. D. Dharmaraju, A. Roy-Chowdhury, P. Hovareshti, and J. S. Baras. Inora-a unified signaling and routing mechanism for qos support in mobile ad hoc networks. *International Conference on Parallel Processing Workshops, Proceedings*, Vancouver, British Columbia, Canada, pp. 86–93, 2002.

56. R. Sivakumar, P. Sinha, and V. Bharghavan. Cedar: A core-extraction distributed ad hoc routing algorithm. *IEEE Journal on Selected Areas in Communications*, 17(8):1454–1465, 1999.

57. R. Sivakumar, P. Sinha, and V. Bharghavan. Core extraction distributed ad hoc routing (cedar) specification, *IETF*, Editor, 1998.

58. Chen Shigang and K. Nahrstedt. Distributed QoS routing with imprecise state information. *7th International Conference on Computer Communications and Networks, Proceedings*. Honolulu, Hawaii, USA, pp. 614–621, 1998.

59. X. Li, W. Jun, and M. Nahrstedt. The enhanced ticket-based routing algorithm. *IEEE International Conference on Communications, ICC*, New York City, USA, vol. 4, pp. 2222–2226, 2002.

60. J. Li and P. Mohapatra. Panda: An approach to improve flooding based route discovery in mobile ad hoc networks. Technical report, CSE, 2002.

61. W.-H. Liao, S.-L. Wang Shu-Ling, J.-P. Sheu Jang-Ping, and Y.-C. Tseng. A multi-path qos routing protocol in a wireless mobile ad hoc network. *Telecommunication Systems*, V19(3):329–347, 2002.

62. C. Perkins and E. M. Royer. Quality of service for ad hoc on-demand distance vector routing, *IETF*, 2001.

63. Hakim Badis. Quality of service for ad hoc optimized link state routing protocol (QOLSR), 2006.

64. A. Juttner, B. Szviatovski, I. Mecs, and Z. Rajko. Lagrange relaxation based method for the QoS routing problem. *INFOCOM 2001. Twentieth Annual Joint Conference of the IEEE Computer and Communications Societies. Proceedings IEEE*, Anchorage, Alaska, vol. 2, pp. 859–868, 2001.

65. K. Bur and C. Ersoy. Ad hoc quality of service multicast routing. *Computer Communications*, 29(1):136–148, 2005.

66. M. Mirhakkak, N. Schult, and D. Thomson. Dynamic quality-of-service for mobile ad hoc networks. *First Annual Workshop on Mobile and Ad Hoc Networking and Computing, MobiHOC*, Boston, Massachusetts, USA, pp. 137–138, 2000.

67. J. Xue, P. Stuedi, and G. Alonso. ASAP: An adaptive QoS protocol for mobile ad hoc networks. vol. 3, pp. 2616–2620, 2003.

68. G.-S. Ahn, A. T. Campbell, S.-B. Lee, and X. Zhang. Insignia, 1999.

STANDARDS AND APPLICATIONS

Chapter 14

WiMAX Architecture, Protocols, Security, and Privacy

S.P.T. Krishnan, Bharadwaj Veeravalli, and Lawrence Wong Wai Choong

CONTENTS

14.1 Introduction

The appeal of worldwide interoperability for microwave access (WiMAX) goes well beyond mobility provided by third generation (3G) and wireless Internet access popularized by Wi-Fi. The standard offers a true broadband connection that supports multiple usage scenarios, including fixed, portable, and mobile access using the same network infrastructure. With WiMAX in place, the applications and services which is now [restricted to stationary locations such as] home and office will become available everywhere. Ubiquitous broadband access will encourage work productivity, personal communications, and entertainment on the go. New services and applications that are specifically suited to mobile usage scenarios will also appear: mobile office, onboard entertainment, mobile search, fleet management, surveillance, and public safety are likely to be the first round stalwarts.

The first WiMAX standard provided stationary or nomadic wireless broadband access and was dubbed as "Fixed WiMAX" standard and was approved in June 2004. It is based on the IEEE 802.16-2004 standard (which revises and replaces IEEE 802.16a and 802.16REVd versions). This technology provides a wireless alternative to the cable modem, digital subscriber lines of any type (xDSL), transmit/exchange (Tx/Ex) circuits, and optical carrier level (OC-x) circuits.

The next WiMAX standard was built on the solid foundation of the fixed WiMAX and is intended to provide true broadband wireless access at vehicular speeds in excess of 120 kmph. This version referred to as "mobile WiMAX", is based on IEEE 802.16e-2005 and was ratified in December 2005. Unlike 802.11b g, which operates in unlicensed 2.4 GHz spectrum, the mobile WiMAX will initially operate in the licensed 2.3, 2.5, 3.3, and 3.4–3.8 GHz spectrum bands with channel sizes ranging from 3.5 to 10 MHz. The only caveat is that the physical layer is not backward compatible with 802.16-2004 and therefore requires new hardware/software solutions.

WiMAX was designed from the ground-up to be an all-Internet Protocol (IP) technology that is optimized for high-throughput, real-time data applications and is not beholden to a legacy infrastructure. Global roaming among WiMAX service providers will allow subscribers to access different networks using the same device and a single, familiar interface.

The 802.16d/e standards have a strong commercial backing to go along with their technical capabilities. The WiMAX Forum [1], a nonprofit industrial consortium that promotes the technologies, has as its goal the certification of inter-operable standards compliant products regardless of vendor. In this regard, the WiMAX forum is following the lead of the Wi-Fi alliance [2], which helped popularize and commercialized the 802.11 technology. Founded in June 2001, the WiMAX Forum as of June 2007 includes 420 member companies.

This chapter is primarily focused on mobile WiMAX as it is the most sought after standard and because it can provide stationary, nomadic, and mobile Internet access. Nevertheless, where appropriate and required, the features of fixed version of WiMAX are referred as well. WiMAX references in the text refer to both variants and things specific to either fixed/mobile will be mentioned as such. In this chapter, we interchangeably use 802.16-2004 to refer to fixed WiMAX and 802.16e-2005 for mobile WiMAX and vice versa, respectively.

The chapter is organized as follows. In Section 14.2, we look at the WiMAX physical layer and the innovations that WiMAX has introduced. In Section 14.3, we delve into the Media Access Control (MAC) layer and in Section 14.4 we note the radio frequency innovations in WiMAX. Sections 14.5 through 14.7 look into more visible side of WiMAX such as the architecture, distinct features, and applications. Section 14.8 details on the competing standards and Section 14.9 concludes this chapter.

14.2   Physical Layer

The lowest layer in any networking stack is the physical layer and it is one of two layers that define the WiMAX standard, the other being the MAC layer. This layering of the networking stack and its isolation and independence of each layer except on its immediate neighboring layers is the cornerstone of the Open Standard Interconnection (OSI) Model, which has kept it relevant as we moved from coaxial cables to Ethernet to Wi-Fi and now to WiMAX.

Fixed WiMAX adopted the orthogonal frequency division multiple access (OFDMA) scheme and mobile WiMAX adopted the S-OFDMA at the physical layer. In radio frequency (RF) transmissions, collision/overlap of packets and retransmission are generally the case that reduce the effective bandwidth of the wireless channel. In WiMAX, the introduction of the cyclic prefix (CP) can completely eliminate inter-symbol interference (ISI) as long as the CP duration is longer than the channel delay spread. The CP is typically a repetition of the last samples of data portion of the block that is appended to the beginning of the data payload. The CP prevents inter-block interference. A perceived drawback of CP is that it introduces overhead, which effectively reduces bandwidth efficiency.

14.2.1   OFDMA

Orthogonal frequency division multiplexing (OFDM) is a multiplexing technique that subdivides the bandwidth into multiple frequency sub-carriers. In an OFDM system, the input data stream is divided into several parallel sub-streams of reduced data rate (thus increased symbol duration) and each sub-stream is modulated and transmitted on a separate orthogonal sub-carrier. The increased symbol duration improves the robustness of OFDM to delay spread.

OFDMA is a multiple-access/multiplexing scheme that provides multiplexing operation of data streams from multiple users thereby providing multiple access onto the downlink (DL) and uplink (UL) sub-channels. OFDMA, however, goes a step further by then grouping multiple sub-carriers into sub-channels. A single client or subscriber station might transmit using all of the sub-channels within the carrier space, or multiple clients might transmit with each using a portion of the total number of sub-channels simultaneously. The use of OFDMA has improved the multipath performance in non-line-of-sight environments, which is typical in an urban or a mobile setup. The minimum frequency-time resource unit of sub-channelization is one slot. The OFDMA symbol structure consists of three types of sub-carriers: (a) data sub-carriers for data transport, (b) pilot sub-carriers for estimation/synchronization, and (c) null sub-carriers for no transmission typically used for guard bands.

14.2.2 Scalable OFDMA

The IEEE 802.16e-2005 wireless MAN amendment introduced the concept of scalable OFDMA (S-OFDMA). S-OFDMA allows a network operator or systems integrator to vary the bandwidth for different users. It supports a wide range of channel bandwidths from 1.25 to 20 MHz to flexibly address the need for various spectrum allocation and usage model requirements worldwide. Additionally, when the distance between a subscriber and the base station (or access point [AP]) increases, it becomes challenging for the subscriber to transmit successfully to the base station at a given power level. Also, for handheld mobile devices, it is often not possible for them to transmit to the base station over long distances and over wide channel bandwidths. The 802.11 channel bandwidth is fixed at 20 MHz. In contrast, applications modeled on 3G principles limit channel bandwidth to about 1.5 MHz to provide longer range. S-OFDMA goes a step further and introduces scalable channel bandwidths with the range 1.25–20 MHz. This flexibility of channel bandwidth is also crucial for cell planning, especially in the licensed spectrum. Superior performance, made possible by the adoption of S/OFDMA multiplexing, gives WiMAX a performance edge in delivering IP data services compared to 3G technologies and thus justifies for the S/OFDMA in mobile WiMAX.

14.2.3 Time Division Duplex

WiMAX performance is further enhanced by the use of time division duplex (TDD), but WiMAX can also support frequency division duplex (FDD) in full/half duplex mode, which dominates in 3G networks. Where FDD keeps the UL and the DL channels separate in frequency, TDD is a less complex and more efficient mechanism that uses a single frequency channel, with UL and DL traffic separated by a guard time. TDD enables adjustment of the DL/UL ratio to efficiently support asymmetric DL/UL traffic, while with FDD, DL and UL always have fixed and generally, equal DL and UL bandwidths. In addition, for IP-based services, the use of a single channel for the UL and the DL makes it substantially less complex and more cost-effective to implement multiple input multiple output (MIMO) and beamforming in WiMAX networks than in code division multiple access (CDMA)-based networks. MIMO and beamforming are expected to bring a substantial improvement in throughput in TDD-based WiMAX networks.

TDD is to be exclusively used for the initial mobile WiMAX profiles for its added efficiency in support of asymmetric traffic and channel reciprocity for better support of link adaptation, easy support of MIMO, and other closed loop advanced antenna systems. Transceiver designs for TDD implementations are less complex and therefore less expensive.

14.3 MAC Layer

The 802.16 standard was developed from the outset for the delivery of broadband services including voice, data, and video. The MAC layer is based on the time-proven Data over Cable Service Interface Specification standard and can support data traffic bursts with high peak rate demand while simultaneously supporting streaming video and latency-sensitive voice traffic over the same channel. The resource allocated to one terminal by the MAC scheduler can vary from a single time slot to the entire frame, thus providing a very large dynamic range of throughput to a specific user terminal at any given time. Furthermore, because the resource allocation information is conveyed at the beginning of each frame, the scheduler can effectively change the resource allocation on a frame-by-frame basis to adapt to the spiky nature of the traffic.

14.3.1   MAC Scheduling

The 802.16-2004 standard relies upon a Grant Request Access Protocol that in contrast to the contention-based access used under 802.11 does not allow data collisions and therefore uses the available bandwidth more efficiently. No collisions means no loss of bandwidth due to data retransmission. All communication, upstream and downstream, is coordinated by the base station. Other characteristics of the fixed WiMAX standard include improved user connectivity, the full support for wireless metropolitan access network (WMAN) service, and robust carrier-class operation.

From its inception, the WiMAX standard was designed to provide WMAN service. The standard keeps more users connected by virtue of its flexible channel widths and adaptive modulation. The standard uses channels narrower than the fixed 20 MHz channels used in 802.11, and therefore can serve lower data-rate subscribers without wasting bandwidth. When subscribers encounter noisy conditions or low signal strength, the adaptive modulation scheme keeps them connected when they might otherwise be dropped. Also, the standards deliver faster data rates at longer distances than the 802.11g standard.

14.4   WiMAX RF Innovations

The physical layers (PHYs) for both 802.11 and 802.16-2004 are designed to tolerate delay spread. Because the 802.11 standard was designed for 100 m, it can tolerate only about 900 ns of delay spread. The fixed WiMAX standard tolerates up to 10 μs of delay spread—more than 1000 times than in the 802.11 standard. In the case of 802.16-2004, the OFDM signal is divided into 256 carriers instead of 64 as with the 802.11 standard. The larger number of sub-carriers over the same band results in narrower sub-carriers, which is equivalent to larger symbol periods. The same percentage of guard time or CP provides larger absolute values in time for a larger delay spread and multipath immunity. Multipath interference and delay spread improve performance in situations where there is no direct line-of-sight path between the base station and the subscriber station. Tolerance to multipath and self-interference with sub-channel orthogonality in both the DL and the UL is inbuilt. The 802.11 standard provides 1/4th of the OFDM options for CP than does the 802.16-2004 standard, which provides 1/32th, 1/16th, 1/8th, and 1/4th, where each can be optimally set. For a 20 MHz bandwidth, the difference between a 1/4 CP in 802.11 standard and 802.16-2004 standard would be a factor of four because of the ratio 256/64. In OFDMA with 2048 fast fourier transform size, the ratio is 32. The inclusion of MIMO antenna techniques along with flexible sub-channelization schemes and Advanced Coding and Modulation enable mobile WiMAX technology to support peak DL data rates up to 63 Mbps per sector and peak UL data rates up to 28 Mbps per sector in a 10 MHz channel.

14.4.1   Smart Antennas

Smart antennas are being used to increase the spectral density (i.e., the number of bits that can be communicated over a given channel in a given time) and to increase the signal-to-noise ratio for WiMAX solutions. Because of performance and technology, the fixed WiMAX standard supports several adaptive smart antenna types. Below we describe some of the popular types.

Receive spatial diversity antennas use more than one receiving antenna. The antennas are placed at least half a wavelength apart to operate effectively. Maintaining this minimum distance ensures that the antennas are incoherent, that is, they will be impacted differently by the additive/subtractive effects of signals arriving by means of multiple paths.

Simple diversity antennas detect the signal strength of the multiple (two or more) antennas attached and switch that antenna into the receiver mode. The likelihood of getting a strong signal is directly proportional to the number of (incoherent) antennas.

Beam-steering antennas shape the antenna array pattern to produce high gains in the useful signal direction or notches that reject interference. High antenna gain increases the signal, noise, and rate. The directional pattern attenuates the interference out of the main beam. Selective fading can be mitigated if multipath components arrive with a sufficient angular separation.

Beamforming antennas as the name indicates shape the RF beam so that the coverage area is targeted and is narrower than what is produced in a unidirectional antenna. This allows the area around a base station to be divided into sectors, allowing additional frequency reuse among sectors.

14.4.2   Other Advanced Features

In addition to the major features listed above, WiMAX also includes the following advanced features. We provide below a summary and for more details the reader is directed to other materials that deal on more advanced engineering.

Mobile WiMAX has introduced the adaptive modulation and coding (AMC), hybrid automatic repeat request (HARQ), and fast channel feedback (CQICH) to enhance coverage and capacity for WiMAX in mobile applications. HARQ provides added robustness with rapidly changing path conditions in high mobility situations. Frequency-selective scheduling and sub-channelization with multiple permutation options, gives mobile WiMAX the ability to optimize connection quality based on relative signal strengths to specific users. Fractional frequency reuse controls co-channel interference to support universal frequency reuse with minimal degradation in spectral efficiency. Optimal trade-off between overhead and latency is provided by 5 ms frame size.

14.5   WiMAX Architecture

The IEEE only defined the Physical (PHY) and MAC layers in 802.16. To deploy operational and successful commercial systems, there is need for support beyond 802.16 (PHY/MAC) air interface specifications. Chief among them is the need to support a core set of networking functions as part of the overall End-to-End WiMAX System Architecture.

The WiMAX Architecture is based on an all-IP Packet-Switched Framework with no legacy circuit switching. An all-IP core places the network on the performance growth curve of general purpose processors and computing devices, often termed Moore's law. The architecture permits decoupling of access architecture (and supported topologies) from connectivity IP service. Network elements of the connectivity system are independent of the IEEE 802.16 radio specifics. It offers the advantage of reduced total cost of ownership.

The architecture features modularity and flexibility to accommodate a broad range of deployment options from small-scale to large-scale (sparse to dense radio coverage and capacity) WiMAX networks that can be deployed in urban, suburban, and rural radio propagation environments. It can operate in license or licensed-exempt frequency bands and in hierarchical, flat, or mesh topologies, and their variant configurations. Coexistence of fixed, nomadic, portable, and mobile usage models is an added feature of the overall architecture.

The architecture includes IP multimedia subsystem (IMs) support. This enables access to third-party Application Service Provider (ASP) hosted applications like voice, multimedia services, emergency services, lawful interception, and mobile telephony using Voice-over-IP (VoIP). Support for interfacing with various inter-working and media gateways permitting delivery of incumbent/legacy services translated over IP (e.g., SMS over IP, MMS, WAP) to WiMAX access networks and delivery of IP broadcast and multicast services over WiMAX access networks are now possible.

Inter-working and roaming is another key strength of the network architecture with support for a number of deployment scenarios. Inter-working with existing wireless networks such as third generation partnership project (3GPP) (2) or existing wire-line networks such as digital subscriber line (DSL) will enable current broadband operators to deploy WiMAX in a complementary way. Global roaming across WiMAX operator networks, including reuse of access credentials, authentication, authorization, and accounting (AAA), and consolidated billing, is being implemented. A variety of user authentication credential formats such as username/password, digital certificates, subscriber identify module (SIM), universal SIM (USIM), and removable user identify module (RUIM) will be supported.

14.5.1 *WiMAX Profiles*

802.16-2004 addresses the entire sub-11GHz frequency range; therefore, there is an inherent need for a number of different solutions, or profiles to use the vernacular of the WiMAX Forum. Presently, the WiMAX Forum has identified at least five profiles for 802.16-2004 that allow the technology to accommodate different frequency bands, channel bandwidths, and duplexing schemes (TDD/FDD).

Similarly, the mobile technical group (MTG) in the WiMAX Forum is developing the mobile WiMAX system profiles that will define the mandatory and optional features of the IEEE standard that are necessary to build a mobile WiMAX compliant air interface that can be certified by the WiMAX Forum. Release-1 Mobile WiMAX profiles will cover 5, 7, 8.75, and 10 MHz channel bandwidths for licensed worldwide spectrum allocations in the 2.3, 2.5, 3.3, and 3.5 GHz frequency bands. The first mobile WiMAX system profiles were released in February 2006. Mobile WiMAX based on the 802.16e-2005 enables WiMAX systems to address portable and mobile applications in addition to fixed and nomadic applications.

14.6 WiMAX Distinct Features

14.6.1 *Security*

The security features in mobile WiMAX are best in class. Privacy and Key Management Protocol version 2 (PKMv2) is the basis of mobile WiMAX security as defined in 802.16e. This protocol manages the MAC security using PKM-REQ/RSP messages. PKM EAP authentication, traffic encryption control, Handover key exchange, and multicast/broadcast security messages are all used in this protocol.

By default, in Wi-Fi and also in Ethernet, the user always authenticates to the infrastructure and not vice versa. This leads to spoofing of the AP by malicious users. Mobile WiMAX supports device and user authentication using IETF EAP by providing support for credentials that are SIM-based, USIM-based, digital certificate or username/password-based. Corresponding Extensible Authentication Protocol-Subscriber Identity Module (EAP-SIM), Extensible Authentication Protocol Authentication and Key Agreement (EAP-AKA), EAP over Transport Layer Security (EAP-TLS), or EAP-MSCHAPv2 authentication methods are supported through the Extensible Authentication Protocol (EAP). Key deriving methods are the only EAP methods supported.

AES-CCM is the cipher used for protecting all the user data over the mobile WiMAX MAC interface. The keys used for driving the cipher are generated from the EAP authentication. A traffic encryption state machine that has a periodic key temporal encryption key (TEK) refresh mechanism enables sustained transition of keys to further improve protection. In Wi-Fi, management frames are always sent as plaintext that enabled eavesdroppers to at least know who is talking to whom even if data encryption is used. In WiMAX, control and management frames are protected using advanced encryption standard (AES)-based CMAC, or MD5-based HMAC schemes.

Wi-Fi (WLAN) was designed from the ground-up to be a wireless extension to last mile access solutions such as DSL or cable. It was designed to support stationary users or users at pedestrian speeds. WiMAX was designed to support WMAN and therefore needs to support users at vehicular speeds. In this scenario, secure/fast hand-over between APs is a major issue. A 3-way handshake scheme is supported by mobile WiMAX to optimize the re-authentication mechanisms for supporting fast handovers. This mechanism is also useful for the prevention of any man-in-the-middle attacks.

Security protocol optimizations for fast handovers, data integrity, replay protection, confidentiality, and nonrepudiation all use appropriate key lengths and are a part of mobile WiMAX.

14.6.2 WiMAX Quality of Service

The WiMAX specifications has inbuilt quality of service (QoS) mechanisms. With high data rates, asymmetric DL/UL capability, fine bandwidth resource granularity, and a flexible allocation mechanism, differentiated levels of QoS—coarse-grained (per user/terminal) or fine-grained (per service flow per user/terminal)—admission control are feasible. Mobile WiMAX can meet QoS requirements for a wide range of data services and applications. The standard thus enables wireless Internet service providers to provide service level agreement (SLA) for customers who require it and to tailor service levels on traffic type as well. For example, the standard can guarantee high bandwidth to business customers or low latency for voice and video applications, while providing only best-effort and lower-cost service to residential Internet users.

In the mobile WiMAX MAC layer, QoS is provided via service flows. This is a unidirectional flow of packets that is tagged with a particular set of QoS parameters and applies to both DL and UL data streams. Mobile WiMAX supports a wide range of data services and applications with varied QoS requirements. Additionally, sub-channelization and Media Access Protocol (MAP, Medium Access Protocol)-based signaling schemes provide a flexible mechanism for optimal scheduling of space, frequency, and time resources over the air interface on a frame-by-frame basis. Table 14.1 summarizes the different classes of services flows in WiMAX.

The Mobile WiMAX MAC scheduling service is designed to efficiently deliver broadband data services including voice, data, and video over time-varying broadband wireless channel.

In Wi-Fi, there is no scheduler running in AP and therefore there is no master for the radio channel, whereas in WiMAX the scheduler is located at each base station to enable rapid response to traffic requirements and channel conditions. The data packets are associated to service flows with well-defined QoS parameters in the MAC layer. Multiple UL bandwidth request mechanisms, such as bandwidth request through ranging channel, piggyback request, and polling are designed to support UL bandwidth requests.

The MAC supports frequency-time resource allocation in both DL and UL on a per-frame basis. The resource allocation is delivered in MAP messages at the beginning of each frame. Therefore, the resource allocation can be changed frame-by-frame in response to traffic and channel conditions.

Table 14.1 Different Classes of WiMAX Client Traffic

QoS Category	Applications	QoS Specifications
UGS Unsolicited grant service	VoIP	Maximum sustained rate Maximum latency Tolerance Jitter tolerance
rtPS Real-time polling service	Streaming audio or video	Minimum reserved rate Maximum sustained rate Maximum latency tolerance Traffic priority
ErtPS Extended real-time polling service	Voice with activity detection (VoIP)	Minimum reserved rate Maximum sustained rate Maximum latency tolerance Jitter tolerance Traffic priority
nrtPS Non-real-time polling service	File Transfer Protocol (FTP)	Minimum reserved rate Maximum sustained rate Traffic priority
BE Best-effort service	Data transfer, Web browsing, etc.	Maximum sustained rate Traffic priority

Additionally, the amount of resource in each allocation can range from one slot to the entire frame. The fast and fine granular resource allocation allows superior QoS for data traffic.

The MAC scheduler handles data transport on a connection-by-connection basis. Through frequency-selective scheduling, the scheduler can allocate mobile users to their corresponding strongest sub-channels. The frequency-selective scheduling can enhance system capacity with a moderate increase the in channel quality indicator overhead in the UL.

14.6.3 Mobility Management

The WiMAX Architecture has extensive capability to support mobility and handovers. It will include vertical or inter-technology handovers, for example, to Wi-Fi, 3GPP, 3GPP2, DSL, or Multiple System Operator (MSO, Multiple Systems Operator) when such capability is enabled in multimode mobile station (MS). Within this framework, and as applicable, the architecture is expected to accommodate MS with multiple IP addresses and simultaneous IPv4 and IPv6 connections, and support roaming between network service providers (NSPs).

Mobile WiMAX also supports seamless handoff to enable the MS to switch from one base station to another at vehicular speeds without interrupting the connection. There are three handoff methods supported within the 802.16e standard—hard handoff (HHO), fast base station switching (FBSS), and macro diversity handover (MDHO). Of these, the HHO is mandatory while FBSS and MDHO are two optional modes. The WiMAX Forum has developed several techniques for optimizing HHO within the framework of the 802.16e standard. These improvements have been developed with the goal of keeping layer 2 handoff delays to less than 50 ms to ensure real-time applications such as VoIP perform without service degradation. Flexible key management schemes assure that security is maintained during handover.

14.6.4 Power Management

Battery life is arguably the most critical issue for mobile devices. Mobile WiMAX supports two modes for power-efficient operation—sleep mode and idle mode. Sleep mode is a state in which the MS conducts pre-negotiated periods of absence from the serving base station air interface. These periods are characterized by the unavailability of the MS, as observed from the serving base station, to DL or UL traffic. Sleep mode is intended to minimize MS power usage and minimize the usage of the serving base station air interface resources. The sleep mode also provides flexibility for the MS to scan other base stations to collect information to assist handoff during the sleep mode.

The idle mode provides a mechanism for the MS to become periodically available for DL broadcast traffic messaging without registration at a specific base station as the MS traverses an air link environment populated by multiple base stations. The idle mode benefits the MS by removing the requirement for handoff and other normal operations and benefits the network and base station by eliminating air interface and network handoff traffic from essentially inactive mobile stations while still providing a simple and timely method (paging) for alerting the MS about pending DL traffic.

14.7 WiMAX in Real Life

This section explores WiMAX deployment scenarios, applications, and products.

14.7.1 Scenarios

WiMAX can be deployed either in a broadband wireless access or back-haul transport scenario. The following are some of the deployment options:

Access: Wireless DSL, wide-area area hot spot, rural broadband access
Transport: Mobile back-haul, Wi-Fi back-haul, Self-Backhaul

WiMAX is a suitable technology for a wireless DSL service, that is, a high-speed wireless Internet connection comparable to today's DSL and cable offerings. It is suitable for new operators to compete with established DSL/cable operators in cities and enables green field operators to deploy wide area broadband wireless access in suburban and rural areas where the cost of laying cables is prohibitively expensive, thus making fixed broadband access a nonviable option. Finally, it could be deployed as a hot spot solution due to the range being much wider than that of Wi-Fi.

WiMAX was primarily designed for carrying IP/Ethernet traffic. It is technically possible to use WiMAX systems for transporting Time Division Multiplexing (TDM) links (e.g., E1 and T1). Circuit emulation (CE) is used to carry TDM links (mobile traffic) as IP over Ethernet. This traffic is then mapped to the WiMAX MAC and OFDMA physical layer. The main pretext of introducing wireless in the last-mile configuration is to reduce the cabling cost to each end-user premises and systems. However, the upstream traffic from the AP is still currently carried through xDSL or cable infrastructure. WiMAX introduces a new option for carrying this upstream back-haul traffic. This enables the Internet service providers to deploy the APs at strategic locations in urban, suburban, and rural areas and upstream linked by WiMAX technology. By this way, even the cost of cabling and time lag for rolling-out new services become nonexistent. Similarly, when WiMAX is deployed as an access solution, a part of the bandwidth can be reserved to carry the upstream traffic in a transport mode thereby creating a hybrid WiMAX deployment. Figure 14.1 pictorially describes the deployment scenarios in both access and transport mode for WiMAX.

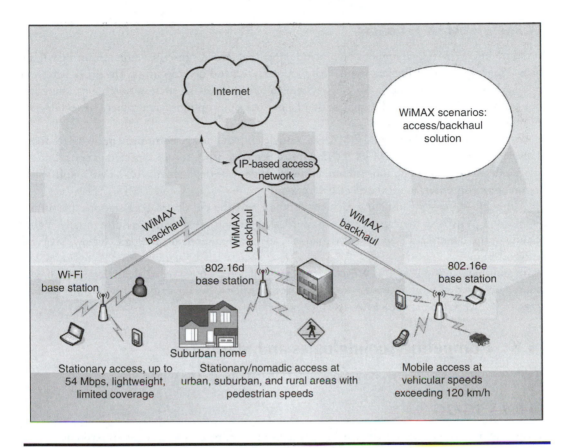

Figure 14.1 **WiMAX applications and deployment model.**

14.7.2 WiMAX Applications

VoIP is expected to be one of the most popular WiMAX applications. Its value proposition is immediate to most users. With a data connection plan, VoIP calls can be received or placed at a very low or, in some cases, no additional cost. WiMAX will provide full support for VoIP traffic, thanks to QoS functionality and low latency.

Broadcast is another potential WiMAX application. Multicast and broadcast service (MBS) supported by mobile WiMAX combines the best features of DVB-H, MediaFLO, and 3GPP E-UTRA and satisfies the requirements such as high data rate and coverage using a single frequency network, flexible radio resource allocation, low mobile device power consumption, low channel switching time, and support of data-casting in addition to audio and video streams.

Vertical applications like surveillance, public safety, connectivity to remote devices, inventory tracking, fleet management, and educational services can also be supported by mobile WiMAX networks with little or no incremental cost to network operators.

In general, mobile WiMAX applications can be classified based on instantaneous bandwidth requirements, latency specifications, and jitter tolerance. One method of categorization could be multiplayer interactive gaming, VoIP and video conference, streaming media, Web browsing and instant messaging, and media content downloads.

14.7.3 WiMAX Products

Initial mobile WiMAX equipment will include notebook-based subscriber units (mini PCMCIA cards, PCI Express, PCI Express mini, USB modules, etc.) and desktop units. The introduction of mobile devices with embedded WiMAX systems-on-chips (SOCs), such as notebooks, the Ultra mobile PC (UMPC), personal digital assistants, phones, smart phones, and other wireless devices are expected to follow soon after.

The integration of Wi-Fi and WiMAX in a single chipset and the commitment by device manufacturers to incorporate a WiMAX interface into their new products are expected to contribute to an even deeper cost reduction for subscriber units. Wide scale deployments of WiMAX and Wi-Fi in notebook computers are also expected very soon.

Although the cost of installing each base station and the density of base stations are similar for 3G and WiMAX, the capacity that a WiMAX base station provides is substantially higher because of the use of OFDMA in wider channels, advanced antenna technologies like MIMO and beamforming.

In summary, in the WMAN, WiMAX is an excellent complement to other wireless technologies that are designed to work in the LAN (Wi-Fi) or that offer wider coverage but with more limited capacity (GSM, CDMA, WCDMA, EV-DO).

14.8 Competing Technologies and Standards

In this section, we briefly look at current competing technologies to WiMAX.

14.8.1 3G/2G

Second- and third-generation cellular infrastructure is optimized to carry circuit switched voice traffic and is not designed to cope with the growing amount of traffic generated by high-speed and real-time applications.

WiMAX will coexist and interwork with existing and emerging technologies, both wired and wireless. Even though it can support VoIP, WiMAX will not replace or compete with 2G or 3G technologies for voice services. Cellular networks provide the extensive coverage that circuit-switched voice services require and which the WiMAX infrastructure is not designed to support. Third-generation networks cover many urban and suburban areas, but they will not be able to offer sufficient capacity or throughput for data applications. Similarly, WiMAX and Wi-Fi are complementary and are expected to be incorporated in dual-mode chipsets in mobile devices, as WiMAX provides wider coverage, while Wi-Fi is better suited for high-throughput, indoor LAN applications.

WiMAX is a next-generation technology that will facilitate the cellular operators' transition to all-IP networks. Cellular networks are also moving toward an IP core with the long-term evolution and system architecture evolution (SAE) efforts, but this activity is in its early stages with service expected to be rolled out in 2010. WiMAX fully supports IMS2 and its 3GPP2 counterpart, multimedia domain (MMD).

Figure 14.2 shows the range comparisons between WiMAX and the competing standards and technologies in use today.

14.8.2 WiBro

Another standard worth mentioning is WiBro (wireless broadband). WiBro is a South Korean initiative to establish a homegrown wireless technology, much like China's Time Division-Synchronous

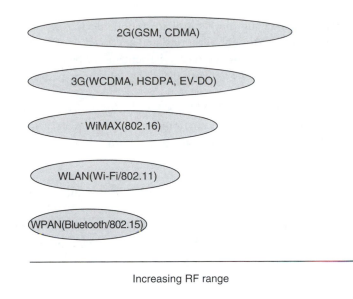

Figure 14.2 WiMAX RF range in comparison with other current wireless technologies.

Code Division Multiple Access (TD-SCDMA). Specifically, WiBro is a TDD-based system that operates in a 9 MHz radio channel at 2.3 GHz with OFDMA as its access technology. According to its proponents, WiBro supports users traveling at speeds up to 120 km/h and peak user data rates of 3 Mbps in the DL (UL = 1 Mbps) and 18 Mbps of peak sector throughput in the DL (UL = 6 Mbps). Average user data rates are expected to be in excess of 512 kbps, and with the cell radius limited to 1 km, it will largely be deployed in densely populated areas. Initially, WiBro was perceived as a portable solution, even though it could support mobile users, as the technology did not support seamless cell handoffs. With WiBro now included within the 802.16e umbrella, making it another WiMAX profile, there is a desire to introduce vehicular mobility, or near seamless handoffs.

The first commercial portable/mobile application for WiMAX-certified products is expected to take place in Korea with the launch of WiBro services. Products for WiBro services operate in the licensed 2.3 GHz frequency band with an 8.75 MHz channel bandwidth. This initial product launch will use single input, single output antenna configurations and support mobile speeds greater than 60 km/h.

14.8.3 *Hiperman*

An alternative standard to 802.16a was the European Telecommunications Standards Institute's (ETSI) BRAN HA (broadband radio access networks HiperAccess) or high-performance radio metropolitian area network (HiperMAN). ETSI has two specifications, HiperAccess, which operates above 11 GHz and HiperMAN for below 11 GHz. This has since been consumed into the WiMAX standard.

14.9 Conclusion

WiMAX meets all the requirements for mobile Internet access. It supports multiple handoff mechanisms, ranging from hard handoffs (with break-before-make links) to soft handoffs (with make-before-break links), power-saving mechanisms for mobile devices, advanced QoS and low latency for improved support of real-time applications, and advanced AAA functionality.

Mobile WiMAX is the technology that best meets the demand for personal broadband services. It is based on a next-generation all-IP core network that offers low latency, advanced security, QoS, and worldwide roaming capabilities.

The advanced performance of mobile WiMAX is largely due to its use of OFDMA, a multiplexing technique well suited to multipath environments that gives network operators higher throughput and capacity, great flexibility in managing spectrum resources, and improved indoor coverage.

OFDMA has clearly emerged as the technology of choice for next-generation mobile networks. 3GPP has incorporated OFDMA in its long-term evolution (LTE) specification and the third generation partnership project two (3GPP2) is moving in the same direction. WiMAX has a two-to-three year time advantage over LTE, which is still in its early stages of development. Furthermore, LTE is expected to use OFDMA only in the DL, with single-carrier FDMA employed in the UL. This is likely to impact the UL throughput from mobile devices and result in lower spectrum efficiency in comparison to WiMAX.

Bibliography

For further information on WiMAX technical and business opportunities the readers are referred to other online materials as cited in references [3–12]. All of the materials, as of this writing, are available online and the readers are encouraged to use Google search engine for locating them.

Abbreviations

2G	second generation
3G	third generation
3GPP	third generation partnership project
3GPP2	third generation partnership project two
AAA	authorization, authentication, and accounting
AES	advanced encryption standard
ASP	application service provider
ASP	average selling price
CBC-MAC	cipher block chaining with message authentication code
CCM	counter with CBC-MAC
CDMA	code division multiple access
CMAC	Cipher based message authentication code
CP	cyclic prefix
DL	downlink
DSL	digital subscriber line
DVB	digital video broadcast
DVB-H	digital video broadcast-hand held
EAP	Extensible Authentication Protocol
ETSI	European Telecommunications Standards Institute
EV-DO	evolution data optimized

FDD   frequency division duplex
FDMA   frequency division multiple access
GSM   global system for mobile communication
HiperMAN   high performance radio metropolitan area network
IEEE   Institution for Electrical and Electronics Engineers
IMS   IP multimedia subsystem
IP   Internet Protocol
LAN   local area network
LTE   long-term evolution
MAC   Media Access Control
MIMO   multiple input multiple output
OFDM   orthogonal frequency division multiplexing
OFDMA   orthogonal frequency division multiplexing access
PCI   peripheral component interconnect
PCMCIA   Personal Computer Memory Card International Association
PHY   physical layer
QoS   quality of service
RF   radio frequency
S-OFDMA   scalable-orthogonal frequency division multiplexing access
TDD   time division duplex
TDM   time division multiplexing
UP   Uplink
USB   universal serial bus
VoIP   Voice over Internet Protocol
WCDMA   wideband code division multiple access
Wi-Fi   wireless fidelity
WiBro   wireless broadband
WiMAX   worldwide interoperability for microwave access
WMAN   wireless metropolitan area network

REFERENCES

1. http://www.wimaxforum.org/home/.
2. http://www.wi-fi.org/.
3. Mobile WiMAX—Part 1: A Technical Overview and Performance Evaluation.
4. WiMAX Forum Plugfest white paper.
5. WiMAX, Wikipedia.
6. WiMAX: The critical wireless standard.
7. Understanding Wi-Fi and WiMAX as metro-access solutions.
8. WiMAX opportunities and challenges in a wireless world.
9. WiMAX—Copper in the air.
10. Mobile WiMAX: A performance and comparative summary.
11. WiMAX capacity white paper.
12. WiMAX presentations.

Chapter 15

Detailed DSRC-WAVE Architecture

Yasser Morgan, Mohamed El-Darieby, and Baher Abdulhai

CONTENTS

15.1   Motivation and Rationale

Large urban areas in North America and in many parts of the world are suffering unprecedented and soaring congestion problems. Conventional solutions in the form of building more roads are neither desirable nor feasible in most cases. Hence, innovative technological advances that enhance the efficiency of existing transportation infrastructure can play a vital role in combating transportation problems. The rationale, simply put, is to use the available capacity more efficiently before expansions can be justified.

Undoubtedly, seamless transportation is the key to our societal and economical sustainability. The efficient, safe, cost-effective movement of people and goods is so fundamental to our daily lives that we tend to take the transportation system for granted until it fails in some way. Despite this, modern societies have under-invested in public infrastructure of all types for decades. Nowhere is this more the case than in transportation. Therefore, it is imperative that modern societies upgrade their transportation systems to remain competitive in the global market and to maintain the high quality of life and social well-being that we rightly prize so highly.

Congestion is typically the most visible transportation issue for both the general public and decision makers. And certainly congestion is a huge problem in large cities and a growing problem in smaller cities. Regardless of the size of the city or metropolitan area, it is almost always the case that transportation infrastructure is failing to keep pace with growth in demand. It costs businesses billions of dollars annually in lost productivity, and it generates considerable frustration, inconvenience, and stress among urban travelers who are attempting to move about the urban area during the course of their daily activities. Congestion, however, is but one of many issues/impacts of importance with respect to the transportation system. Others include serious detrimental impact on safety, the environment, trade, urban form, and the well-being of our social fabric.

A renewed worldwide recognition, however, exists of the importance of investing in transportation infrastructure, planning, and systems operations to successfully maintain an efficient, cost-effective transportation system in support of economic and social well-being. Transportation systems fail when transportation demand (amounts of people of goods movement) exceeds capacity, recurrently during peak periods or nonrecurrently during incidents and unusual events. Transportation remedies generally attempt to reduce the demand to capacity ratio and can be generally grouped into three major categories: (1) building new roads and transit capacity, (2) managing demand, and (3) operating existing capacity more effectively [3].

Among the three major categories, building new roads and transit capacity has been the traditional approach since the dawn of the twentieth century. However, this approach is failing to keep pace with demand growth due to obvious economical, environmental, and sustainable disadvantages. Demand management, on the other hand, is a group of strategies that attempt to ease demand pressure to the extent possible. Examples include congestion pricing, high occupancy vehicle (HOV) lanes, staggered work hours to spread the peak, telecommuting, and land-use strategies to name a few. Demand management, due to its restricting nature, is usually opposed by end users and is hence feared to some extent by decision makers. Operational improvements of the road and transit systems, also known as supply management, have been receiving increasing attention over time as they promise efficient use of capacity and hence relieve the pressure to build more. These include centralized control via transportation systems management centers, incident management strategies, event management strategies, real-time provision of information and guidance to travelers and stake holders, real-time system control of system components such as freeways and intersections, among others. Many of these strategies heavily rely on technology including advances in digital computing and communications, under the field of intelligent transportation systems (ITS). Numerous evaluations have shown that the cost of ITS programs generally outweigh the cost. In addition, ITS deployment typically causes minimal disruptions, unlike major construction projects. Therefore, ITS have gained tremendous attention over the past decade or so.

The heart of ITS undoubtedly lies in gathering, communicating, and using system information in real-time to improve the real-time control and management of the system. In doing so, the intention is to reduce delay due to congestion, and also reduce other unwanted system externalities such as accidents and pollution. Developing such real-time information and control systems involves a wide variety of interconnected research and development problems, the nature of which can be technical (sensors, communications, computing hardware, automated vehicle technology, etc.), methodological (control systems theory, traffic flow theory, artificial intelligence, very large-scale database management and other software requirements, simulation methods, image processing, etc.), or behavioral (driver–vehicle interactions, traffic controller–control display interactions, user decision making in response to information and other stimuli, etc.).

On the side of sensors and data communication, the state of the practice is dominated by the use of point detectors for surveillance, and wire-line communication networks for data and information transmission. In almost all large metropolitan areas, major freeways and arteries are covered by pavement-embedded induction loop detector stations to measure traffic volumes and speeds. Gathered information is typically aggregated over 20–30 s then transmitted over copper or fiber optic wire lines to the nearest operations center. This approach, although dominant at the moment, is losing appeal due to detector reliability issues and the cost of building and maintaining the detector network, not to mention traffic stream disruptions during construction and maintenance. Modern off-road detector technology has improved significantly over the past decade resulting in new and more mature detector types based on radar, ultrasound, infrared, acoustic, and other technologies. Wireless communication is also gaining grounds because of several advantages over wire-line communication including cost, maintenance, and data capacity advantages. Among the rapidly emerging communication technologies is dedicated short range communication (DSRC). DSRC systems are being designed to provide short-range, wireless links to transfer information between vehicles and roadside units, other vehicles, or portable roadside units. DSRC is anticipated to be essential to many ITS applications that improve traveler safety, decrease traffic congestion, facilitate the reduction of air pollution, and help conserve fossil fuels. Examples of such information transfer include traffic light control, traffic monitoring, traveler alerts, automatic toll collection, traffic congestion detection, emergency vehicle traffic signal pre-emption, and electronic inspection of moving trucks through data transmissions with roadside inspection facilities.

DSRC is a block of unlicensed spectrum in the 5.850–5.925 GHz band allocated by the United States Federal Communications Commission (US-FCC) to be used for enhancing safety and productivity of the transportation systems. The North American development of the DSRC standards has been going on since early 1990s. The American Society for Testing and Materials (ASTM) formed a standardization committee called E17.51 that adopted the development of the DSRC standards [12]. The ASTM-E17.51 started by building over the legacy 902–928 MHz band and continued till late 2004. The E17.51 community realized the need to switch to the IEEE 802.11 generic architecture, and therefore established five standard streams within the IEEE framework, namely IEEE P 802.11p, IEEE 1609.1/.2/.3/.4 and, two complementary projects IEEE 1609.0/.5 ([2,5–8], consecutively).

Beside the North American research on DSRC, similar efforts evolved from the European Committee for Standardization (CEN) realizing the importance of developing public transportation safety framework and DSRC standards. In response, CEN formed a technical committee (TC-278) [16] that held its kickoff meeting as early as July 1991. By the mid 2000s, both TC-278 and ASTM-E17.51 decided to adopt the 5.9 GHz spectrum instead of the 915 MHz. Unfortunately, the current American, European, and Japanese standards are incompatible without signs of unification. The first ACM Workshop on Vehicular Ad Hoc Networks organized in October 2004 played a major role in attracting researchers to the short-range communication domain.

DSRC is a medium range wireless communication service (<1 km) designed to support the proposed ITS and public safety applications as illustrated in the VII initiative [17]. DSRC is capable of delivering 27 mbps data rate by using a two way line-of-sight radio, which is significantly lower in cost compared to cellular, WiMAX, or satellite communications. Furthermore, the DSRC applications lend themselves, naturally, to localized road communications, which lead to a much desired distributed decentralized deployment. DSRC operates in a stringent environment that requires fast hand off to maintain the connection with speeding vehicles at all times, strict quality of service (QoS) committed to predefined threshold delays for special safety messages, minimal use of power when transmitting periodical information, and maintaining privacy and anonymity of roaming users in addition to many other environmental challenges. DSRC is also vulnerable to deployment problems due to the short-range nature. The deployment of massive number of units is inevitable, which require expensive and hard-to-build infrastructure. An alternative solution is to build the DSRC network vertically over cellular or WiMAX networks. As illustrated in this chapter, building DSRC over cellular or WiMAX requires, relatively, limited redesign to the DSRC and significantly major work on resolving mobility issues.

The chapter is organized as follows; Section 15.2 introduces the DSRC to the unfamiliar reader and briefly explains motives behind the DSRC initiative in statistical terms as part of the overall ITS paradigm. Section 15.3 identifies the most common categories for DSRC applications. Section 15.4 defines the commonly used terms and building blocks then identifies the wireless access in a vehicular environment (WAVE) reference model from a layered perspective, and defines important over-the-air frame formats.

Sections 15.5 through 15.7 elaborate on the detailed characteristics of vehicular communications and other wireless interoperability and mobility issues. For instance, Section 15.5 identifies WAVE Media Access Control (MAC) services such as WAVE QoS elements, then investigates the channel coordination Function (CCF) of the 1609.4 and extends that to synchronization tolerance issues. The section then identifies the different WAVE communication types. Section 15.6 examines the WAVE network layer services and covers the possible use of 802.11 portal functions to simplify mobility routing in addition to the use of IPv6 neighbor cashe. Section 15.7 visits the management of WAVE services in terms of persistence, service registration, discovery, security, and defining relevant management policies. Section 15.8 briefly visits the major DSRC security concerns. Finally Section 15.9 concludes and comments on the research.

15.2   Overview

The early work on DSRC standards presented beacon-based approaches similar to those in Refs. [13,15]. This approach was abandoned quickly as the complexity of the system became apparent and the amount of data loaded on beacons burdened the system resource. Alternatively, continuous research led to adding significant enhancements to the mobile units' intelligence. This chapter describes the added value to the DSRC standard by diverting from a beacon-based approach.

The DSRC networks are designed to provide public safety and other network services for roaming vehicles throughout North America. The DSRC network is built over two basic units; road-side unit (RSU) and on-board unit (OBU). The RSU is, typically, a stationary unit that connects roaming vehicles to the access network, which, in turn, is connected to a larger infrastructure or to a core network. The OBU is, typically, a network device fixed in a roaming vehicle and is connected to both the DSRC wireless network and to an in-vehicle network. This simple architecture is illustrated in Figure 15.1.

The wireless connection between RSU and OBU is based on DSRC standards suite (Figure 15.2), and is called WAVE. The cones shown in smaller dots in Figure 15.1 represent the RSU communication zones while the ellipse represents the radio range of the OBU. As OBUs move between communication zones, vehicles exchange information with the roadside; in addition, vehicles use the same WAVE media to communicate with each other.

The communication zone covered by each IEEE 802.11p RSU is limited to a maximum of 1 km diameter and uses a 5.9 GHz radio transmission. OBUs are expected to join the WBSS (WAVE basic

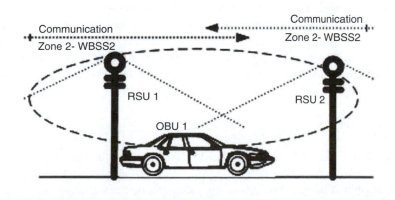

Figure 15.1   OBU roaming between two RSUs.

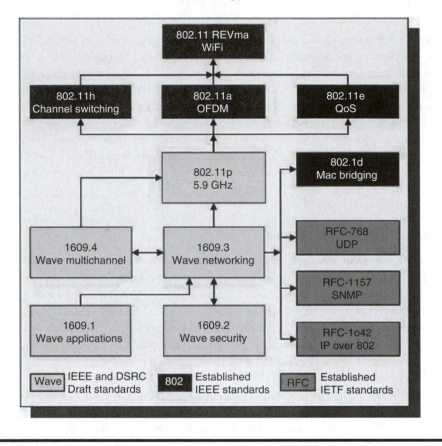

Figure 15.2 Simplified DSRC standards suite.

service set) of the RSU, exchange information, and may leave within very limited time. The use of the term closest here is a reasonable approximation and the term is defined in more detail later. The limited lifetime of an OBU within specific WBSS communication zone imposes hard requirements on the design of the DSRC standards suite and on the nature of future DSRC applications as described later. DSRC networks use WAVE Short Messages Protocol (WSMP) to exchange safety information between vehicles and roadside or just between vehicles.

WAVE devices employ an architecture that supports a predefined control channel (CCH) and multiple service channels (SCHs). The CCH is used to transmit WSMP and to announce WAVE services, while a carefully selected SCH is used for application interactions and data transmissions. The specific designations of these channels and the specification of the PHY are defined in IEEE 802.11p.

15.3 DSRC Applications

There are eight groups identified as DSRC-based ITS applications including:

1. Travel and traffic management
2. Maintenance and construction operations
3. Public transit management

4. Electronic payment
5. Commercial vehicle operations
6. Emergency management
7. Advanced vehicle safety systems
8. Information management

These applications also be categorized based on public or private used radio, based on range (less than 15, 15–100, 100–400, and 400–1000 m), based on vehicle type (all vehicles, buses, trains, heavy trucks, and emergency vehicles), or based on safety measures (public safety, private safety, warnings, and commercial applications).

Examples within these groups include electronic toll collection, transit vehicle signal priority, fuel payment, parking payment, electronic commerce, traffic information, management of public transportation and commercial vehicles, fleet management, work zone warning, road condition warning, weather information, and border crossing clearance.

Currently, public safety applications are the focal point for governments and government-driven researches. Car manufacturers are more focused on applications enhancing driver experience in-vehicle applications like safe lane change, and driver–vehicle–road interaction. The highway electronics industry is focused on other issues like toll collection, parking management, and processing of electronic payment. The vehicular infrastructure integration VII [17] is a consortium of different interest groups focused on the interoperability of DSRC and non-DSRC applications.

15.4 WAVE: Basic Architecture

The following subsections illustrate the major building blocks of the DSRC networks, identify a reference model from a layered perspective, and then illustrate new frame formats that WAVE uses for over-the-air communications.

15.4.1 DSRC Building Blocks

The DSRC initiative has derived the development of the suit of standards illustrated in Figure 15.2. The arrows in Figure 15.2 represent the standards dependencies. Figure 15.2 is simplified to show the relation between DSRC and other major standards. For instance, because 802.11p depends ultimately on 802.11, the 802.11p follows the 802.11 Carrier Sense Multiple Access with Collision Avoidance (CSMA/CA) mechanisms. This chapter focuses on the IEEE 802.11p [5], IEEE 1609.3 [2], and IEEE 1609.4 [6]. Obviously both 1609.1 [7] and 1609.2 [8] depend highly on 1609.3/.4 and on IEEE 802.11p. The details of both 1609.1/.2 remain vulnerable to changes in 1609.3/.4. It is expected that 1609.3/.4 will incur changes in the near future. The entire DSRC standards suite is scheduled to be released in late 2007 as a trial-use standards; and likely be subject to future changes and amendments. Therefore, we are focused on the core part of the DSRC standards suite, which is seen in IEEE 802.11p, IEEE 1609.3, and IEEE 1609.4.

As illustrated in Figure 15.2, the IEEE 1609.1 covers toll applications that use a resource manager to maintain compatibility with the 915 MHz legacy standards. However, the DSRC system provides a generic platform for a variety of applications that are not too specific in nature. The 1609.2, on the other hand, is focused on securing WAVE communications. Although the subject is relevant to this chapter and to our research interests, we decided to cover DSRC security in a separate article due to its volume and complexity.

WAVE devices such as RSU and OBU are expected to be implemented using two types of radio devices. The first is a single-channel WAVE device that exchanges information or listens to only one radio frequency (RF) channel at a time (commonly inaccurately called single-channel device). The second is a multi-channel WAVE device that exchanges on one channel while at least actively listening on a second channel (commonly inaccurately called multi-channel device).

To allow interoperability and accommodate the limited capabilities of single-channel devices, a synchronization procedure is required to ensure that all WAVE devices monitor or utilize CCH at common time intervals. Both CCH and SCH intervals are uniquely defined with respect to an absolute external time reference like the coordinated universal time (UTC) [14]. UTC is commonly provided by global positioning system (GPS) systems; however, the time base accuracy requirements for WAVE devices are sufficiently lenient that MAC management frames containing UTC time estimates from other devices can be used as well. Once a group of WAVE devices are synchronized, each single-channel WAVE device must meet the requirement by monitoring the CCH during the specified CCH intervals.

15.4.2 Reference Model

From a layered perspective, WAVE devices employ almost common ISO/OSI (International Organization for Standardization/Open Systems Interconnection) stack as displayed in Figure 15.3. The figure illustrates a common communication stack and identifies the related standard for both data and management planes. Figure 15.3 also shows service access point (SAP) for each entity.

In Figure 15.3, blocks highlighted in blue represent the focus of the WAVE standards. Blocks highlighted in dark blue represent WAVE amendments to other standards. The green blocks in the top right represent common WAVE interfaces with established standards outside the DSRC. The WAVE applications are not presented in Figure 15.3, but applications use common interface with the classical UDP/TCP stack.

From a networking perspective, a WAVE communication stack shown in Figure 15.3 consists of the following:

1. Data plane: It contains the communication protocol stacks and the hardware used for data delivery. The data plane carries traffic primarily generated by, or destined to, applications. For simplicity, this chapter assumes that all the WAVE protocol entities shown in Figure 15.3 reside in a single physical device, but this need not be the case.
2. Management plane: It performs system configuration, control, and maintenance functions. Management functions may employ the data plane services to pass management traffic between devices. The WAVE management entity (WME) is a general collection of WAVE management services. Note that the WME provides its management interface to all data plane entities, including WSMP, though this is not explicitly illustrated in Figure 15.3. The WAVE security entity provides management of data encryption mechanisms and key management. The WSM also takes part in enforcing security policies besides monitoring traffic patterns and responding to possible attacks [2].

WSMP packets may require special services like being transmitted using a particular power or data rate. These unique requirements impose challenges to the WAVE-MAC layer and below. The MAC and PHY layers must test the contents of each packet to adjust radio power and data rate before each packet transmission.

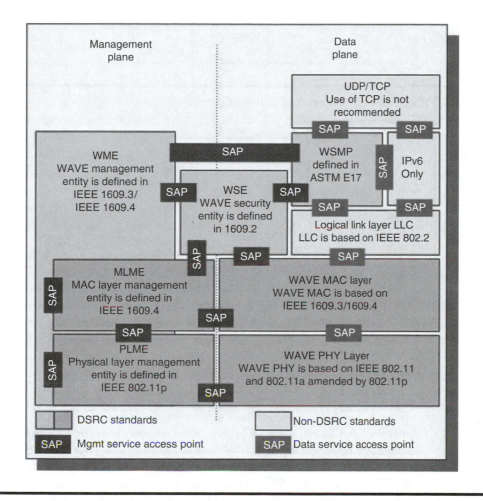

Figure 15.3 WAVE device communication stack.

15.4.3 Over-the-Air Frame Formats

The subtle differences between the WAVE environment and classical WiFi resulted in limited differences in used frame formats. These differences are illustrated hereafter.

15.4.3.1 EtherType

The IEEE 802.2 header carries only one of two EtherType values in a WAVE environment. A value of 0x86DD is used when the frame carries an IPv6 packet and a value of 0x88DC is used when the frame carries a WSMP packet. Figure 15.4 illustrates the IEEE 802.2 frame passed between logical link control (LLC) and MAC layers in a WAVE environment [2].

15.4.3.2 WSMP Format

The WSMP header and frame format is illustrated in Figure 15.5. The WAVE Architecture employs both IPv6 and WSMP stacks simultaneously.

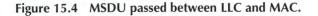

802.2 Header	IPv6 or WSMP Header	IPv6 or WSMP Data

Figure 15.4 MSDU passed between LLC and MAC.

This format permits a WSMP application to control the physical parameters used to transmit a specific WSMP packet.

15.4.3.3 WAVE Service Information Element

The WAVE Architecture defines a WAVE announcement action frame, which carries a WAVE service information element (WSIE), described in Figure 15.6. The WSIE carries the WAVE service announcement (WSA) and timing information. The WAVE announcement action frame format is described in Figure 15.6 and is used to carry WSIE, which communicates essential timing information used to synchronize WAVE devices in addition to carrying the WSA [7].

15.4.3.4 WAVE Service Advertisement Format

The WAVE Architecture defines WSA to carry essential announcement over the air as part of the WSIE. The WSA is secured and is inserted between security header and trailer.

Figure 15.7 shows the WSA format. Control fields are shown in dark blue and fixed field lengths are shown above each field. Information fields are shown in light green and optional fields in tan color. All contents field are bitmaps representing used fields. All length fields may equal to zero if required. The WSA is composed of a WAVE version field, a provider service table (PST), and WAVE routing advertisement. Figure 15.7 shows the details and contents of each field. The details of Figure 15.7 provide important implementation in detail.

15.5 WAVE MAC Services

The following subsections illustrate the MAC layer services in WAVE communications. The focus is on WAVE QoS and MAC queues operations, channel coordination between different WAVE devices, and describing the different types of WAVE communications.

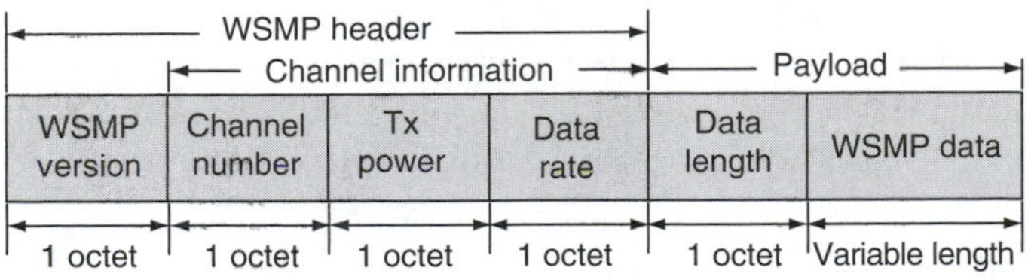

Figure 15.5 WSMP header format.

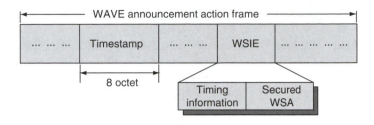

Figure 15.6   WSIE format within the WSA.

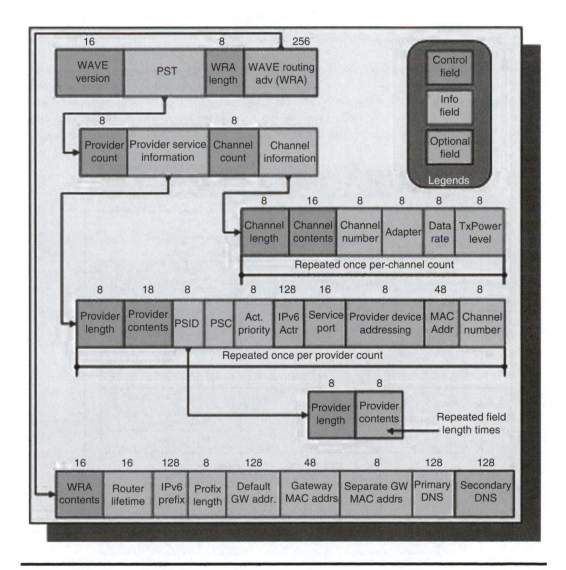

Figure 15.7   WSA format.

15.5.1   WAVE and 802.11p QoS

As illustrated in Figure 15.2, the WAVE and 802.11p follow the 802.11e EDCA QoS paradigm. Referring to Figures 15.4 and 15.5, one of the major additions in DSRC networks is the transmission of WSMP packets using an in-packet specified power, data rate, or channel number. The WAVE design extrapolates the 802.11e EDCA Architecture as illustrated in Figure 15.8 to service both IPv6 and WSMP packets. For simplicity, Figure 15.8 ignores management and signaling traffic.

The reference architecture for WAVE MAC is distinguished from the 802.11e Architecture by implementing access category queues on a per-channel basis and by the addition of CCF. The CCF, in a sense, is implemented using a channel router and a channel selector.

15.5.1.1   Channel Selector

The channel selector carries out multiple decisions as to when monitor specific channel, what are the set of legal channels at a particular point in time, and how long the WAVE device monitors and utilizes a specific channel. The channel selector decides also to drop data if it is supposed to be transmitted over an invalid channel, for example, if a channel no longer exists. The set of policies enforced by channel selector can be fairly complex. Those policies are defined and communicated

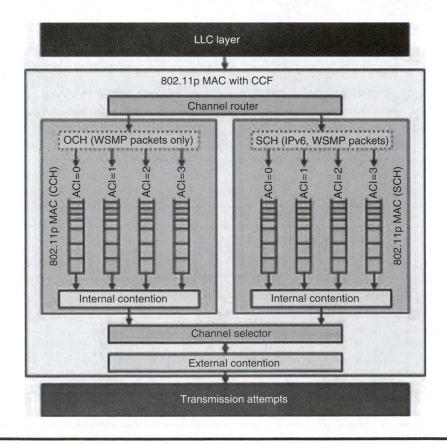

Figure 15.8   WAVE MAC-QoS Architecture.

to the channel selector through the WME as indicated in Figure 15.3. The 1609.4 provides the detailed Management Information Base (MIB) identifying these policies.

15.5.1.2 Channel Routing

The channel router detects the arrival of a WSMP datagram by checking the EtherType field of the 802.2 header. The channel router then forwards the WSMP datagram to the correct queue based on the channel identified in the WSMP header and based on packet priority. If the WSMP datagram is carrying an invalid channel number, the packet is discarded without issuing any error to the sending application.

The IP datagram transmission is slightly different. Before initializing IP data exchanges, the IP application registers the transmitter profile with the MAC layer management entity (MLME) (see Figure 15.3). The transmitter profile contains the SCH number, power level, data rate and the adaptable status of power level, and data rate. When IP datagram is passed from the LLC to the channel router, the channel router routes the datagram to a data buffer that corresponds to the current SCH. At any given time, there is only one active transmitter profile in a WAVE device. If the transmitter profile indicates a specific SCH that is no longer valid, the IP packet is dropped and no error message is generated.

15.5.1.3 User Priority

The concept of priority can be used in various ways. Applications have an application priority level, which is used by networking services to help decide which application gets preferred access to the communication services, i.e., which application's WBSS to announce/join in case of a conflict. A different concept is the priority assigned to network data traffic. The lower layers use a separate MAC transmission priority to prioritize packets for transmission on the medium. IP packets are assigned the MAC priority associated with the traffic class of the generating application. The MAC priority for WSM packets is assigned by the generating application on a packet-by-packet basis.

The general architecture of prioritized access for data transmission on one channel is shown in Figure 15.8. Upon the arrival of a datagram to the channel router, it forwards the datagram to the appropriate channel and data queue. The appropriate priority queue is selected by mapping the user priority (UP as defined in transmitter profile) to an access category index (ACI). The channel selector schedules data for external contention by de-queuing priority queues based on their ACI. The channel selector also configures and confirms the media use of desired channel information.

It is imperative that end-to-end QoS requires more than MAC QoS. The reader is encouraged to augment the MAC QoS mechanism described here with intra-domain mechanisms described in Ref. [11] in addition to cross-domain mechanisms described in Ref. [10] to shape a comprehensive QoS solution.

15.5.2 Channel Coordination

WAVE devices are expected to be implemented as either single-channel WAVE devices that exchange information or listen to only one radio channel at a time, or as multi-channel WAVE devices that exchange information or listen to at least two channels at a time. In a WAVE environment, single-channel and multi-channel WAVE devices may also decide to remain on the CCH ignoring the SCH, but cannot ignore the CCH. Thus, the WAVE environment identifies a time

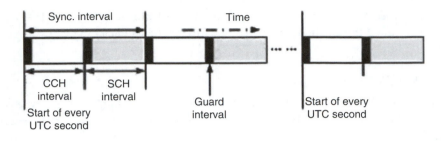

Figure 15.9   WAVE synchronization timeline.

period called the CCH interval (CCHI), and all WAVE devices that are members of the same WBSS are scheduled to listen and utilize the CCH during the common (CCHI).

Figure 15.9 shows both the CCHI and the SCH Interval (SCHI). WAVE devices may decide to ignore the SCH and remain on the CCH even during the SCHI. Therefore, it is imperative that WAVE devices within the same WBSS sustain accurate synchronization over time.

When a WAVE device joins a WBSS, it listens to the CCH until it receives the WSA, which contains both CCHI and SCHI. The values of CCHI and SCHI can be formed as follows:

$$\text{Sync. Interval} = \text{CCHI} + \text{SCHI} \tag{15.1}$$

At the beginning of each scheduled channel interval, a guard interval (GI), shown in Figure 15.9, is used to account for variations in timing inaccuracies among different devices. WAVE devices cannot transmit during the GI. To prevent devices from attempting to transmit simultaneously at the end of a GI, a medium busy is declared during the GI so that all transmission attempts are subject to a random back off at the start of each channel interval.

At the scheduled channel interval, the prioritized MAC activities on the previous channel are suspended, and the prioritized access activities on the current channel starts or resumes if they were suspended. While the CCF prevents packets from being transmitted on the incorrect RF channel.

A good implementation would consider avoiding transmission at scheduled GI. For instance, if the expected MAC Service Data Unit (MSDU) transmission time is more than the time left before the next GI, then the transmission can be avoided and the MSDU can be buffered for next channel interval. Optimizing CCF is complex and requires explicit research on this topic.

15.5.2.1   Channel Coordination Function

A WAVE device can only be a member of one WBSS at a time. The IEEE 802.11 active scanning is prohibited in 802.11p. Multi-channel WAVE devices monitor CCH at all times, but single-channel WAVE devices must implement CCF to monitor CCH at all common CCHI.

The WAVE standards assume the availability of external accurate common time reference. WAVE devices utilize the IEEE 802.11 timestamp field (TSF) illustrated in Figure 15.6 along with the timing information field of the WSIE as inputs in estimating UTC time. The IEEE 802.11 TSF has a length of 8 octets, and is part of the WAVE announcement action frame. The value of the IEEE 802.11 TSF is an integer modulus 2^{64}, incremented in units of microseconds [6].

15.5.2.2 Synchronization Parameters

The coordinated universal time (UTC) is a relative time measure compared to instant 2004-01-01T00:00:00.000000 as in. UTC is available in most GPS devices. A typical GPS-UTC precision, at the time of writing this chapter, is at least 1 PPS (pulse per second) UTC signal (with error 100 nanoseconds) [14]. Experimental trials confirmed that 1 PPS signals are sufficient to synchronize multiple WAVE device timing.

WAVE devices use the UTC time to form TSF. The TSF timer plus any offsets estimated, which resolves to an integer number and incremented in units of microseconds, are used as the WAVE device's internal estimate of UTC time. Single-channel WAVE devices use this UTC estimate to determine when the device exchanges data on SCH. Single-channel WAVE devices continuously monitor the CCH when not synchronized to UTC and when joining a new WBSS. It is important to highlight here that the use of UTC is entirely optional. Other sources of universal time reference can be used as long as the required accuracy is maintained [6].

15.5.2.3 Common Time Base Estimation

Single-channel WAVE devices can synchronize to UTC time by implementing an estimator of UTC time as described before and by implementing an estimator of the standard deviation of the error in the UTC time estimate. Standard deviation is defined as the square root of the second moment of the probability density function of the estimation error. Note that this includes the effect of any biases in the estimate of UTC time. For example, GPS can be used as an input to a simple estimator that changes the GPS time reference to UTC 2004-01-01T00:00.000000, calculates the necessary TSF timer offset value, and sets the standard deviation to that of the standard deviation of the time output of the GPS device under the given operating conditions [14]. A slightly more complex implementation could use information from the timing information field in received WSAs to update an internal estimator of UTC time along with an estimation error variance (standard deviation) as outlined hereafter.

The timing information field depicted in Figure 15.6 is detailed in Figure 15.10. This field holds information that can be used by recipients of WSIE to estimate UTC. The timing information field includes timing capabilities, a TSF timer offset, and TSF timer standard deviation subfields.

The timing capabilities subfield identifies the timing capabilities of the initiator WAVE device in terms of being single/multi-channel, having generated UTC through a GPS, and having continuous time source availability.

The TSF timer offset subfield contains the 2[2032?]s complement integer of the TSF timer offset in microseconds, which when added to the WAVE device's TSF timer value generates the WAVE device best estimate of UTC time at the instant when the first bit of the WSA is transmitted from the WAVE device's antenna connector.

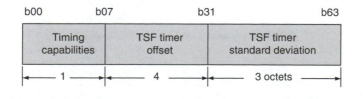

Figure 15.10 Timing information field.

The TSF timer standard deviation subfield is an unsigned integer of number of microseconds. This number represents the estimate of the standard deviation of the UTC error in the WAVE device's estimate of UTC time. When the TSF timer standard deviation subfield is set to its maximum value (224), it indicates that the value of the TSF timer offset subfield is meaningless.

The values of the TSF timer standard deviation and the TSF timer offset are generally a function of various environmental and implementation factors and will vary with time. At startup, the TSF timer offset is set to zero and the TSF timer standard deviation is set to its maximum value to indicate that the TSF timer offset is currently invalid. Once a lower variance source of UTC time becomes available, the new source and its error variance are adopted as initial conditions for the UTC estimator.

15.5.2.4   Synchronization Tolerance

Figure 15.9 illustrated the synchronization timeline in a WAVE environment. Each CCHI and SCHI starts by a GI. Figure 15.11 details the GI to be the sum of synchronization tolerance and the max channel switch time. The synchronization tolerance is defined to be double the 95 percent probability threshold value that determines whether a WAVE device is synchronized to UTC. The max channel switch time is the maximum time the WAVE device takes to change channels.

Because local device clocks at two different devices may drift in opposite directions, WAVE devices are defined to be synchronized to UTC if it complies with the following condition [2]:

$$3 \cdot \text{TSFTimerStdDev} < \text{Sync.Tol.}/2 \tag{15.2}$$

Single-channel WAVE devices that are not synchronized to UTC monitor the CCH continuously, do not offer services on SCH, and do not act as a service provider. If a WAVE device is a member of a WBSS and is processing a transaction on SCH at the time when synchronization is lost, the device simply discontinues the ongoing transaction and reverts to monitoring the CCH.

15.5.3   WAVE Communication Types

WAVE communication services provide data communications over two protocol stacks: IPv6 and WSMP. WSMP is unique to DSRC standards and is designed for use by specialized applications like public safety applications. Applications using WSMP may initiate a WBSS to configure the

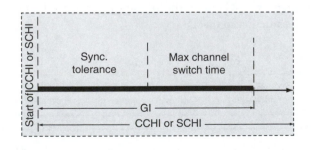

Figure 15.11   Detailed GI.

SCH for their use. But availability of SCH is optional as WSMP can be exchanged on the CCH even in the absence of WBSS.

15.5.3.1   WBSS versus Non-WBSS Operations

Although the use of WBSS is expected to be dominant in DSRC networks, it is not necessarily so. WAVE devices can communicate WSMP messages over WAVE networks without WBSS. A scenario involving WSMP use without WBSS goes like the following.

1. A source WSMP application registers with the WME, and then, composes WSMP data for transmission, and addresses WSMP data to a broadcast MAC address. The MAC reads the required channel information (power level, data rate) from the packet header and set the CCH for transmission using the channel parameters as requested (power level, data rate). Transmission takes place based on internal and media contention.
2. A receiving device accepts the packet and passes it up the communication stack. WSMP stack delivers data to the locally registered application, based on the provider service identifier (PSID) and on the provider service context (PSC). At this point, the receiving application knows the availability and address of the transmitting application, and can continue the exchange on the CCH if desired, using either unicast or broadcast MAC addresses, as appropriate.

When operating with a WBSS, a WAVE device initiates the WBSS at the request of any application running on (through) the same device. The WAVE device initiates the WBSS on the CCH of the provider [2].

15.5.3.2   WAVE Communication Service Types

WAVE supports two types of communication services: persistent and nonpersistent services. The distinction is that a persistent WBSS is announced at each CCH interval (Figure 15.12), and a nonpersistent WBSS is announced only on initiation. A usage of the persistent WBSS would be to offer an ongoing service to any device that arrive into range. A usage of the nonpersistent WBSS would, typically, be to support an on-demand service.

A persistent WBSS is announced periodically, and could be used to support an ongoing service for indefinite duration, such as general Internet access. Persistent WBSS communication service

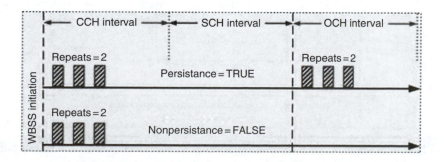

Figure 15.12   WAVE persistent communication service.

resembles normal vehicle operations on the road. A nonpersistent WBSS is announced only on WBSS initiation, and might be used to support a WBSS with limited duration. A practical example of nonpersistent service is garage service where a consumer vehicle can join a private WBSS to download files or upgrade software. Typical examples show stationary units forming a WBSS for a long-term relation [2].

In WAVE communication services, WAVE devices may take the role of either provider or user on a given WBSS. This is determined by the role chosen by the application operating on the device. The provider device generates the announcements that inform other devices of the existence of the WBSS, and the presence of the associated application services.

The user role is assumed by any devices that join the WBSS based on receipt of the announcement. A device may change its role when it participates on a different WBSS. The terms provider and user do not imply any particular behavior of the applications once the WBSS is initiated or joined. A device can be a provider for one service and a user of another.

15.5.3.3 WBSS Initiation and Operations

A WBSS is initiated by WME when a provider application redefines some of the WBSS parameters like the persistence status. The announcing WAVE device forms WSA and WSIE, which includes the PST as described in Figures 15.6, 15.7, and 15.13.

As soon as a user application starts on a WAVE device, it must register all services it may need with the local WME (Figure 15.3). Upon receipt of a WSIE, the receiving WME checks whether the provider application, defined by PSID and PSC in the announcement, is of interest to any locally registered user applications. User applications have the option to be informed of

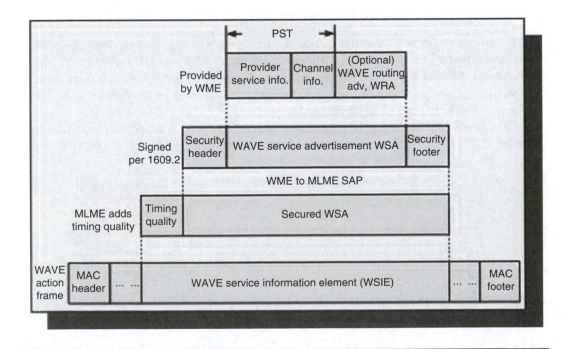

Figure 15.13 Building WAVE announcement.

announcement containing both PSID and PSC match, or containing only PSID match. When a match is found, the WME takes one of two actions, depending on the user application registration parameter. In the simple case, the WME generates the necessary MAC primitives to cause the local device to join the announced WBSS, by tuning the device to the correct SCH at the correct time, and by setting any other lower layer configuration parameters appropriately to support the communications. Alternately, the user application may choose to reconfirm before joining of the announced WBSS. This gives the user application an additional control level allowing any application to decline the participation in a specific service if it has recently accomplished milestone objectives.

Upon the decision to join the announcing WBSS and as soon as the communication parameters are set successfully, the WME sends a notification to the local user application. Subsequently, the user application is free to generate WSMP or IPv6 data packets for transmission on the SCH. Received packets are delivered up the WSMP or IPv6 stack. The WBSS stays in place at the local device until it is terminated.

15.5.3.4   WBSS Termination

Once initiated, a WBSS stays in place at each participating device until locally terminated. WAVE devices may independently decide to leave a WBSS. There is no protocol exchange over-the-air interface to confirm the end of a WBSS. The WME of a WAVE device may decide to issue a request to its MAC layer to leave a WBSS and to inform all the affected applications through a notification (i.e., process WBSS termination) in any of the following situations:

1. All applications indicate the completion of their activities, via a request changing their status.
2. Participation on a conflicting WBSS (e.g., on another channel) is required to support a higher priority application.
3. Lower layer indicates the SCH has been idle for a predefined amount of time, implying an irrecoverable loss of the WBSS.
4. Security credentials associated with the WBSS announcement expire at the user or are determined to be invalid when checked.

15.5.3.5   Changing WBSS Services

In a nonpersistence WBSS, provider application that are registered after issuing the initial WBSS cannot join the on going WBSS services.

On the other hand, persistent WBSS offers different sets of applications over time by altering the announced PST, which is part of WSA (see Figure 15.7), during different CCH interval. To support the dynamic PST feature, the announcement's destination MAC address is constrained to be the broadcast address. Applications come and go from the provider's WBSS as triggered by the application request primitive. The provider WME may also end a persistent WBSS as described in Section 15.5.3.4.

User applications start and stop participation on a persistent WBSS during its existence. The WME maintains a table of active/inactive status of each participating application. The WME of the user WAVE device joins and ends local participation on the WBSS as required based on the user application [2032?] status [2].

15.6   WAVE Network Layer Services

Data packet delivery is not the major focus of the WAVE standards. However, the stringent operational environment of the wireless media and the short time available for devices to communicate impose some limitations on the exchange of control traffic like ICMP. Therefore, the DSRC standards provide an extended view on how services of the network layer can be conducted [2].

Figure 15.14 illustrates a RSU-OBU communication through a WAVE air-link. The long dotted border in Figure 15.14 (box 1) shows the part of the network layer that is defined in the WAVE standards. In the same figure, boxes 2 and 3 highlighted in smaller dots in the top left and right show the stacks that are vendor dependent. Vendor-dependent stacks may vary in implementation but must be able to support the use of multiple MAC header addresses as specified in this section. Box 4 at the bottom of Figure 15.14 shows the external systems that can be connected to any DSRC network in large. For instance, the connection between the in-vehicle data distribution system (the cloud at the right) and any wireless access network can take place through cellular, satellite, or WiMAX technologies. External systems remain out of the scope of this chapter.

15.6.1   Use of Portal Function

The IEEE 802.11 describes the standards for portal function between the wireless network and distribution systems like the right and left clouds in Figure 15.14. The following portal function

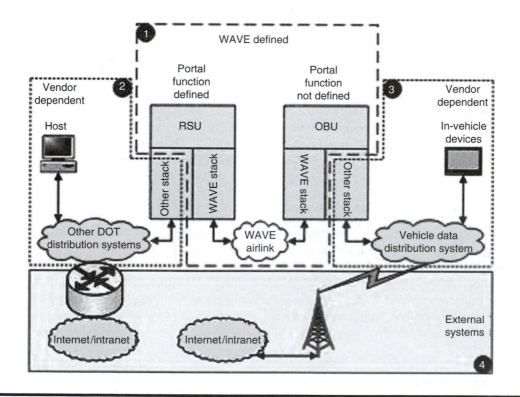

Figure 15.14   RSU–OBU portal function.

is required on the RSU; however, implementers must maintain interoperability with the described portal function if an alternative approach is followed.

The portal function operates at the data link layer of the protocol stack, and thus employs MAC addressing for packet delivery. The following definitions are necessary to describe the portal function mechanism:

1. Source address (SA): Originator of the layer 2 frame.
2. Destination address (DA): End destination of the layer 2 frame. The DA may exist on the air interface, the DA can exist in a distribution system, or the DA may be outside the distribution system. In the last case, the MAC address of the default router is used as a DA instead of the MAC address of the actual end destination.
3. Source distribution system address (FromDS): Indicates frame sent from a distribution system.
4. Target distribution system address (ToDS): Indicates frame sent to a distribution system.

It is important to realize that the use of a distribution system is not required. For instance, the RSU may implement the portal function and can even be the host at the same time. In this discussion, the BSSID equals the MAC address of the RSU for simplicity.

The use of a portal function supported by the utilization of the four MAC addresses in the MAC header provide a simple solution to the complex problem of routing packets in a dynamic environment with significant unit mobility (Table 15.1). The utilization of the four MAC addresses provides efficient use of the scarce wireless media compared with the classical IP layer signaling. This solution, however, is suboptimal. It is possible to elaborate scenarios where it fails. Nevertheless, it serves the purpose of the current DSRC standards, and elaborate solutions to quick mobility and handover can be adopted in future [2].

15.6.2 IPv6 Neighbor Cache

The common IPv6 neighbor discovery, following a multicast-response mechanism as in RFC 2461, is highly discouraged in the WAVE environment due to the amount of control traffic it generates. It is desirable in WAVE implementations to keep traffic on the CCH to minimum.

Table 15.1 Portal Function Use of MAC Addresses

Field	OBU–OBU, RSU–OBU, OBU–RSU	OBU to RSU to Host via DS	Host via to OBU DS to RSU OBU
ToDS	0	1	0
From DS	0	0	1
Addr1 (reciever) MAC	DA	BSSID (RSU)	DA (OBU)
Addr2 (transmit) MAC	SA	SA (OBU)	BSSID (RSU)
Addr3	BSSID	DA (host or router)	SA (host or router)
Addr4	Not used	Not used	Not used

WAVE devices listening to the wireless media learn the MAC and IPv6 address pairs of active devices participating in the same WBSS. Each WAVE device builds its neighbor cache based on the information obtained by merely listening to the media. Because provider WAVE devices must send at least WSA, even in nonpersistent WBSS, user devices are guaranteed to learn the MAC and IPv6 address pairs for all devices providing services. If the provider service is routed, the user host adds an entry to its neighbor cache for the default gateway router address. The DNS server address is announced as part of the WSA (see Figure 15.7), following service provider registration of the IP service announcement. In the same way, the IPv6 prefix and the default gateway address are announced as part of the WSA (Figure 15.7).

15.7   WAVE Services Management

As a general rule, the WAVE environment prohibits unregistered applications from gaining access to WAVE services. This rule has limited exception that is explained later.

A provider device is a WAVE device that has legitimate applications registered to provide WAVE services and is also the initiator of a WBSS. The provider WAVE device is a sender of WSM. Similarly, a user device is a WAVE device that has legitimate applications registered to use WAVE services and is also the joiner of a WBSS. The user WAVE device is a receiver of WSM.

A PST (see Figures 15.7 and 15.13) is a collection of data describing the applications that are registered with, and available through, a WAVE device with related supporting channel information. A PSID is the code number that identifies a service provided by an application. Applications register with the WAVE device as a provider or user using the correct PSID. Each WAVE device builds its PST and maintains it over time.

A provider WAVE device announces its PST as part of its WSA. When a user WAVE device locates the WBSS, it extracts the provider PST from the remote WSA; then, the user WAVE device compares the set of applications available in the provider PST with its own user PST that is available through its WME and contains local user applications demanding service. The user WME generates a new table of applications with matching interests. This new table represents the set of applications that are allowed to gain access to WAVE services.

The mechanism described here, although inaccurate, illustrates how WAVE devices grant application access to WAVE media and simplifies reading. The following details identify the operations in more precise terms.

15.7.1   Application Registration and Removal

All applications register, as a provider or user, with its WAVE device WME before gaining services. Applications registered on the same WAVE device must have a unique PSID. This rule is not limited to only WAVE-applications; it applies even to typical IP-based data access. The only exception is that unregistered WSMP provider application may send WSMP packets. WSMP user application must be registered to receive any WSMP.

The WME maintains a record of all registered applications. Before accepting a new registration, the WME confirms that the new application is unique and has the necessary security credentials to gain the required access to its WAVE media. Applications failing to satisfy the two conditions are denied registration and receive a registration denial message. Otherwise, applications are registered and receive a confirmation message.

In addition to the application registration table, the WME maintains an application status table. The WME uses the application status table to maintain a record of active and inactive applications besides other operational information like notification IPv6 address, port, and priority. Any application can remove its registration by sending a message to the WME. In such a case, the WME removes the application registration entry from the application registration table and from the application status table. If the application status table contains no active applications, the WME ends the current WBSS [2].

15.7.2 WBSS Management

Generally, all WBSS are established on the CCH. However, on the basis of an application request, a WAVE device may establish the WBSS on a SCH, and announce its presence on the CCH for other devices to join that WBSS. All WBSS are initiated based on service priority and the availability of radio resources. A WBSS can be newly established or it can be a modification to the existing persistent WBSS.

A WAVE device can join (or initiate) one WBSS at a time. If a specific user device is not a member of any WBSS, it joins a new WBSS upon determining a match between a locally registered user application and a received WSA that has a registered provider application with the same PSID. The WME takes the decision of joining a specific WBSS and this decision is illustrated in Figure 15.15. However, if the user device is currently a member of a specific WBSS, it joins a new WBSS only if the new WBSS provides a service match of comparatively higher priority. A user device may decide to leave a WBSS if the currently active applications complete, timeout, or lose security credentials.

The provider device transition to a new WBSS and the termination of old WBSS are clearly illustrated in Figure 15.15. It is important here to highlight that the WME checks the availability of the requested radio resources, inspects the WSA security credentials, and validates other

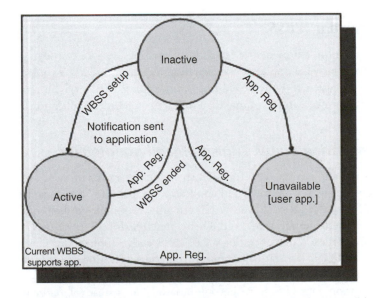

Figure 15.15 WBSS status diagram.

MLME requests before it disrupts the current WBSS services and before seeking transition to a newer WBSS. If the WME detects a failure in the transition to the new WBSS, it stays on the old WBSS and sends an error message to the relevant application indicating the proper reason code [2].

15.7.3 WBSS Join Policies

A WAVE device joins a new WBSS based on a set of policies that are stored and executed by its WME. Those policies are invoked after the WME compares the set of services received from a remote device through the WSA with its own set of desired services. The policies can be illustrated as follows:

1. If no match is found, the WME takes no action; else, the service priority of each matched application is compared with the service priority of each application in the application status table that has an active application status.
2. If there is an active application with higher service priority than any matched application, the WME takes no action; else, if a match is found, and there is no active application or no higher priority active application.
3. If any of the matched applications was registered with the no Confirm-Before-Join flag set, the WME joins the WBSS. In the case of a pre-existing lower priority WBSS, the WME ends participation on that WBSS first. Else, all of the matched applications were registered with the Confirm-Before-Join flag set, the WME joins the WBSS immediately.

As soon as the device joins the new WBSS, the WME changes the status of relevant applications to [2032?] active [2032?]. Multiple applications matching with the same channel number share the same WBSS.

15.7.4 Persistent WBSS

In a persistent operation, a provider device adds, or modifies, service to the ongoing WBSS when the WME receives a provider application registration request. Similarly, a provider application terminates or sends a request to be removed from the ongoing WBSS of the provider device if the device is in persistent mode and has more than one active provider application.

15.7.5 Application Status Transition and Maintenance

The WME uses the application status table to maintain the status of each registered application based on the supporting WBSS status and application availability. The default status of all applications is [2032?] inactive [2032?]. The WME attempts to match the PST of any received WSA with its locally registered applications, and when a match is found, the WME may initiate a join WBSS as specified earlier. While operating on a WBSS, active applications may generate traffic for transmission on the SCH. The state diagram showing the application status transition is illustrated in Figure 15.16.

A user application may choose to become unavailable instead of removing its registration. An unavailable status indicates a user application that is dormant for sometime. One use of the unavailable status is the case when an application waits for an event to take place on a different

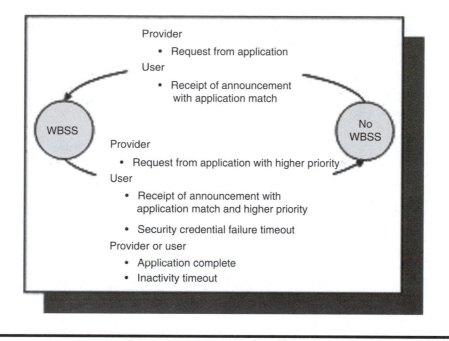

Figure 15.16 Application status diagram.

application before it is retriggered. The WME neither joins a new WBSS nor maintains a current WBSS on behalf of unavailable applications. Simply the WME treats unavailable applications like nonregistered applications.

15.7.6 Channel Activity and Usage Monitor

The MLME monitors channel inactivity and triggers an internal indicator to the WME when the active SCH is idle longer than a prespecified time. The WME assumes that a user device should cease operation on the current WBSS and proceed with WBSS termination.

The MLME also keeps track of all SCH in use by other devices within a listening range. Consequently, the WME can choose to operate on a SCH that is likely to have the least congestion whenever the WME is required to initiate communication on the arbitrary SCH. The WME keeps track of the most recent time at which a WSA was received on each SCH (from the set 174, 176, 180, and 182). When the application requests the best available channel, the WME uses the SCH with the oldest received WSA [2].

15.8 Security Considerations

As indicated before, the IEEE 1609.2 is a part of DSRC standards focused on securing WAVE communications. Because of its complexity and volume, we decided to avoid a lengthy discussion on security mechanisms, preferring to cover this topic in a separate paper. However, it is important to highlight here that both WSA and WSM messages are covered by the 1609.2 message security.

It is also important to indicate that the 1609.2 was developed in parallel with the evolution of other 1609.x and 802.11p standards. Therefore, the designers did not get sufficient time to cover some of the major WAVE environment concerns. For instance, there has been growing concern about violating user anonymity and privacy. Any man-in-the-middle type of attack can listen and maintain a list of the communicating MAC addresses. Knowing the MAC address of a specific user, it is easy to trace the user mobility. One way to counter this attack is by randomizing the user MAC address periodically. However, a solution like this complicates the WAVE standard and it was decided to release the standard for a trial use until these issues are examined.

15.9 Conclusion

This chapter covers the current status of the DSRC standards suite and illustrates the integration of DSRC multiple standards. Most of the DSRC standard suite is expected to be released by mid-2008. The current IEEE 1609.x has been charted for [2032?] trial use [2032?] and the involvement of the research community is highly desired for future DSRC development.

This chapter focused on covering the different mechanisms of the DSRC and how various standards cooperate together to provide a resilient platform for road safety and other DSRC applications. The chapter illustrates detailed WAVE-MAC mechanisms and describes how different WAVE components address mobility issues. Fertile research areas are defined in our series of articles, for example, Refs. [1,9].

Most skeptical reviews on the future of DSRC focus on the DSRC feasibility and viability. Many North American cities have deployed WiFi downtown coverage and used it to provide users with free access to the Internet. In most cases, cities and municipalities paid for the system and the services due to their low cost. Through these pioneering experiences, we learned that the cost of deployment and mounting of devices accounts for more than 80 percent of the overall cost. Typically, WiFi devices account for about 7 percent. The cost of WiFi-based infrastructure is expected to go lower in the coming decades. Furthermore, SCH can be utilized for commercial use, which attracts businesses to pay for part of the infrastructure cost in return for utilizing local SCH for commercial purposes. In conclusion, we expect the DSRC networks to dominate the wireless evolution within the next decade. Much research is needed to improve DSRC security, as well as communication and service layers.

This chapter is fundamental in combining the various technologies used to provide short-range vehicular communication and defines the use of each technology in addition to defining interoperability issues. It presents a strong base for researchers investigating solutions in this domain by identifying common terms and interoperability mechanisms.

REFERENCES

1. Y. Morgan and H. Roshdy, Novel issues in DSRC vehicular communication radios, Submitted to the *IEEE-VT Magazine*, April 2007.
2. IEEE P1609.3 SWG et al., Wireless access in vehicular environments (WAVE) networking services, IEEE P1609.3 D23, February 2007.
3. Mallet W. et al., Surface transportation congestion: Policy and issues, Report for US Congress, Congressional Research Services, 2007.
4. Industry Canada, Proposed spectrum utilization policy, technical and licensing requirements to introduce dedicated short-range communications-based intelligent transportation systems applications in the band 5850-5925 MHz, Report on DSRC, 2007.
5. IEEE 802.11p SWG et al., Draft amendment to standard for information technology-telecommunications and information exchange between systems-local and metropolitan networks-specific

requirements-Part 11: Wireless LAN medium access control (MAC) and physical layer (PHY) specifications: Ammendment: Wireless access in vehicular environments, IEEE 802.11p D1.2, September 2006.

6. IEEE P1609.4 SWG et al., IEEE P1609.4 trial-use standard for wireless access in vehicular environments (WAVE)-multi-channel operation, IEEE P1609.4 D09, August 2006.

7. IEEE P1609.1 SWG et al., IEEE P1609.1 trial-use standard for wireless access in vehicular environments (WAVE) resource manager, IEEE P1609.1 D17, July 2006.

8. IEEE P1609.2 SWG et al., IEEE P1609.2 trial-use standard for wireless access in vehicular environments-security services for applications and management messages, IEEE P1609.2 D07, April 2006.

9. R. Bera, J. Bera, S. Sil, S. Dogra, N. Sinha, and D. Mondal, Dedicated short-range communications (DSRC) for intelligent transport system, in the 2006 IFIP International Conference on Wireless and Optical Communications Networks, April 2006.

10. Y. Morgan and T. Kunz, PYLON-lite: QoS model for gateways to ad-hoc network, Journal for the special issue on Recent Advances in Wireless Networks and Systems, *Computers and Electrical Engineering*, 32(1–3), 68–87, October 2005.

11. Y. Morgan and T. Kunz, Enhancing SWAN QoS model by adopting destination based regulation (ESWAN), in the Proceedings of the 2nd WiOpt-04 Conference for Modeling and Optimization in Mobile Ad-hoc and Wireless Networks, pp. 112–121, Cambridge, UK, March 2004.

12. ASTM E2213-03, Standard specification for telecommunications and information exchange between roadside and vehicle systems-5 GHz band dedicated short range communications (DSRC) medium access control (MAC) and physical layer (PHY) specifications, ASTM-E17.51 for DSRC, September 2003.

13. C. Cseh, Architecture of the dedicated short-range communications (DSRC) protocol, in the Proceedings of the IEEE 48th Vehicular Technology Conference VTC-98, vol. 3, pp. 2095–2099, May 1998.

14. International Radio Consultative Committee (CCIR), Report 517, Standard Frequency and Time-Signal Emissions: Detailed Instructions by SWG-7 for the Implementation of Recommendation 460 Concerning the Improved Coordinated Universal Time (UTC) System, January 1972, XIIth Plenary Assembly CCIR, New Delhi, India 1970, Geneva, International Telecommunication Union, 1970, Vol. III, 258a–258d; reprinted in Time and Frequency: Theory and Fundamentals, U.S. Monograph 140, Washington, D.C., U.S. Govt. Printing Office, 1974, 32–35. Reprinted by Geneva, International Telecommunication Union, Radio Comm. Bureau 1998.

15. J. Kaltwasser and M. Pietschmann, Transactions via short range communication links-analytical evaluation of communication parameters, in the Proceedings of the IEEE 44th Vehicular Technology Conference VTC-94, vol. 2, pp. 1351–1354, June 1994.

16. J. Kossack and B. McQueen, European progress towards standardization in ATT communications, in the IEEE Proceedings of Vehicle Navigation and Information Systems Conference, pp. 307–311, Ottawa, Canada, October 1993.

17. Vehicular Infrastructure Integration VII url: http://www.its.dot.gov/vii/.

Chapter 16

Supporting Heterogeneous Services in Ultra-Wideband-Based WPAN

Kuang-Hao Liu, Lin Cai, and Xuemin (Sherman) Shen

CONTENTS

Ultra-wideband (UWB) is a promising technology for short-range bandwidth-demanding wireless communications. Several portable devices equipped with UWB transceivers can autonomously form a so-called wireless personal area network (WPAN) to conveniently exchange bulky data or high-quality multimedia streams. Because the wireless medium is shared by multiple users with different quality-of-service (QoS) requirements, fulfilling the diverse needs of user applications and achieving efficient utilization of the scarce wireless resource are important and challenging issues. This chapter is devoted to effective resource allocation for UWB-based WPAN. We start by identifying the deficiency of the existing scheduling mechanisms when they are applied to UWB networks, mainly because they do not take UWB specific features into account. By coupling the physical characteristics of UWB with the resource allocation process, we propose practical scheduling algorithms for different design objectives, namely throughput maximization for homogeneous traffic and QoS provisioning for heterogeneous traffic, respectively. Simulations are used to demonstrate the effectiveness and efficiency of the proposed algorithms.

16.1   Introduction

Recent advances in the semiconductor industry have boosted the integration of various radio technologies in a single device to provide seamless wireless broadband access (WBA) [1]. For portable devices with limited power supply, UWB is a promising technology for ubiquitous connectivity in home/personal space. Devices equipped with UWB transceivers can provide several hundred megabits per second transmission rate in very low-power consumption. As a result of high-speed wireless connectivity, a wide set of multimedia applications such as high-definition television, music sharing, wireless classroom, etc., can be carried out over UWB devices with high display quality and sustainable battery life.

To allow fast network formulation among users with QoS support, the IEEE 802.15.3 working group has defined the Medium Access Control (MAC) mechanisms for WPAN, by which a number of devices can exchange high-volume data with each other free from pre-established infrastructure [2]. Among a set of devices in the vicinity, one of them is elected as the piconet coordinator (PNC), responsible for channel access control and radio resource allocation. Unlike other centralized solutions such as the base station in cellular networks or the access point in wireless local area networks (WLANs), the PNC is not restricted to be a fixed facility, facilitating dynamic piconet configuration. In addition, users in WPAN can communicate in a peer-to-peer fashion that greatly relieves the load of PNC, which needs to relay the user traffic in the legacy Bluetooth-based WPAN.

The IEEE 802.15.3 was originally designed for narrowband wireless systems without leveraging any salience of UWB technology. For instance, as an extreme form of spread-spectrum systems, several UWB links within the communication range can concurrently transmit without destroying each other. This is significantly different from narrowband systems where transmissions within one-hop communication range disrupt ongoing links therein. By strategically selecting concurrent transmission links, the network throughput can be improved to serve more high bandwidth-demanding users, an important factor in determining the success of future UWB networks in dense environments, such as trains/buses, classrooms, libraries, and so on.

Lacking of explicit considerations of the physical properties of UWB, existing solutions remain considerable margin to be further improved [3]. This chapter attempts to link the physical characteristics of UWB communications with the Media Access Control (MAC) Protocol design. We explore the physical properties of UWB and demonstrate that the network throughput can be significantly increased by properly scheduling concurrent UWB links. Because the optimal scheduling for concurrent transmissions is shown to be NP-hard, efficient scheduling algorithms with different design objectives are proposed. In some circumstances, users use the same type of service with large bandwidth demand, for example, file sharing. It is desirable to increase the network throughput and accommodate more users in the vicinity. The first design objective is thus to maximize the aggregate throughput, given a user constellation. Furthermore, users may launch different applications with heterogeneous bandwidth demands. For such a scenario, the resource allocation strategy aiming at throughput maximization may benefit certain users while starve others at the same time. How to provide QoS support for heterogeneous traffic types is deemed another important design objective. Extensive simulations have been conducted to evaluate the performance improvement and implementation complexity of the proposed algorithms.

The remainder of this chapter is organized as follows. Section 16.2 introduces the basic principle of UWB communications, and Section 16.3 overviews the existing IEEE 802.15.3 standard for WPAN. Section 16.4 presents the related work on scheduling problems for UWB networks. Section 16.5 shows how the special property of UWB can be used to enhance the MAC design. The result serves as a guideline for Section 16.6, where practical scheduling algorithms aiming throughput maximization are proposed. Section 16.7 addresses the issue of QoS support for heterogeneous traffic types. Section 16.8 evaluates the performance and computation complexity of the proposed algorithms. Concluding remarks and future research directions are presented in Section 16.9. Table 16.1 summarizes the important symbols for handy reference.

16.2 UWB PHY

To leverage the advantage of UWB in networking design, we first review the fundamental properties and limitations of UWB. UWB technology encompasses a broad range of signal forms and design approaches. The original UWB systems, the so-called impulse radio (IR), reveal different traits from the traditional narrowband communication systems. In IR-UWB, the transmitter sends data by a series of extremely short (0.1–1.5 ns), low duty-cycle pulses using a carrier-free modulation. The corresponding receiver then decodes the information by listening to an a priori known pulse sequence sent by the transmitter. IR technology has been widely used in radar and military applications due to its high spatial resolution, immunity to passive interference, low probability of detection, etc. [1]. On the other hand, the low duty-cycle property of IR-UWB limits the achievable data rate up to several tens of megabits per second. Modern UWB systems employ different techniques such as direct-sequence spread-spectrum (DS-SS) [4] or orthogonal frequency division multiplexing (OFDM) [5] to achieve high data rate and spread the signal over the wide spectrum. To prevent interfere with existing radios, stringent emission rules have been imposed on UWB transmissions that restrict the communication range of UWB signals. Therefore, UWB technology is most suitable for short-range wireless communications with high bandwidth demand.

The inherent features of UWB communications offer new design options, for example, the wide bandwidth of UWB signal immediately translates to accurate ranging capability. As a form

Table 16.1 Summary of Important Symbols

Symbol	Definition
$R_i^{\mathbb{T}}$ $(R_i^{\mathbb{C}})$	achivable data rate of flow i using TDMA (concurrent transmissions)
α	cross-correlation among concurrent transmissions
$p_t(i)$ $(p_r(i))$	transmit (received) power of flow i
η	one-sided PSD of background noise
I_0	one-sided PSD of interference
γ	path-loss exponent
D	exclusive region size
G (G')	network conflict graph (with half-duplex constraint)
$U(r_i)$	utility of flow i with average data rate r_i
$\tilde{U}(\cdot)$	noisy version of $U(\cdot)$
$\overline{U}(\cdot)$	sample mean of $U(\cdot)$
$|V|$	number of active flows
κ^*	optimal scheduling for one slot
$U_i(s)$	instant utility of flow i in slot s
c_i	revenue/importance parameter of flow i
$\rho_i^{(s)}$	utility control parameter of flow i in slot s

of spread-spectrum, concurrent UWB transmissions can coexist without destroying the reception of each other. This is fundamentally different from narrowband systems, where the concurrent transmissions are inhibited to protect the ongoing links. Allowing concurrent UWB transmissions implies that additional time-multiplexing gain can be expected to increase the network throughput. However, as the bandwidth is large but not infinite, there still exists the near–far problem (i.e., strong signal swamping weaker signal at the receiver) if concurrent transmissions are not properly managed. Section 16.5 addresses this issue.

16.3   MAC Protocols in IEEE 802.15.3

This section briefly reviews the current WPAN standard—IEEE 802.15.3. This standard defines a superframe structure as shown in Figure 16.1. Each superframe starts with a beacon period (BP) for network synchronization and control message broadcast, followed by a contention access period (CAP) and a channel time allocation period (CTAP). In the CAP, devices send their access requests to the PNC using the Carrier Sensing Multiple Access/Collision Avoidance (CSMA/CA) mechanism. According to the CSMA/CA mechanism, the transmitter is required to first sense that the medium is idle for a random period. Only if the medium is idle after that period shall the device

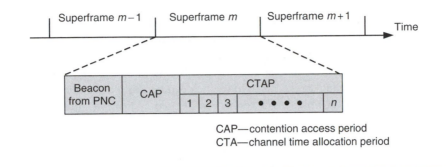

Figure 16.1 **Superframe structure defined in IEEE 802.15.3 MAC Protocol.**

start its transmission. However, the extremely low-power spectral density (PSD) of UWB communications may challenge the efficiency of carrier-sensing functionality* for detecting channel activities.

In CTAP, time is slotted and the PNC allocates time slots for both isochronous streams and asynchronous data traffic simply using the time division multiple access (TDMA) discipline. However, TDMA is inefficient for UWB networks in terms of channel utilization. Taking a DS-UWB network with a 480 Mbps data rate, for example, the time to transmit a packet of 1024 bytes is only 17.06 μs. If the acquisition time for each packet is 15 μs (the default preamble length in Ref. [4]), neglecting other timing components and overheads, the channel efficiency, i.e., the fraction of time used for actual data transmission, is reduced to only 53.2 percent. Similarly, for MB-UWB, the channel efficiency is only 40.75 percent, considering the overhead of the physical and MAC layers specified in Ref. [5]. Because the time-exclusive mechanism of TDMA leads to low channel utilization, it becomes challenging to provide services to multiple users in a dense environment using IEEE 802.15.3 MAC.

16.4 Related Work

Considerable research work on UWB networks has appeared in the literature. This section reviews previous work closely related to our study. In Ref. [6], the cross-layer design issue for UWB ad hoc networks was investigated. Aiming at throughput maximization, the cross-layer optimization problem is explored based on simple network topologies. The paper suggested that by implementing an exclusive region around each receiver, the optimal flow rate can be attained, and the radius of exclusive region is relevant to the achievable flow rate. However, how to determine a proper exclusive region size is untouched. A joint scheduling and power allocation problem for UWB networks was formulated in Ref. [7]. Through mathematical modeling, they showed that for UWB networks, the throughput improvement resulted from power allocation is very limited, although the implementation of power allocation is beneficial to manage interference and reduce power consumption. Thus, sophisticated power allocation may not be necessary if the design goal is to maximize the network capacity. Different from the previous work where certain power control/allocation mechanisms have been deployed to reduce power consumption and interference, this chapter focuses on increasing the network throughput and maintaining the processing

---

* In 802.11a WLAN, the transmit PSD is 2.5 mW/MHz ≈ 3.98 dBm/MHz, whereas it is −41.3 dBm/MHz for UWB according to the Federal Communication Commission's regulations.

load of PNC as low as possible. Thus, we seek the solution when transmission power is regulated by a simple rule: either zero or the maximum allowable level according to the FCC emission mask for UWB communications. Although our strategy does not directly reduce power consumption, the proposed algorithm can significantly ease the load of PNC and improve network throughput. In addition, we derive the optimal exclusive region which can be easily determined according to the underlying propagation environment. The results can be used as a simple guideline for fast UWB network planning.

The ultra-wide bandwidth of UWB inherently allows multiple concurrent transmissions. This physical property has been extensively studied at the signal level [8,9]. It has been shown that the average symbol error rate in DS-UWB is a function of UWB monocycle and channel behavior [10]. With a properly selected format of monocycle, the nature of time-hopping (TH) IR, one of the variations of UWB systems, allows several users to access to network at the same time without the need of user synchronization. This property facilitates ad hoc networking where distributed channel access can be implemented at each node according to the locally monitored multi-access interference level in the vicinity [10,11]. Under this principle, techniques such as chip discrimination [12] and adaptive coding [13] have been proposed to further improve the UWB system capacity. However, the aggregated throughput may be even worse than one-by-one transmission in a dense environment. Thus, an effective scheduling algorithm is required to control the potential interference, which is the main focus of this chapter.

Furthermore, supporting heterogeneous traffic is deemed an important requirement for UWB networks. Previous research has proved that an optimal multi-service scheduling problem is NP-hard [14]. The most existing work on multi-service scheduling lies in fixed bandwidth allocation [15,16], which is not the case in wireless networks. In this chapter, we propose a meta-heuristic algorithm to solve the optimal scheduling problem for heterogeneous traffic.

16.5 MAC Enhancement by Concurrent Scheduling

To exploit the inherent capability of UWB to allow concurrent transmissions, this section derives the sufficient conditions to ensure that scheduling concurrent transmissions results in a higher aggregate throughput than when using TDMA. The result serves as a guideline for developing resource allocation strategies for UWB networks.

16.5.1 System Model

The network of interest consists of multiple users where each user communicates with others via peer-to-peer links. The PNC in WPAN is responsible for accepting the resource request upon which the scheduling decision is made and distributing to the network members. Unlike other centralized systems, the PNC in such a peer-to-peer network does not route the traffic for peers and is not capable of learning the instantaneous channel gain of each link. Alternatively, the average channel quality can be predicted using the link distance information provided by the accurate ranging capability of UWB. An additive white Gaussian noise (AWGN) channel is assumed for each link with available bandwidth W, one-sided noise PSD of η, and that of interference I_0, respectively. Shannon's upper bound gives the achievable throughput R by $W \log_2(1 + SINR)$ bits per second (bps), where $SINR = P_r/(\eta + I_0)W$ with p_r being the received power. In the wideband regime, i.e., $W \to \infty$ [17],

$$R \approx \frac{p_r}{\eta + I_0} \log_2 e \qquad (16.1)$$

Aiming at low computational overhead, sophisticated power control is not considered. All transmissions use the same transmit power p_t, which is the maximum value allowed by the UWB spectrum mask. A path-loss propagation model is assumed to estimate the average received power. For link i, the received power is given by $p_r(i) = \alpha p_t d_i^{-\gamma}$, where d_i is the sender–receiver distance of the ith link, α characterized the cross-correlations among concurrent transmissions, and γ is the path-loss exponent. Both parameters α and γ are assumed constant.

16.5.2 Exclusive Region

The exclusive region is defined as the region around each receiver in which no concurrent transmissions are allowed. A large guard zone helps limit the aggregate interference by inhibiting the nearby dominant interferers, at the cost of reduced spatial reuse. For extreme cases, i.e., the guard zone is infinitely large or small, the scheduling discipline reduces to TDMA (no concurrent transmission) or all-at-once (all links transmit concurrently), respectively. Thus, we consider TDMA as the benchmark, and the following proposition derives the sufficient condition that concurrent UWB transmissions are preferable to TDMA.

PROPOSITION 16.1
Without loss of generality, consider the scheduling problem of scheduling N links with the scheduling cycle of N time slots. The link throughput, resulted from concurrent transmissions with the exclusive region of radius D, is larger than that from TDMA when

$$I_{j,i} \leq \eta \quad j \neq i \tag{16.2}$$

where $I_{j,i}$ represents the interference from link j's source to link i's destination.

PROOF Denote $R_i^{\mathbb{T}}$ and $R_i^{\mathbb{C}}$ the throughput of link i under TDMA and concurrent transmissions, respectively, during one scheduling cycle consisting of N slots. With TDMA scheduling, the achievable throughput of link i is

$$R_i^{\mathbb{T}} = k' p_r(i)/\eta = k' \alpha p_t d_i^{-\gamma}/\eta \tag{16.3}$$

where $k' = \log_2 e$ is a scaling factor. For concurrent transmissions, the achievable throughput of link i is

$$R_i^{\mathbb{C}} = \frac{Nk' p_r(i)}{\eta + \sum_{j \neq i} I_{j,i}} = \frac{Nk' \alpha p_t d_i^{-\gamma}}{\eta + \sum_{j \neq i} I_{j,i}} \tag{16.4}$$

where $I_{j,i}$ is the interference from link j's source to link i's destination with distance $d_{j,i}$. Let D be the distance such that $I_{j,i}$ equals N_0. If all interferers are at least D distance away from the receiver of link, i.e., $d_{j,i} \geq D$, it implies $I_{j,i} \leq \eta$ for all $j \neq i$. The resultant link throughput with concurrent transmissions is

$$R_i^{\mathbb{C}} > \frac{Nk' \alpha p_t d_i^{-\gamma}}{\eta + (n-1)\eta} = \frac{k' \alpha p_t d_i^{-\gamma}}{\eta} = R_i^{\mathbb{T}} \tag{16.5}$$

Proposition 16.1 immediately leads to the following theorem:

THEOREM 16.1

Denote by D the radius of exclusive region, as defined in Equation 16.2. Any two links i and j can transmit concurrently in favor of aggregate throughput maximization if the following condition holds:

$$d_{i,j} > D \text{ and } d_{j,i} > D \tag{16.6}$$

The minimal radius of exclusive region that ensures $I_{j,i} \leq \eta$ depends on the cross-correlations of UWB communications and the background noise level, and it is independent of the link length. The above condition implies that scheduling concurrent UWB communications is preferable to TDMA transmissions so long as all interferers are outside the exclusive regions of other receivers.

16.6 Efficient Scheduling Algorithms for Throughput Maximization

Theorem 16.1 suggests that we can improve the aggregate throughput by properly allowing concurrent transmissions. However, scheduling concurrent transmissions to maximize the aggregate throughput is known to be an NP-hard problem [18]. Even with brute force search could induce $\mathcal{O}(N^N)$ complexity, where N is the number of active links. We propose two heuristic algorithms that achieve high aggregate throughput with polynomial-time complexity.

Algorithm 1: Proportional Allocation Algorithm (PaA)

> **Input:** $i := 1$; $\Phi := \{1, \ldots, N\}$
> **repeat**
> $\quad$ $G_i := \emptyset$;
> $\quad$ randomly choose a flow f from **UA**;
> $\quad$ $G_i \leftarrow S_i \cup \{f'\}$;
> $\quad$ $UA \leftarrow UA\backslash\{f\}$;
> $\quad$ **for** *any flow f' other than f* **do**
> $\quad\quad$ **if** $f' \notin ER_l$ **&** $l \notin ER_{f'}$ $\forall l \in G_i$ **then**
> $\quad\quad\quad$ $G_i \leftarrow G_i \cup \{f'\}$;
> $\quad\quad$ **if** $f' \in UA$ **then**
> $\quad\quad\quad$ $\Phi \leftarrow UA\backslash\{f'\}$;
> $\quad$ $i \leftarrow i+1$
> **until** $(UA := \emptyset) \vee (i > K)$;
> $\omega \leftarrow i$;
> **for** $i = 1$ *to* ω **do**
> $\quad$ allocate $\lfloor K \cdot \frac{|G_i|}{\sum_{j=1}^{\omega} |G_j|} \rfloor$ slots to flows in G_i

Let Φ denote the flow set, initiated by $\Phi := \{1, \ldots, N\}$. The first algorithm, called PaA, determines the concurrent transmission flows using the following rule. Denote G_i as a set containing the concurrent transmission flows, which is an empty set initially. The algorithm starts by randomly choosing one flow from Φ, and then moves it to G_i. Then it randomly picks another flow from Φ and copies it into G_i if the condition (Equation 16.6) holds for all flows in G_i. In the algorithm, ER_f represents the exclusive region of a flow f. Repeat this step until all flows are checked and we have generated a concurrent transmission set G_i. The above procedure is repeated until either Φ becomes empty or the number of generated set G_i equals the number of time slots K. Suppose there are a total of ω sets $(G_1, G_2, \ldots, G_\omega)$ generated, then each flow in G_i will be allocated with the number of slots proportional to $|G_i|$, which is the size of G_i. The PaA has a computational complexity of $\mathcal{O}(N^2)$. It can be observed that in PaA (1) each flow will belong to at least one group, and thus will be assigned at least one slot; (2) in each group, the exclusive-region condition is strictly maintained; (3) the time slots allocated to each group are proportional to the number of flows that can transmit concurrently in that group. The rationale is that, considering the multiuser network as a conflict graph, the maximum network throughput can be achieved when each slot allocation is a maximum weighted independent set (MWIS),* where the vertex weight is set to the link throughput. To maximize the aggregate throughput, PaA allocates more slots to the flow group with larger size, which is hypothesized close to the MWIS.

The second algorithm is called a repeating allocation algorithm (RaA). Let ϕ_f be the number of slots being allocated to flow f. For a slot i, RaA will assign it to a group of flows S_i according to the following rule. RaA first randomly chooses a flow with minimal ϕ_f with ties broken arbitrarily, and adds it to S_i. Then, RaA adds other flows that do not conflict with any flows in S_i. This procedure is repeated for all time slots. The RaA has a higher computational complexity of $\mathcal{O}(KN^2 \log N)$, and we anticipate it should be more fair in terms of the number of slots assigned to each flow, because RaA essentially follows the max–min fairness discipline. The performance of these two algorithms will be evaluated via simulations in Section 16.8.

Algorithm 2: RaA

Input: $\phi_f = 0$; $S_i := \emptyset$
for $i = 1$ to K **do**
    $f^* \leftarrow \arg \min_f \{\phi_f\}$;
    $S_i \leftarrow S_i \cup \{f^*\}$;
    $\phi_{f^*} \leftarrow \phi_{f^*} + 1$;
    for any $flow$ $other$ $than$ f^* **do**
        if $f \notin ER_l$ **&** $l \notin ER_f$ $\forall l \in S_i$ **then**
            $S_i \leftarrow S_i \cup \{f\}$;
            $\phi_f \leftarrow \phi_f + 1$;

16.7 QoS Support for Heterogeneous Traffic

Both PaA and RaA algorithms aim at throughput maximization. In the presence of heterogeneous traffic with diverse QoS requirements, we need to modify the problem formulation to provide differentiated services. For instance, real-time video applications generally need a certain minimum bandwidth to initiate the application while higher bandwidth can help to improve the display quality. Data services such as file transfer or Web browsing have no strict requirement on bandwidth but a higher bandwidth is preferable. To satisfy different resource demands of heterogeneous applications, it is important that the resource allocation mechanism takes traffic characteristics into account

---

* An independent set in a graph is the set of vertices such that no two vertices in the independent set share the same edge. Associating each vertex a weight, the independent set with the maximum total weights is the maximum weighted independent set.

to achieve efficient resource utilization. In this context, the notion of utility function is useful to quantitatively characterize the user satisfaction level in response to the amount of bandwidth. Various utility functions have been proposed to characterize different traffic types (e.g., log-function for inelastic traffic, sigmoidal function for elastic traffic, etc.) [16,19–21]. Therefore, the resource allocation problem can be formulated as a utility optimization problem, which jointly considers the physical property of UWB communications and user satisfaction levels. The objective is to maximize the total utility under the constraint of fairness among users with heterogeneous applications. This section is organized as follows. We first briefly introduce the utility functions for characterizing different traffic types. Then we discuss the optimization problem and the proposed utility-based scheduling algorithm.

16.7.1 Utility Functions

The utility functions considered in this chapter are general and nondecreasing functions with values within [0,1]. Traffic types are classified into three classes. Class-1 includes the applications that need a fixed amount of bandwidth continuously available for the lifetime of the connection (e.g., constraint-based routing [CBR] voice applications), which is characterized by a step function:

$$U(r) = \begin{cases} 1 & \text{if } r \geq r_{\min} \\ 0 & \text{if } r < r_{\min} \end{cases} \quad (16.7)$$

where $r_{\min}$ is the minimum bandwidth requirement for a connection. Class-2 applications can adapt to the allocated bandwidth to certain extent (e.g., compressed video streams) described by the following sigmoidal function:

$$U(r) = 1 - e^{-\frac{br^2}{a+r}} \quad (16.8)$$

where the parameters a and b can be adjusted to determine the shape of $u(r)$, as shown in Figure 16.2a. Class-3 applications are most flexible to the available bandwidth. Most TCP-controlled data applications belong to this class. Because there is no minimum rate requirement for such traffic class, the utility can be modeled by the following utility function:

$$U(r) = \begin{cases} 1 & \text{if } r \geq r_{\max} \\ \sin^{\tau}\left(\frac{\pi}{2} \cdot \frac{r}{r_{\max}}\right) & \text{if } r < r_{\max} \end{cases} \quad (16.9)$$

where parameter τ controls the shape of $u(r)$, as shown in Figure 16.2b. Readers may refer to Ref. [15] for more discussions on different utility functions. Generally, the choice of utility functions may affect the efficiency of resource allocation, and leads to different degrees of complexity to the utility-maximization problem [22].

16.7.2 Utility Optimization Problem

Let $U(r_i)$ represent the utility of flow i with the achievable data rate r_i. We first consider a general formulation, where the objective is to maximize aggregate utility over all users. For UWB networks, several users may be scheduled in the same time slot to exploit the time-multiplexing gain as we

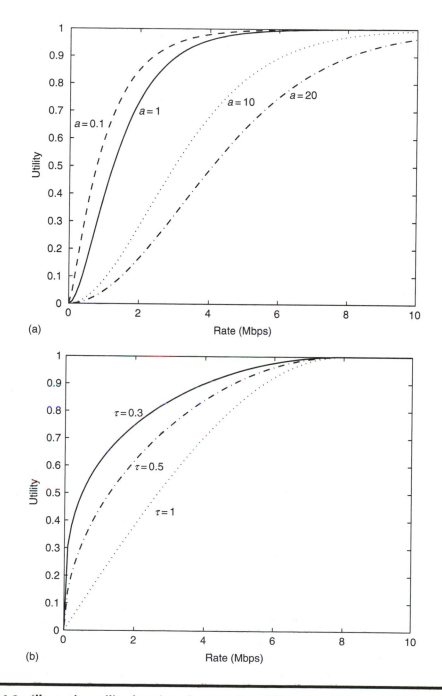

Figure 16.2 **Illustrative utility functions for Class 2 and Class 3 traffic. (a) Utility functions for Class 2, $b = 1$. (b) Utility functions for Class 3, $r_{max} = 8$ Mbps.**

have discussed. Because the concurrent transmissions introduce a multi-user interference (MUI) that decreases the achievable data rate of individual flow per time slot, certain constraints must be applied such that the aggregate utilities are maximized. To describe the impact of MUI due to concurrent transmissions, we can model the network as a conflict graph $G = (V, E)$, where the vertex set V represents the set of flows and E the set of edges. Two vertexes in G are connected if they are not allowed to transmit in the same time slot according to the exclusive-region condition indicated in Equation 16.6. The connected vertexes (i.e., flows) are called neighbors. Let $N(i)$ represent the neighbors of a vertex $i \in V$. Define the following binary variables: $x_i^s = 1$ if flow i is allocated in slot s; otherwise $x_i^s = 0$. Likewise, $y_{ij}^s = 1$ if node j is the recipient of node i in slot s; otherwise $y_{ij}^s = 0$. Now the scheduling problem for a single slot can be formulated as a nonlinear mixed-integer programming (NLIP) problem:*

$$(\mathcal{P}) \quad \max \sum_{i=1}^{|V|} x_i^s U(r_i) \tag{16.10}$$

$$\text{s.t.} \quad x_i^s \sum_{j \in N(i)} x_j^s = 0, \quad \forall i = 1, \ldots, |V| \tag{16.11}$$

$$\sum_{i \in T} y_{ij}^s + \sum_{k \in R} y_{jk}^s \leq 1, \quad \forall_j \in \mathcal{R} \tag{16.12}$$

$$r_i = \frac{K p_t d_i^{-\gamma}}{\eta + x_j^s \alpha p_t \sum_{j=1, j \neq i}^{|V|} d_{ji}^{-\gamma}}, \quad i = 1, \ldots, |V| \tag{16.13}$$

Given the number of active flows $|V|$ and the current slot s, the flow set $\{x_i^s : i = 1, \ldots, |V|\}$ that maximizes the total utilities defined in Equation 16.10 under the constraints given by Equations 16.11 through 16.13 is the optimum of $(\mathcal{P})$. The constraint given by Equation 16.11 ensures that any flow scheduled in slot s is conflict free. The constraint given by Equation 16.12 restricts that a node can communicate with at most one node in a slot, where T represents the set of senders scheduled in the same slot, and $\mathcal{R}$ is the set of corresponding receivers. The constraint given by Equation 16.13 specifies the flow data rate. Note that $(\mathcal{P})$ can be extended as a superframe utility optimization problem by modifying Equation 16.10 as $\max \sum_{s=1}^{S} \sum_{i=1}^{|V|} x_i^s U(r_i)$, where S is the total number of slots in a superframe.

Problem $(\mathcal{P})$ is equivalent to the MWIS problem, which can be interpreted as follows. We incorporate the constraint given by Equation 16.12 into the graph G by adding an edge between two nodes in V if the corresponding flows have common transmitting or receiving nodes, and denote the resulting graph as G'. Let $\mathcal{K}$ denote the collection of independent sets in G'. Because the superset $\mathcal{K}$ satisfies the constraints given by Equations 16.11 and 16.12, finding the optimal solution of problem $(\mathcal{P})$ is equivalent to finding the set with maximum weight in $\mathcal{K}$, denoted by $\kappa^* \in \mathcal{K}$ in G'. In other words,

$$\kappa^* = \arg \max_{\kappa \in \mathcal{K}} \sum_{i=1}^{|\kappa|} U(r_i) \tag{16.14}$$

---

* Problem $(\mathcal{P})$ is NLIP because $U(r_i)$ is nonlinear, x_i is integer, and r_i is continuous.

where $U(r_i)$ corresponds to the weight of vertex $i \in V$ and the slot index s has been dropped such that κ^* represents the optimal scheduling for one slot. As finding MWIS is known to be NP-hard, there is no polynomial-time algorithm to solve ($\mathcal{P}$).

16.7.3 Utility-Based Scheduling Algorithm

By associating each user a utility function according to its traffic class, the utility-maximization formulation described in Section 16.7.2 can be used to derive the optimal scheduling for heterogeneous traffic. Here, the utility $U(r_i)$ depends on the achievable data rate r_i. Because r_i is a function of propagation path loss, inaccurate distance estimation may lead to suboptimal scheduling decisions. The assumption of perfect distance information can be relaxed by the discrete stochastic approximation as follows. Let $\tilde{U}(\cdot)$ denote the noisy version of $U(\cdot)$, i.e., $\tilde{U}(\cdot)$ contains errors due to noisy distance information. Rewrite Equation 16.14 as $\kappa^* = \arg \max_{\kappa \in \mathcal{K}} \tilde{U}(\kappa)$, where $\tilde{U}(\kappa) = \sum_{i=1}^{|\kappa|} \tilde{U}(r_i)$. Subsequently, approximate $\tilde{U}(\kappa)$ by $\mathbb{E}\big[\tilde{U}_m(\kappa)\big]$, where $\mathbb{E}[\cdot]$ is the expectation operator, which yields

$$\kappa^* = \arg \max_{\kappa \in \mathcal{K}} \sum_{i=1}^{|\kappa|} \tilde{U}(r_i) \approx \arg \max_{\kappa \in \mathcal{K}} \mathbb{E}\big[\tilde{U}(\kappa)\big] \tag{16.15}$$

If the objective function is unimodal, optimization techniques such as golden-section search or gradient-based approaches may be used to find the maximum of the nonlinear function $\mathbb{E}\big[\tilde{U}(\kappa)\big]$. Unfortunately, the optimization objective function is not unimodal. In addition, deriving the distribution of $\tilde{U}(\kappa)$ is very difficult, if not impossible, because $\tilde{U}(\kappa)$ is combinatorial, depending on the elements in κ. In situations where the objective function is difficult to derive analytically, discrete approximation is an applicable technique to solve the optimization problems with uncertainties. Let $\{\tilde{U}_m(\kappa), m = 1, 2, \ldots\}$ represent the sequence of noisy utilities associated with set $\kappa \in \mathcal{K}$, where $\tilde{U}_m(\kappa)$ is obtained from different distance estimations. Furthermore, $\overline{U}(\kappa) = 1/M \sum_{m=1}^{M} \tilde{U}_m(\kappa)$ is the sample mean of $\{\tilde{U}(\kappa)\}$. By the strong law of large numbers, $\overline{U}(\kappa)$ converges almost surely (a.s.) to $\mathbb{E}\big[\tilde{U}(\kappa)\big]$. Together with the finiteness of the set $\mathcal{K}$, it is implied that

$$\arg \max_{\kappa \in \mathcal{K}} \overline{U}(\kappa) \to \arg \max_{\kappa \in \mathcal{K}} \mathbb{E}\big[\tilde{U}(\kappa)\big] \text{ a.s.}$$

Therefore, instead of using one biased utility value for solving the optimization problem ($\mathcal{P}$), we take the series $\{\tilde{U}_m(\kappa)\}$ in approximating the noisy objective function to avoid trapping into a local optimum [23]. To solve the aforementioned discrete stochastic optimization problem, we propose a meta-heuristic called exclusive-region global search algorithm (ER-GSA) [24]. On the basis of GSA [25], ER-GSA relies on a random sequence generated during the algorithm iterations to efficiently find the optimum. The resulting random sequence is a Markov chain, where each state represents a point in the solution space that has been visited by the algorithm. In each iteration, the transition of the Markov chain is determined by comparing the objective value of the current state and that of a random chosen point from the solution space. We use the following notations in the algorithm. At the mth iteration, κ_m is the current state ($\kappa_m \in \mathcal{K}$), $W_m(\kappa)$ is the number of times the algorithm has visited state κ, and κ_m^* is the state that the algorithm has visited the most till the mth iteration.

During each iteration m, a new subset κ_m' is randomly selected (**step 2**), and the sample mean of the utility function value is computed as $\overline{U}_m(\kappa_m') = 1/|\kappa_m'| \sum_{i=1}^{|\kappa_m'|} \tilde{U}_m(\kappa_i)$, $\kappa_i \in \kappa_m'$. The above method, called variable-sample approach because the sample size is varied depending on the size of the randomly chosen point κ'_m in each iteration, improves the algorithm convergence compared to original GSA. The sample mean of κ'_m is then compared with that of the current point κ_m (**step 3**). The sequence $\{\kappa_m, m = 1, 2, \ldots\}$ is a Markov chain on the state space $\mathcal{K}$. If the current state is considered better than the newly selected one, i.e., $\overline{U}_m(\kappa_m) > \overline{U}_m(\kappa'_m)$, the algorithm proceeds to **step 5**, updating the best subset κ^*_m according to $W_m(\kappa)$. Because the optimal subset κ^*_m has higher probability of generating a larger utility function value, the associated $W_m(\kappa^*)$ is thus accumulated faster. To accelerate the convergence to the global optimum, a local enhancement in **step 4** using the concept of exclusive region can refine the chosen subset such that the optimal subset can be located faster. In particular, the set ER_l contains those flows within the exclusive region of flow l, and S_i denotes the ith flow in S. The conditions $(S_i \notin ER_l)$ and $(l \notin ER_{S_i})$ imply that flows S_i and l are allowed to transmit concurrently, according to Equation 16.6. The size of the exclusive region can be determined according to the strength of background noise and the pathloss exponent [24]. It can be seen that, the GSA algorithm assumes any two points are neighbors, and only improved moves are accepted. Consequently, the sequence $\{\kappa^*_m\}$ converges to the global optimum. By tracking the counter $W_m(\kappa)$, the algorithm can further avoid false retaining on a local optimum due to the biased objective function value. For more details about the algorithm convergence and implementation issues, readers may refer to Ref. [24].

Algorithm 3: ER-GSA

Step 1 Randomly select an initial user subset $\kappa_0 \in \mathcal{K}$ and let $\kappa^*_0 = \kappa_0$. Set $W_0(\kappa_0) = 1$ and $W_0(\kappa) = 0$ for all $\kappa \in \mathcal{K}\setminus\{\kappa_0\}$. Calculate $\overline{U}_0(\kappa_0)$. Let $m = 0$, and go to step 2.
Step 2 Randomly select another user subset $\kappa'_m \in \mathcal{K}\setminus\{\kappa_m\}$. Compute the corresponding $\overline{U}_m(\kappa'_m)$ using the variable-sample method. Go to step 3.
Step 3 If $\overline{U}_m(\kappa_m) > \overline{U}_m(\kappa'_m)$, let $\kappa_{m+1} = \kappa_m$, and go to step 5. Otherwise, go to step 4.
Step 4 Sort $\overline{U}_m(\kappa'_m)$ in descending order. Denote s_i the i-th flow in the sorted set, and S' an empty set.
for $i = 1$ to $|\kappa'_m|$ **do**
$\quad$ **if** $s_i \notin ER_l$ $\quad$ **&** $\quad$ $l \notin ER_{s_i}$ $\quad$ $\forall l \in S'$ **then**
$\quad\quad\lfloor S' = S' \cup \{s_i\}$
If $\overline{U}_m(S') > \overline{U}_m(\kappa'_m)$, let $\kappa'_m = S'$.
Let $\kappa_{m+1} = \kappa'_m$ and go to step 5.
Step 5 Let $m = m + 1$, $W_m(\kappa_m) = W_{m-1}(\kappa_m) + 1$, and $W_m(\kappa) = W_{m-1}(\kappa)$ for all $\kappa \in \mathcal{K}\setminus\{\kappa_m\}$. If $W_m(\kappa_m) > W_m(\kappa^*_{m-1})$, then let $\kappa^*_m = \kappa_m$. Otherwise, let $\kappa^*_m = \kappa^*_{m-1}$. Go to step 2.

16.7.4 Utility Update

The utility-maximization formulation can be further modified such that the scheduling decision satisfies the fairness criteria in the longterm. We employ a simple rule based on weighted fair queuing (WFQ) [26], which normalizes the instantaneous utility of flow i in slot s, denoted by $U_i(s)$, to the total utility this flow has obtained, $\sum_{t=1}^{s-1} U_i(t)$. In addition, users can be discriminated based on the cost of bandwidth usage. Accordingly, a control parameter denoted as $\rho_i^{(s)}$ for flow i in slot s is defined as

$$\rho_i^{(s)} := \frac{c_i}{\left(\sum_{t=1}^{s-1} U_i(t) + \epsilon\right)^s} \qquad (16.16)$$

where
   c_i is a predefined parameter representing the revenue contribution or importance of a particular user
   $\epsilon > 0$ is a small nominal constant to avoid zero denominator. Consequently, the weighted utility for flow i in slot s is given by

$$U(r_i) = \rho_i^{(s)} U_i(s) \qquad (16.17)$$

where $U_i(s)$ is a function of r_i and the traffic class of each flow, as discussed in Section 16.7.3. The scheduling policy given by Equation 16.17 has the following properties: (1) the weighted utility being an exponential function of slot index s ensures the flows with less sum utility a higher priority; (2) a flow is opportunistically scheduled if it has higher utility value in the current slot than others; and (3) different level of protections to traffic classes can be achieved by adjusting the parameter c_i.

16.8   Performance Evaluation and Discussion

In this section we present some numerical results to demonstrate the performance of the proposed resource allocation algorithms, namely the RaA, PaA, and ER-GSA, for UWB networks. Three metrics are used to study the performance of our proposed solutions: the cumulative utilities, the minimum utility among flows, and the fairness support. The complexity and efficiency issues are also discussed.

16.8.1   Experimental Setting

The simulated network consists of 20 nodes uniformly distributed in a square area of $10 \times 10 \text{ m}^2$. Each sender arbitrarily chooses another node as the receiver, forming ten peer-to-peer communication flows. The data rate of each flow is estimated as $R = k \cdot W \log_2(1 + \text{SINR})$, where $0 < k \leq 1$ reflects the efficiency of the transceiver design; $W = 500$ MHz; the power spectrum density of the transmission and noise are -41 and -114 dBm/MHz, respectively; and the pathloss exponent is set to 4. The distance between two nodes $\tilde{d}$ is modeled by $d = \tilde{d} + \delta$ where d is the actual distance and δ is the estimation error modeled as a normal distributed random variable according to Ref. [27, Eq.(15)], i.e., $\delta \sim \mathcal{N}(0, \sigma^2)$ where $\sigma^2 = 0.05$. The cross-correlation of the target signal and the interfering signals is assumed to be 0.1. The considered three traffic classes and their utility functions are listed in Table 16.2. Each superframe contains ten slots. The size of exclusive region, denoted as d_{ER}, is set to 2 m, except in Section 16.8.2 where we vary the size of exclusive region to study its impact on the aforementioned three performance metrics.

16.8.2   Utility versus Fairness

In a multiuser environment, it is important that the scheduling algorithm can maximize the aggregate utility and maintain fair resource allocation among competing flows. We compare the proposed

Table 16.2 Traffic Characteristics in Simulations

Traffic Class	B.W. Requirement	Utility Function
I	1 Mbps	$\begin{cases} 1 & \text{if } r \geq 1 \\ 0 & \text{if } r < 1 \end{cases}$
II	1–20 Mbps	$1 - e^{\frac{-0.03r^2}{20+r}}$
III	0–250 Mbps	$\begin{cases} 1 & \text{if } r \geq 250 \\ \sin^{0.3}\left(\frac{\pi}{2} \cdot \frac{r}{250}\right) & \text{if } r < 250 \end{cases}$

scheduling algorithms using three performance metrics: (1) the total utility of all flows; (2) the minimum per-flow utility among all flows; and (3) the Jain's fairness index [28]. Each point in Figures 16.3 through 16.5 indicates the result at the end of the 10th slot. The 95 percent confidence interval from ten different random topologies is plotted as error bars. Other parameters follow the default setting defined at the beginning of this section.

16.8.2.1 Total Utility versus Fairness

Utility maximization and fair allocation are known to be two conflicting objectives. For instance, if we always choose the flows with small transceiver distance and high data rate to transmit and neglect those with low data rate, we can achieve high overall throughput and utility, but some flows may be starving. Our proposed solution is to maximize the total utility under the fairness constraint. In Figure 16.3a, the total utility generated by ER-GSA, PaA, RaA, and TDMA among ten Class-3 flows are shown. For PaA and RaA, the total utility varies significantly when different sizes of the exclusive region are used. ER-GSA achieves comparable total utility as that achieved by PaA and RaA with the best d_{ER}. On the other hand, ER-GSA maintains much higher level of fairness than PaA and RaA, as shown in Figure 16.3b. TDMA maintains the fairness in terms of the number of time slots allocated to each flow, but its fairness index in terms of the utility and the total utility are much lower (about 58 percent less) than that achieved by the other three algorithms.

We further evaluate the performance with three traffic classes. To measure fairness for heterogeneous traffic, the Jain's fairness index is computed as $\sum_{i=1}^{N} \left[u(i)/c(i) \right]^2 / \{ N \cdot \sum_{i=1}^{N} \left[u(i)/c(i) \right]^2 \}$, where $u(i)$ and $c(i)$ are the utility and the corresponding weighting factor of flow i. The total utility and the fairness index at the end of the 10th slot are shown in Figure 16.4a and b, respectively. For total utility, ER-GSA achieves about 90 percent total utility of that achieved by PaA or RaA, but the latter two fail to maintain fairness in the presence of multi-class traffic. Together with the results in the single-class case, we conclude that ER-GSA can maintain a good trade-off between utility maximization and fairness, and PaA and RaA can achieve high network throughput with less computation time.

16.8.2.2 Minimum Utility

The maximum aggregate utility is often achieved by maximizing certain users' utility while starving others. Hence, it is important to consider the performance of the worst user. Figure 16.5a and b shows the utility of the worst user in single-class and multi-class scenarios, respectively. It can be

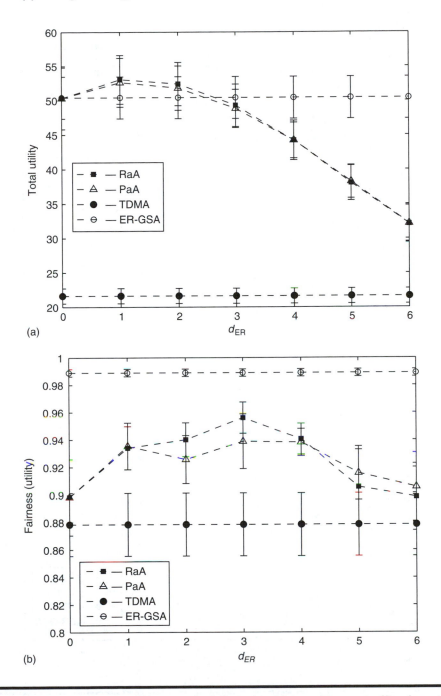

(a)

(b)

Figure 16.3 Comparisons on different scheduling algorithms with one traffic class. (a) Total utility obtained by different scheduling algorithms. (b) Utility fairness obtained by different scheduling algorithms.

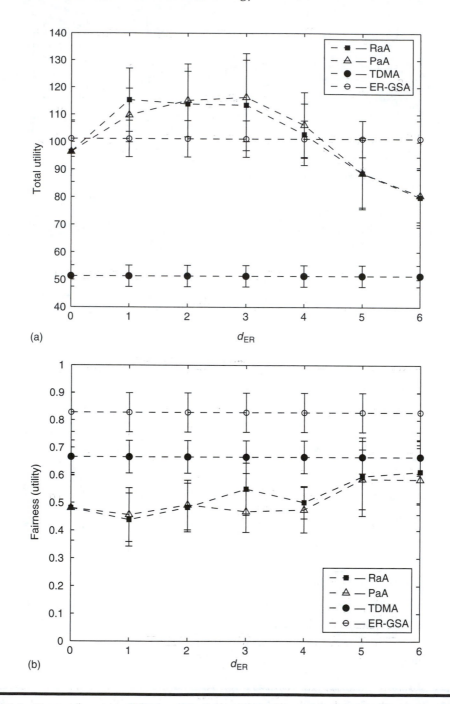

(a)

(b)

Figure 16.4 Comparisons on different scheduling algorithms with three traffic classes. (a) Total utility obtained by different scheduling algorithms. (b) Utility fairness obtained by different scheduling algorithms.

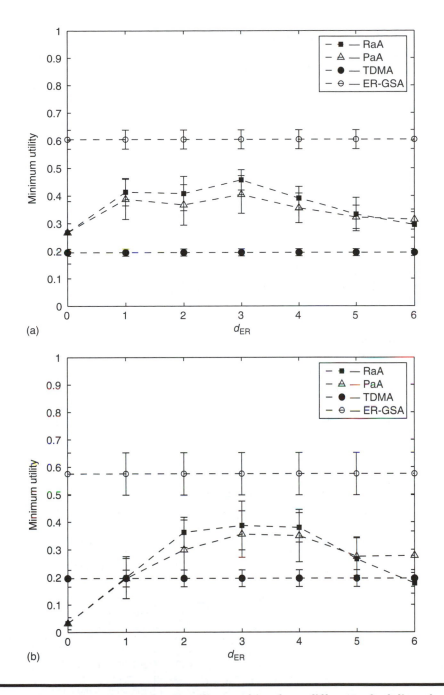

Figure 16.5 Comparisons on minimum utility resulting from different scheduling algorithms. (a) Minimum utility resulting from different scheduling algorithms: one traffic class. (b) Minimum utility resulting from different scheduling algorithms; three traffic classes.

seen that the minimum utility offered by ER-GSA is much higher than that of the other algorithms. Note that in RaA, the user with minimum number of assigned slots is chosen at each scheduling iteration. Thus, RaA performs slightly better than PaA by increasing the minimum utility. Using TDMA not only results in low aggregate utility but also tends to starve the worst user.

16.8.3 Algorithm Efficiency and Stability

To measure the algorithm robustness against estimation errors, define the stability factor ξ as [25]

$$\xi = \frac{\sum_i (m'_i - m_i)}{M - m_1}$$

where

$(m'_i - m_i)$ represents the number of iterations that the algorithm stays at the optimal point consecutively because the optimal point is reached at the ith time

M is the total number of iterations.

The smaller value of ξ implies that the algorithm is more sensitive to noise. The variable-sample path random search (SPRS) algorithm proposed in Ref. [29] increases the sample size adaptively to overcome the impact of noise. Thus, the SPRS algorithm will stay at the optimal point once it is reached if the sample size grows sufficiently fast, at the cost of prohibitively high computational load.

We compare the stability and computation cost of all algorithms being discussed. All algorithms are coded in C language and executed on a Pentium-4 2.8 GHz CPU. We schedule 10 flows for 10 slots, and repeat the algorithm 10 runs where each run contains 1000 iterations. Table 16.3

Table 16.3 Execution Time and Stability ξ

Algorithm	GSA	ER-GSA	SPRS	RaA	PaA
Execution time (ms)	2.63	4.28	290.89	0.438	0.114
	2.61	3.44	362.06	0.560	0.110
	2.47	3.42	136.66	0.594	0.115
	2.58	3.43	206.73	0.706	0.078
	2.29	3.80	379.75	0.579	0.121
	2.66	3.52	261.06	0.589	0.076
	2.63	3.32	351.34	0.570	0.158
	2.58	3.30	325.87	0.595	0.122
	2.55	3.68	313.74	0.601	0.105
	2.71	3.34	309.51	0.593	0.106
Stability factor ξ (percent)	73	84	100	—	—

gives the execution time and stability factor ξ corresponding to each algorithm.* It can be seen that SPRS has the best stability property, but it is not feasible to implement it for real-time scheduling. Considering the typical superframe length of 65–90 ms, the complexity of the other four algorithms are acceptable. On the other hand, a sufficient number of iterations are needed to ensure the convergence of ER-GSA and GSA. A simple rule is to let it be at least the size of the entire solution space, so each point can be checked once statistically. More complicated termination rules, such as performing a paired t-test after a certain number of iterations to decide the termination of the algorithm (as in SPRS), may introduce excessive computation overhead, and thus they are not recommended.

16.9   Conclusion and Open Issues

UWB is a promising wireless technology that can provide seamless connection in short-range communications. This chapter has addressed the resource allocation problem for UWB-based WPAN, considering both the physical characteristics of UWB and the users' requirements in WPANs. The resource allocation problem has been mathematically formulated, and shown to be NP-hard. To maximize the aggregate throughput, two heuristic algorithms called RaA and PaA leveraging the inherent property of UWB have been proposed. To maximize the aggregate utility with fairness constraint, a meta-heuristic algorithm tailed for UWB networks called ER-GSA has been proposed to solve the optimization problem. Several performance metrics, such as aggregate utility, fairness, and algorithm complexity have been evaluated via simulations. It has been demonstrated that, for single traffic class, the proposed RaR, PaR, and ER-GSA scheduling algorithms can provide more than 200 percent utility gain over TDMA. For multiple traffic classes, the ER-GSA algorithm can maximize the total utility under the constraints of both inter-class fairness and intra-class fairness.

Some issues may need further investigations. As mentioned in Section 16.5, without the support of inter-piconet signaling, links that associate with different piconets may interfere with each other, due to the co-channel interference (CCI). Recently, the CCI issue has also been studied for UWB-based piconets [30,31]. How to reduce the interference among simultaneously operating piconets deserves further study. On the other hand, the MAC protocols considered in this chapter rely on a centralized PNC to coordinate the resource allocation. An alternative solution is a fully distributed MAC, where each network entity makes its own decision on resource allocation with or without the cooperation among users. Although distributed mechanisms eliminate the need of a centralized controller leading to a more flexible operation and avoiding the inter-piconet interference, resource sharing strategies become more challenging. How to jointly achieve QoS provisioning and power saving is another critical problem for UWB networks. In addition, many potential applications of UWB, such as vehicular communications, biomedical sensor networks, etc., can be envisioned in the near future [32,33], and how to efficiently manage radio resources for these applications will be an interesting topic.

---

* Because the heuristic RaA and PaA are not designed to ensure optimum, we do not consider their stability performance.

REFERENCES

1. R. J. Fontana, Recent system applications of short-pulse ultra-wideband (UWB) technology, *IEEE Trans. Microwave Theory Tech.*, 52(9), 2087–2104, 2004.

2. IEEE standard part 15.3: Wireless Medium Access Control (MAC) and Physical Layer (PHY) Specifications for High Rate Wireless Personal Area Networks WPANs, IEEE Std 802.15.3-2003, September 2003.

3. X. Shen, W. Zhuang, H. Jiang, and J. Cai, Medium access control in ultra-wideband wireless networks, *IEEE Trans. Veh. Technol.*, 54(5), 1663–1677, 2005.

4. IEEE 802.15 WPAN High Rate Alternative PHY Task Group 3a. DS-UWB Physical Layer Proposal. IEEE P802.15-04/0137r4, January 2005.

5. IEEE 802.15 WPAN High Rate Alternative PHY Task Group 3a. Multiband OFDM Physical Layer Proposal, September 2004.

6. B. Radunovic and J.-Y. Le Boudec, Optimal power control, scheduling, and routing in UWB networks, *IEEE J. Select. Areas Commun.*, 22(7) 1252–1270, 2004.

7. J. Cai, K.-H. Liu, X. Shen, J. W. Mark, and T. D. Todd, Power allocation and scheduling for ultra-wideband wireless networks, *IEEE Trans. Veh. Technol.*, 57(2), 1103–1112, 2008.

8. M. Win and R. Scholtz, Ultra-wide bandwidth time-hopping spread-spectrum impulse radio for wireless multiple-access communications, *IEEE Trans. Commun.*, 48(4) 679–689, 2000.

9. V. S. Somayazulu, Multiple access performance in UWB systems using time hopping vs. direct sequence spreading, in Proceedings of IEEE WCNC'02, Orlando, Florida, pp. 522–525, 2002.

10. M.-G. D. Benedetto, L. De Nardis, M. Junk, and G. Giancola, UWB^2: Uncoordinated, wireless, baseborn medium access for UWB communication networks, *Mobile Netw. Appl.*, 10(5) 663–374, 2005.

11. F. Cuomo, C. Martello, A. Baiocchi, and F. Captriotti, Radio resource sharing for ad hoc networking with UWB, *IEEE J. Select. Areas Commun.*, 20 1722–1732, 2002.

12. W. M. Lovelace and J. K. Townsend, Adaptive rate control with chip discrimination in UWB networks, in Proceedings of IEEE Ultrawideband Systems Technology (UWBST), Reston, Virginia, pp. 195–199, 2003.

13. R. Merz, J. Widmer, J.-Y. Le Boudec, and B. Radunovic, A joint PHY/MAC architecture for low-radiated power TH-UWB wireless ad hoc networks, *Wireless Commun. Mobile Comput.*, 5(5) 567–580, 2005.

14. C. Lee, J. Lehoczky, R. Rajkumara, and D. Siewiorek, On quality of service optimization with discrete QoS options, in Proceedings of the 5th IEEE Real-Time Technology and Application Symposium, pp. 276–286, Vancouver, Canada, June 1999.

15. V. Rakocevic, J. Griffiths, and G. Cope, Analysis of bandwidth allocation schemes in multiservice IP networks using utility functions, International Teletraffic Congress (ITC17'), Salvador da Bahia, Brazil, December 2001.

16. N. Lu and J. Bigham, Utility-maximization bandwidth adaptation for multi-class traffic QoS provisioning in wireless networks, in Proceedings of 1st ACM International Workshop on Quality of Service & Security in Wireless and Mobile Networks (Q2SWinet'05), Montreal, Canada, pp. 136–143, 2005.

17. J. G. Proakis, *Digital Communication*. 4th edition, McGraw-Hill, New York, 2001.

18. K.-H. Liu, L. Cai, and X. Shen, Performance enhancement of medium access control for UWB WPAN, in Proceedings of IEEE Globecom'06, San Francisco, CA, December 2006.

19. M. Dianati, X. Shen, and S. Naik, Cooperative fair scheduling for the downlink of CDMA cellular networks, *IEEE Trans. Veh. Technol.*, 56(4), 1749–1760, 2007.

20. Y. Cao and V. Li, Utility-oriented adaptive QoS and bandwidth allocation in wireless networks, in Proceedings of IEEE ICC'02, New York, pp. 3071–3075, 2002.

21. J.-W. Lee, R. R. Mazumdar, and N. B. Shroff, Downlink power allocation for multiclass DS-CDMA wireless networks, in Proceedings of IEEE INFOCOM'02, New York, pp. 1480–1489, 2002.

22. J.-W. Lee, R. R. Mazumdar, and N. B. Shroff, Non-convex optimization and rate control for multi-class services in the Internet, *IEEE/ACM Trans. Netw.*, Issue 4, 13, 827–840, 2005.

23. G. C. Pflug, *Optimization of Stochastic Models: The Interface between Simulation and Optimization*, Kluwer Academic Publishers, Dordrecht, Boston, 1996.
24. K.-H. Liu, L. Cai, and X. Shen, Multi-class utility-based scheduling for UWB networks, *IEEE Trans. Veh. Technol.*, 57(2), 1176–1187, 2008.
25. S. Andradottir, A global search method for discrete stochastic optimization, *SIAM J. Optimization*, 6(6), 513–530, 1996.
26. A. Demers, S. Keshav, and S. Shenker, Analysis and simulation of a fair queueing algorithm, in Proceedings of SIGCOMM'89, Austin, TX, pp. 1–12, 1989.
27. B. Denis, J.-B. Pierrot, and C. Abou-Rjeily, Joint distributed synchronization and positioning in UWB ad hoc networks using TOA, *IEEE Trans. Microwave Theory Tech.*, 54(4), 1896–1911, 2006.
28. R. Jain, A. Durresi, and G. Babic, Throughput fairness index: An explanation, ATM Forum Document Number: ATM Forum/990045, February 1999.
29. T. Homem-De-Mello, Variable-sample methods for stochastic optimization, *ACM Trans. Model. Comput. Simul. (TOMACS)*, 13(2), 108–133, 2003.
30. N. Kumar, S. Venkatesh, and R. M. Buehrer, A spread-spectrum MAC protocol for impulse-radio networks, in Proceedings of IEEE VTC'05, Dallas, Texas, pp. 665–669, September 2005.
31. P. Gong, P. Xue, J. S. Lee, and D. K. Kim, Performance enhancement of an MB-OFDM based UWB system in multiple SOPs environments, in Proceedings of 1st International Symposium on Wireless Pervasive Computing, Phuket, Thailand, January 2006.
32. H. M. Jafari, W. Liu, S. Hranilovic, and M. J. Deen, Ultrawideband radar imaging system for biomedical applications, *J. Vacuum Sci. Technol. A: Vacuum, Surfaces, and Films*, 24(3), 752–757, 2006.
33. I. Gresham, A. Jenkins, R. Egri, C. Eswarappa, N. Kinayman, N. Jain, R. Anderson, F. Kolak, R. Wohlert, S. P. Bawell, J. Bennett, and J.-P. Lanteri, Ultra-wideband radar sensors for short-range vehicular applications, *IEEE Trans. Microwave Theory Tech.*, 52(9), 2105–2122, 2004.

Chapter 17

New UMA Paradigm: Class 2 Opportunistic Networks

Zill-E-Huma Kamal, Leszek Lilien, Ajay Gupta, Zijiang Yang, and Manish Kumar Batsa

CONTENTS

Opportunistic networks can be divided into class 1 opportunistic networks that use opportunistically only communication resources, and class 2 opportunistic networks, called oppnets in short, that can use opportunistically all kinds of resources—computation, sensing, actuation, storage, including communication capabilities. After introducing oppnets, we discuss basic oppnet ideas and operations, followed by the control flow in oppnets. Next, we present a standard implementation framework for oppnets, named oppnet virtual machine (OVM), and discuss a list of primitives needed for the control center (CC), seed and oppnet helpers, and lites, followed by a categorization and description of various oppnet applications and scenarios.

Following this comprehensive discussion on oppnets, we present the design and implementation of a small-scale oppnet experiment, named MicroOppnet. MicroOppnet, originally developed as a proof of concept, is now being extended to serve as a test bed for experimentation and pilot implementations of oppnet architectures and their components. We present control flow of the application with respect to the OVM primitives discussed and presented earlier.

We conclude with a summarization of our work and some open issues in class 2 opportunistic networks.

17.1 Introduction

17.1.1 Pervasive Technologies and Unlicensed Mobile Access Technology

Simply stated, pervasive technologies are those that enable easy information access and processing for anyone from anywhere at any time [1]. Sun MicroSystems provides a more integrated definition of pervasive computing as the convergence of computers, communication, consumer electronics, content and services, where devices interact in an hierarchical manner over two underlying layers—service and standards [1]. The service layer establishes infrastructure for computing, communication, content, access, etc., whereas the standards layer allows information and application exchange (e.g., standards include Java, XML, HTML, etc.) [1]. Hansmann et al. [1] postulate that pervasive computing realizes four fundamental paradigms:

- Decentralization: through heterogeneous distributed environments/systems with a dynamic network of relationships
- Diversification: instead of one device offering all functionalities, multiple diverse devices that are best capable of meeting the requirements of the specific purpose
- Connectivity: communication standards must be integrated on a global basis of interoperability
- Simplicity: easy to use

Unlicensed mobile access (UMA) technology promotes the use of unlicensed spectrum technologies such as WiFi/WLAN/802.11b/g and Bluetooth (BT) as part of an integrated communication network. Currently, use of UMA technology is being promoted for 3G mobile phones and data standards [2]. UMA, also known as generic access network (GAN), is a telecommunication system that allows seamless integration of the cellular network with unlicensed wireless technologies. Under the GAN system, if the dual-mode phone detects the presence of a compatible unlicensed wireless spectrum, say a WLAN, then it establishes an Internet Protocol (IP) connection, such that Voice-over-IP (VoIP) calls can be placed and data can be transferred via the underlying cheaper UMA technology. When a compatible UMA technology is not available, the phone operates in the carrier's specific cellular network. A GAN/UMA system headset can operate in four modes: cellular network only, cellular network where available and WiFi otherwise, WiFi where available and cellular network otherwise, and WiFi only [2].

The UMA technology exhibits all the fundamental characteristics of a pervasive technology. It is a decentralized, diverse, seamless connection, and simple for the user.

Currently, UMA is used by cellular network operators to promote connectivity. We would like to discuss a newer computing paradigm termed opportunistic networks. These not only utilize UMA technology for connectivity but also leverage other resources, such as computational power, sensing capabilities, storage, etc., to meet the requirements for a specific purpose.

17.1.2 Opportunistic Networks and Their Classes

In what we call class 1 opportunistic networks, opportunism is quite restricted, usually limited to opportunistic connectivity, that is, establishing communications when devices are within each other's range. In contrast, we proposed a new paradigm and a new technology called class 2 opportunistic networks, or oppnets, to enable both an opportunistic growth of networks and opportunistic use of resources gained by this growth [3]. (It evolved from our initial idea of opportunistic sensor networks [4]).

Effectively, oppnets leverage their capabilities by exploiting the wealth of resources available on all kinds of pervasive devices that are within their reach—crossing communication and hardware and software barriers. Use of oppnets can be envisioned in numerous applications ranging from military to emergency response to mundane domestic applications.

Class 1 opportunistic networks provide low-level UMA, while class 2 in addition provides services that it is able to discover (also via true discovery, not just a directory lookup). Opportunistic data dissemination techniques [5–7] might be considered class 1.5 opportunistic networks.

17.2 Class 2 Opportunistic Networks: New Paradigm for UMA

17.2.1 Basic Oppnet Ideas and Operations

This section shows basic oppnet ideas and operations [3,8,9].

17.2.1.1 Seed Oppnets and Oppnet Helpers

Seed oppnets: Each oppnet starts as a seed oppnet—a set of nodes employed together at the time of the initial network deployment (Figure 17.1). The seed is predesigned, and can be viewed as a network in its own right. It might be very small, in the extreme consisting of a single node.

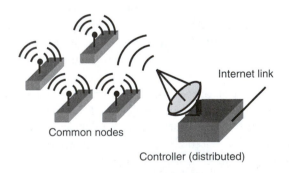

Figure 17.1 Seed oppnet. (From L. Lilien, A. Gupta, and Z. Yang, Proceedings of the 1st International Workshop on Next Generation Networks for First Responders and Critical Infrastructure (NetCri07), New Orleans, Louisiana, April 2007; L. Lilien, Z. H. Kamal, V. Bhuse, and A. Gupta, K. Makki et al. (Ed.), *Mobile and Wireless Network Security and Privacy*, Springer Science+Business Media, Norwell, Massachusetts, 2007; Z. H. Kamal, A. Gupta, L. Lilien, and Z. Yang, *Conference: Proceedings The 3rd International Conference on Collaborative Computing: Networking Applications and Worksharing*, White Plains, New York, November 12–15, 2007.)

A subset of seed nodes constitutes a distributed CC. CC can grow admitting other nodes, and can shrink expelling any of its nodes. Admitted notes are called helpers. We can have both regular helpers and lightweight helpers or lites (such as a smoke detector). Lites are oppnet-enabled, equipped with inexpensive, simple means of standard oppnet communications. In this way, even lites can be triggered to operate in the oppnet mode when needed and commanded to do so by CC regular helpers, but not lites, and can discover and may admit other helpers.

Summarizing, a node belongs to one of the four categories: (1) CC nodes; (2) seed nodes, which really are the seed nodes that are not CC nodes; (3) helpers, which really are the regular helpers that are not lites; and (4) lites.

Potential helpers and their discovery: In general, the set of potential helpers for oppnets is very broad, including communication, computing, and sensor systems, both wired and wireless, both freestanding and embedded. As pervasive computing progresses, the candidate pool will continue increasing dramatically, in infrastructures, buildings, vehicles, appliances, etc.

More densely populated areas will have, in general, a denser coverage by potential helpers. Thus, it will be easier to leverage capabilities of an oppnet in such areas. This is a desirable property: more resources become available in areas with a possibility of more human victims and more property damage.

Before a seed oppnet can grow, it must discover its own set of potential helpers available to it. In addition to a mere lookup of a previously prepared information (e.g., a directory), which is often referred to as discovery, we mean also much more challenging true discovery. True discovery could involve an oppnet node scanning the spectrum for signals or beacons, and collecting enough information to contact their senders.

Candidates, helpers, and utilizing helpers: Those potential helpers that are considered promising and are contacted by an oppnet become its candidate helpers or candidates. Candidates admitted into an oppnet become its actual helpers.

Oppnets can utilize resources of helpers to significantly enhance their capabilities. This has the form of leveraging of all kinds of resources and skills (provided by smart or intelligent software) that new helpers bring with them. In this way, oppnets obtain a lot of help effectively and efficiently (even for free in emergency situations, as discussed later).

Use of helper functionalities can be innovative in at least two ways. First, oppnets are able to exploit dormant capabilities of their helpers. For instance, even entities with no obvious sensing capabilities can be used for sensing: (1) a desktop can sense its user's presence at the keyboard; (2) a smart refrigerator monitoring opening of its door can sense presence of potential victims at home in a disaster area. As another example, the water infrastructure sensornet (sensor network) with multisensor capabilities, which is positioned near roads, can be directed to sense vehicular movement, or the lack thereof.

Second, helpers might be used in novel combinations of existing technologies, as in the following scenario [9]. A seed oppnet is deployed in a metropolitan area after an earthquake. It finds many potential helpers, and integrates some of them into an expanded oppnet. One of the nodes of the expanded oppnet, a surveillance system, looks at a public area scene with many objects. The image is passed to an oppnet node that analyzes it, and recognizes one of the objects as an overturned car (Figure 17.2). Another node decides that the license plate of the car should be read. As the oppnet currently includes no image analysis specialist, a helper with such capabilities is found and integrated into the oppnet. It reads the license plate number. The license plate number is used by another newly integrated helper to check in a vehicle database whether the car is equipped with the OnStar communication system [10]. If it is, the appropriate OnStar center facility is contacted, becomes a helper, and obtains a connection with the OnStar device in the car. The OnStar device in

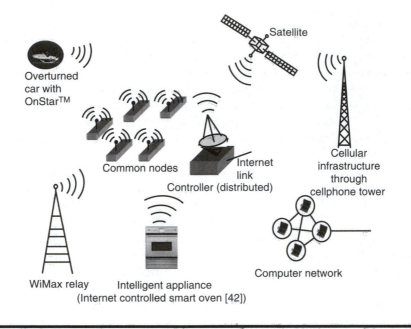

Figure 17.2   Expanded oppnet. (From L. Lilien, A. Gupta, and Z. Yang, Proceedings of the 1st International Workshop on Next Generation Networks for First Responders and Critical Infrastructure (NetCri07), New Orleans, Louisiana, April 2007; L. Lilien, Z. H. Kamal, V. Bhuse, and A. Gupta, K. Makki et al. (Ed.), *Mobile and Wireless Network Security and Privacy*, Springer Science+Business Media, Norwell, Massachusetts, 2007; Z. H. Kamal, A. Gupta, L. Lilien, and Z. Yang, *Conference: Proceedings The 3rd International Conference on Collaborative Computing: Networking Applications and Worksharing*, White Plains, New York, November 12–15, 2007.)

the car becomes a helper and is asked to contact BANs (body area networks) on and within bodies of car occupants. Each BAN available in the car becomes a helper and reports on the vital signs of its owner. The reports from BANs are analyzed by scheduling nodes that schedule the responder teams to ensure that people in most serious condition are rescued sooner than people who can wait for help longer. (Please note that with the exception of the BAN link that is just a bit futuristic—its widespread availability could be measured in years not in decades—all other node and helper capabilities used in the scenario are already quite common.)

17.2.1.2   Growth of Seed Oppnet into Expanded Oppnet

A seed oppnet grows into an expanded oppnet after admitting new helpers. For example, the expanded oppnet in Figure 17.2 admitted these helpers: (1) a computer network, contacted via a wired Internet link; (2) a cell phone infrastructure (represented by the cell phone tower), contacted via oppnet's cell phone peripheral; (3) a satellite, contacted via a direct satellite link; (4) a home area network, contacted via an intelligent appliance (e.g., a refrigerator) with a wireless link; (5) a microwave network, contacted via a microwave relay; (6) BANs of occupants of an overturned car, contacted via OnStar.

Helpers are either invited or ordered to join. In the former case, contacted candidates are free to either join or refuse the invitation. In the latter case, they must accept being conscripted in the spirit of citizens called to arms (or suffer the consequences of going absent without official leave (AWOL)).

17.2.1.3 Asking or Ordering Helpers and Oppnet Reserve

Ordering candidate helpers to join may seem controversial, and requires addressing. First, it is obvious that any candidate can be asked to join in any situation.

Second, any candidate can be ordered to join in life-or-death situations. It is an analogy to citizens being required by law to assist with their property (e.g., vehicles) and labor in saving lives or critical resources.

Third, some candidates can always be ordered to become helpers in emergencies. They include many kinds of computing and communication systems serving police, firefighters, the National Guard, etc. Also, the federal/local governments can make some of their systems available upon an order from an emergency preparedness and response (EPR) oppnet.

The category of systems always available on an order of an EPR oppnet includes systems that volunteer—actually, are volunteered by their owners. In an analogy to Army, Air Force, and other reserves, they all can be named collectively as the oppnet reserve. Individually they are oppnet reservists. As in the case in the human reserves, volunteers sign up for oppnet reserve for some incentives, be they financial, moral, etc. Once they sign up, they are trained for an active duty: facilities assisting oppnets in their discovery and contacting them are installed on them. For example, a standard OVM software is installed on them (see Section 17.2.3.) The training makes reservists highly prepared for their oppnet duties.

Oppnet reserve is not necessary for the oppnet paradigm but is very helpful for at least two reasons. First, oppnet reservists in an incident area increase the pool of candidates that can be ordered—rather than asked—by an oppnet to join it. Second, having trained reservists (e.g., OVM-equipped ones) significantly simplifies discovery of candidates. Specifically, it facilitates finding by an oppnet the very first contact in an incident area, which is always most difficult. Once a reservist joins an oppnet, reservist's own contacts become easy next-wave contacts for the oppnet.

We have assumed that at least one reservist survives an incident. With numerous reservists in practically every area of the country—the more reservists the more densely populated is an incident area—we are practically guaranteed that some reservists will survive (and some of the reservists' contacts will survive).

By employing helpers working for free as volunteers or conscripts, opportunistic networks can be extremely competitive economically in their operation. Full realization of this crucial property requires determining the most appropriate incentives for volunteers and enforcements for conscripts.

One more issue needs be addressed. Integrating helpers by oppnets could have unintended consequences such as disruptions of operations of life-support and life-saving systems, traffic lights, utilities, PTSN and cell phones, the Internet, etc. [4,11].

To protect critical operations of oppnets and of helpers joining an oppnet, oppnets must obey the following principles:

- Oppnets must not disrupt critical operations of potential helpers. In particular, they must not take over any resources of life-support and life-saving systems.
- For potential helpers running noncritical services, risk evaluation must be performed by an oppnet before they are asked or ordered to join the oppnet. This task may be simplified by potential helpers identifying their own risk levels, according to a standard risk level classification.
- Privacy and security of oppnets and helpers must be assured, especially in the oppnet growth process.

17.2.2 Control Flow in Oppnets

The control flow in oppnets, that is, the basic sequence of oppnet operations is shown in Figure 17.3. Oppnets first deploy a seed oppnet (see Figure 17.3), which may be viewed as a pretty typical ad hoc network. It self-configures, and then works to detect foreign devices or systems using all kinds of communication media—including wired Internet, WiFi, cell phones, RFIDs, satellites, etc. At this stage, oppnets start to differ from typical networks.

Detected systems are identified and evaluated for their usefulness and dependability as candidates for joining the oppnet. Best ones are invited to the expanded oppnet. A candidate can accept or reject the invitation (but it might be ordered to join during disaster response operations). Upon accepting the invitation, a candidate is admitted into the oppnet, becoming its helper. The resources of the helper are integrated with the oppnet, and oppnet's tasks can be offloaded to or distributed among this and all other helpers (collaborative processing).

A decentralized command center—either augmenting human operators or fully autonomous—presides over the operations of the oppnet throughout its life. If the oppnet needs more resources to achieve its goal, the process repeats, and once the goal of the oppnet has been achieved, the helpers are restored and released.

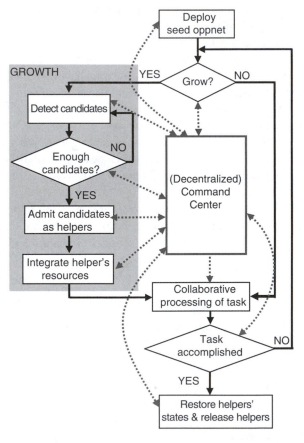

Figure 17.3 Basic operations of an oppnet. (From Z. H. Kamal, A. Gupta, L. Lilien, and Z. Yang, *Conference: Proceedings The 3rd International Conference on Collaborative Computing: Networking Applications and Worksharing*, White Plains, New York, November 12–15, 2007.)

17.2.3 Oppnet Virtual Machine: Standard Implementation Framework for Oppnet Applications

The OVM is a part of the oppnet. The goal of OVM project is to propose a standard for implementation of oppnets. The standard will facilitate implementations from different software vendors and will assure their interoperability.

OVM will allow developing and marketing standard library routines and application programming interfaces (APIs) to be used for implementing all kinds of oppnet-based applications. OVM will not only facilitate application development but will also assure interoperability among different oppnet implementations and third-party oppnet products.

A list of goals for OVM includes the following:

■ Design an application programming interface
  – Semantics of the interface should be language independent
    • Allow convenient C/C++/Java bindings for the interface
  – Provide extensions that allow greater flexibility
  – Can be implemented on many vendor platforms
  – Can be used in a heterogeneous environment
■ Allow efficient communication
  – With uniformed data format
  – Assume a reliable communication interface: the user need not cope with communication failures. Such failures are dealt with by the underlying communication subsystem

The OVM primitives are intended for use by all those who want to write programs in C/C++/Java/C++/C# for oppnet seeds or oppnet-enabled devices. This includes individual application programmers, manufactures of hardware devices, and creators of environments and tools. To be attractive to this wide audience, the standard must provide a simple, easy-to-use interface. However, this standard does not specify

■ Program construction tools
■ Debugging facilities
■ Support for task management
■ Underlying mechanism for communication

Oppnet features that are not included in the OVM standard can always be offered as extensions by specific implementations.

The Appendix includes a detailed list of primitives, divided into categories. It is summarized in Tables 17.1 through 17.4. The procedures for oppnet CC, seeds, helpers, and lites have prefixes "CTRL_," "SEED_," "HLPR_," and "LITE_," respectively.

Separate primitives for the four node classes help preventing situations when a node attempts to play a role of a node from another node class. The two main advantages of having distinct primitive classes are

■ Better security. Seed nodes have higher clearance level than helpers, which in turn have higher clearance level than lites. (Within each class, clearance sublevels can be defined.) Extra class-based layers in security mechanisms facilitate addressing security concerns more efficiently.
■ Resource savings. Most helpers and lites have quite limited resources. By knowing the limitations of the roles they can play, we can install on them only the relevant partial virtual

Table 17.1   Partial List of OVM Primitives for CC Nodes

Name of the Primitive	Functions of the Primitive
CTRL_initiate	Initiate oppnet
CTRL_terminate	Terminate oppnet
CTRL_command	Send commend to seed nodes

Source: From L. Lilien, A. Gupta, and Z. Yang, Proceedings of the 1st International Workshop on Next Generation Networks for First Responders and Critical Infrastructure (NetCri07), New Orleans, Louisiana, April 2007.

Table 17.2   Partial List of OVM Primitives for Seed Nodes

Name of the Primitive	Functions of the Primitive
SEED_scan	Scan communication spectrum to detect devices that could become candidate helpers
SEED_discover	Discover candidate helpers with a specific communication mechanism
SEED_listen	Receive and save messages in buffer
SEED_validate	Verify the received command
SEED_isMember	Checks if a device is already an oppnet node (oppnet member)
SEED_evaluateAdmit	Evaluate a device and admit it into oppnet if the device meets criteria for admittance
SEED_sendTask	Send a task to other oppnet device
SEED_delegateTask	Delegate a task that requires a permission from the delegating entity
SEED_release	Release a helper when no longer needed
SEED_processMsg	Process a message from buffer
SEED_report	Report information to control center/coordinator
SEED_update	Update a device in the oppnet with new expectations
SEED_receiveTask	Receive task from control center or another seed
SEED_wait	Wait for a certain amount of time before taking another action
SEED_barrier	Block the caller until all devices specified in the input parameter have called it

Source: From L. Lilien, A. Gupta, and Z. Yang, Proceedings of the 1st International Workshop on Next Generation Networks for First Responders and Critical Infrastructure (NetCri07), New Orleans, Louisiana, April 2007.

Table 17.3 Partial List of OVM Primitives for Helpers

Name of the Primitive	Functions of the Primitive
HLPR_isMember	Test if a helper is already a member of oppnet
HLPR_joinOppnet	Join oppnet
HLPR_scan	Scan communication spectrum to detect devices that could become candidate helpers (regular or lites)
HLPR_discover	Discover candidate helpers with a specified communication mechanism
HLPR_validate	Verify the received command
HLPR_switchMode	Switch between helpers' regular application and oppnet application
HLPR_report	Send information/data to specified device
HLPR_selectTask	Select a task from the task queue to execute
HLPR_listen	Receive message and save it
HLPR_evaluateAdmit	Evaluate a candidate helper and admit it into oppnet if it meets criteria defined by oppnet
HLPR_runApplication	Execute application indicated by authorized oppnet seed or helper node
HLPR_release	Release a helper (unless delegated a release task, a helper H can release only helpers admitted by H)
HLPR_processMsg	Process a message from buffer
HLPR_sendData	Send information/data to specified authorized oppnet node
HLPR_leave	Inform a seed that the caller will quit oppnet
HLPR_strongTask	Respond to the request sent from device and express the willingness to join oppnet. By accepting this task, the device will abort previous task
HLPR_weakTask	Respond to the request sent from device and express the willingness to join oppnet. By accepting this task, the device will put the task in a queue
HLPR_assignStrongTask	Assign tasks to a device. If accepted, the task will interrupt the previous task at the device
HLPR_assignWeakTask	Assign tasks to a device. If accepted, the task will be queued

Source: From L. Lilien, A. Gupta, and Z. Yang, Proceedings of the 1st International Workshop on Next Generation Networks for First Responders and Critical Infrastructure (NetCri07), New Orleans, Louisiana, April 2007.

Table 17.4 Partial List of OVM Primitives for Lites (Lightweight Helpers)

Name of the Primitive	Functions of the Primitive
LITE_isMember	Test if a lit is already a member of oppnet
LITE_joinOppnet	Join oppnet
LITE_validate	Verify the received command
LITE_switchMode	Switch between lites' regular application and oppnet application
LITE_report	Send information/data to specified device
LITE_selectTask	Select a task from the task queue to execute
LITE_listen	Receive message and save it
LITE_runApplication	Execute application indicated by authorized oppnet seed or helper node
LITE_processMsg	Process a message from buffer
LITE_sendData	Send information/data to specified authorized oppnet node
LITE_leave	Inform a seed that the caller will quit oppnet
LITE_strongTask	Respond to the request sent from device and express the willingness to join oppnet. By accepting this task, the device will abort previous task
LITE_weakTask	Respond to the request sent from device and express the willingness to join oppnet. By accepting this task, the device will put the task in a queue

Source: From Z. H. Kamal, A. Gupta, L. Lilien, and Z. Yang, *Conference: Proceedings The 3rd International Conference on Collaborative Computing: Networking Applications and Worksharing*, White Plains, New York, November 12–15, 2007.

machines. For example, a lite will not be burdened with the tasks of discovering other helpers or lites, thus eliminating the need to install on it OVM components needed for scanning, discovery, etc.

17.2.3.1 Using OVM Primitives in Oppnet Application Scenarios

In this section, we show examples that how a programmer can develop oppnet applications based on the primitives defined in the preceding sections.

The following simple scenario illustrates an oppnet application. In a natural disaster area, one priority is to find survivors caged in houses and cut off by earthquake, hurricane, or flooding. After the oppnet seed is deployed, the oppnet will expand by admitting helpers and lites. Among others, the motion sensors embedded in BT-equipped smoke detectors will become lites. If any such lite detects any movement, data will be transmitted to oppnet coordinators. The sequence chart of such a scenario is displayed in Figure 17.4. It shows how seed nodes obtain information from a lite via a helper. (The lite runs a small motion detection application.)

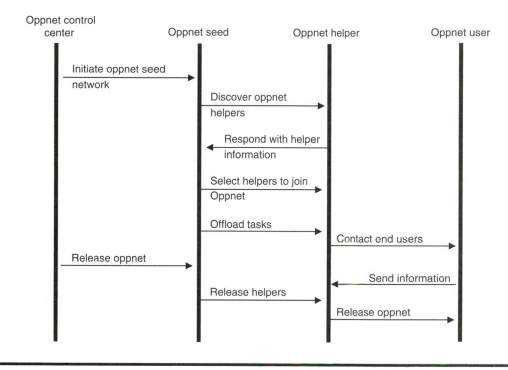

Figure 17.4 A sequence chart for an oppnet OVM use scenario. (From L. Lilien, A. Gupta, and Z. Yang, Proceedings of the 1st International Workshop on Next Generation Networks for First Responders and Critical Infrastructure (NetCri07), New Orleans, Louisiana, April 2007.)

The pseudocodes for nodes in the oppnet scenario illustrated in Figure 17.4 are shown in Figures 17.5 through 17.7. Oppnet control center is a reactive system that responds to human commands and processes data reported from oppnet seeds. (Due to space constraints, the pseudocodes have been significantly simplified to serve illustrative purposes only.)

The nodes of an expanded oppnet, including seed nodes, helpers, and lites, keep listening to commands from the oppnet's (distributed) control centers or other authorized nodes (e.g., a helper can accept tasks from another helper). When a command C is received by a node, the node first verifies C. The verification may include (1) checking sender's access rights, (2) checking security level of C, and (3) estimating resources needed to carry out C. The command will be executed if it passes the checks. Oppnet helpers and lites perform their daily activities until they are called upon to join an oppnet, in which case they switch to the defined emergency mode. When a command is received by a node, the node will also verify if it has the ability to execute the task. For example, a command that requires connection via Wi-Fi cannot be run by a device with only BT connectivity.

17.2.4 Example Oppnet Applications and Use Scenarios

Oppnets, as most nontrivial technologies, can be malevolent—deployed to harm humans, their artifacts, and technical infrastructure they rely upon [3]. Opportunistic networks are an example of developing application-oriented specializations of ad hoc networks [6].

Opportunistic networks can be extended to opportunistic systems by providing application-oriented primitives above the network layer. This section discusses the various oppnet application and scenarios [9].

```
repeat on command received from control center or other
authorized device
        SEED_validate(command);

        switch (command)
         case "scan":
            SEED_scan(…);

         case "BT (Bluetooth) discover":
            SEED_discover(BT,…);
            SEED_listen(…);

            for each responding BT device D do
               if (not SEED_isMember(D,…))
                  SEED_evaluateAdmit(D,…);
               end if
            end for

            if need more BT helpers
               for each H in subset of regular helpers do
                  SEED_delegateTask(H,"get BT helpers",…)
               end for
            end if
         case "send tasks":
            for each H in subset of helpers do
               SEED_sendTask(H, command,…);
            end for
         case "report":
            for each message M in buffer do
               SEED_processMessage(M);
            end for

            SEED_report(…);
         ...
        end_switch
    end_repeat
```

Figure 17.5 Pseudocode for seed nodes in the scenario. (From L. Lilien, A. Gupta, and Z. Yang, Proceedings of the 1st International Workshop on Next Generation Networks for First Responders and Critical Infrastructure (NetCri07), New Orleans, Louisiana, April 2007.)

17.2.4.1 Characteristics of Oppnet-Based Applications

Use of oppnets is most beneficial for applications or application classes characterized by the following properties:

- It can start with a seed
- It requires high interoperability
- It uses highly heterogeneous software and hardware components
- It can benefit significantly from leveraging diverse resources of helpers
- It is able to maintain persistent connectivity with helpers once it is established

The Standard Implementation Framework, OVM, for oppnets [8] (mentioned above) will facilitate creating oppnet-based applications by providing a standard set of primitives. The primitives for use by application components will, for example, facilitate discovering potential helpers, integrating them, and releasing them when they are not needed any more.

```
repeat on command received from control center or other
                authorized device

        HLPR_validate (command);

        switch (command)

         case "join oppnet":
            HLPR_switchMode (…);
            HLPR_joinOppnet (…);

         case "detect motion":
            HLPR_runApplication (motion,…);
            HLPR_sendData (…);

         case "get BT (Bluetooth) helpers":
            HLPR_scan (BT,…);
            HLPR_discover (BT,…);
            HLPR_listen (…);

            for each responding BT device D do
                if (not HLPR_isMember (D,…))
                    HLPR_evaluateAdmit (D,…);
                    HLPR_report (…,BT, D);
                end if
            end for

         case "report":
            for each message M in buffer do
                HLPR_ processMessage (M,…);
            end for

            HLPR_report (…);

         case "leave oppnet":
            HLPR_leaveOppnet (…);
            HLPR_switchMode (…);
         ...

        end_switch
    end_repeat
```

Figure 17.6   Pseudocode for helpers in the scenario. (From L. Lilien, A. Gupta, and Z. Yang, Proceedings of the 1st International Workshop on Next Generation Networks for First Responders and Critical Infrastructure (NetCri07), New Orleans, Louisiana, April 2007.)

17.2.4.2   Example Oppnet Application Classes

We can envision numerous applications and application classes that can be facilitated by oppnets. Some of them are described next [11,13].

17.2.4.2.1   Emergency Applications

We see important applications for oppnets first of all (not exclusively) in all kinds of emergency situations, for example, in response to man-made or natural disasters in the area of homeland security [12]. Oppnets can significantly improve effectiveness and efficiency of such operations.

For predictable disasters (like hurricanes), seed oppnets can be put into action and their buildup started (or even completed) before the disaster, when it is still much easier to locate and invite other nodes and clusters into the oppnet. The first invited helpers could be the sensornets deployed for structural damage monitoring and assessment in buildings, roads, and bridges.

```
repeat on command received from control center or other
authorized device

        LITE_validate(command);

        switch (command)
         case "join oppnet":
            LITE_switchMode(…);
            LITE_joinOppnet(…);

         case "detect motion":
            LITE_runApplication(motion,…);
            LITE_sendData(…);

         case "report":
            for each message M in buffer do
               LITE_ processMessage(M,…);
            end for

            LITE_report(…);

         case "leave oppnet":
            LITE_leaveOppnet(…);
            LITE_switchMode(…);
         ...

        end_switch
    end_repeat
```

Figure 17.7 Pseudocode for lites in the scenario.

17.2.4.2.2   Home/Office Oppnet Applications

Oppnets can benefit home/office applications by utilizing resources within the domestic/office environment to facilitate mundane tasks. Consider contrast between the two scenarios for viewing a visual message on a PDA in a living room. Without an oppnet-based software, personal digital assistant (PDA) has to present the message using the miniscule PDA screen and its substandard speakers. With an oppnet-based software, PDA (now being a single-node seed oppnet) can quickly find helpers: a TV monitor and an audio controller for hi-fi speakers available in the living room. PDA can ask these helpers to join, and integrate them into an expanded oppnet. The expanded oppnet, now including three nodes (the PDA, the TV monitor, and the audio controller), can present the visual message on high-quality devices.

A similar scenario can be realized in mobile ad hoc networks (MANETs) [14] but with much more programmer's efforts because MANETs do not provide high-level application-oriented primitives to simplify implementation. Only oppnets do [8].

17.2.4.2.3   Benevolent and Malevolent Oppnet Applications

Oppnets, as most nontrivial technologies, can be malevolent—deployed to harm humans, their artifacts, and technical infrastructure they rely upon. Invited nodes might be kept in the dark about the real goals of their host oppnets. Specifically, good guys could be cheated by a malevolent oppnet and believe that they will be used to benefit users. Similarly, bad guys might be fooled by a benevolent oppnet into believing that they collaborate on objectives to harm users, while in fact they would be closely controlled and participate in realizing positive goals.

On the negative side, home-based opportunistic networks could be the worst violators of individual's privacy, if they are able to exploit point coordinators (PCs), cell phones, computer-connected security cameras, embedded home appliance processors, etc.

17.2.4.2.4   Predator Oppnets

To counteract malevolent oppnets threats, predator networks that feed on all kinds of malevolent networks—including malevolent oppnets—can be created. Using advanced oppnet capabilities and primitives, they can detect malevolent networks, plant spies (oppnet helpers) in them, and use the spies to discover true goals of suspicious networks.

Their analysis must be careful, as some of the suspicious networks might actually be benevolent ones, victims of false positives. Conversely, intelligent adversaries can deploy malevolent predator networks that feed on all kinds of benevolent networks, including benevolent oppnets.

17.2.4.3   Example Oppnet Application Scenarios

We now discuss two example oppnet application scenarios: a benevolent one and a malevolent one. Both rely on some reconfiguration capabilities of non-opportunistic (regular) sensornets.

17.2.4.3.1   Benevolent Oppnet Scenario: Citizens Called to Arms

A seed oppnet is deployed in the area where an earthquake occurred. It is an ad hoc wireless network with nodes much more powerful than in a typical ad hoc network (more energy, computing and communication resources, etc.). Once activated, the seed tries to detect any nodes that can help in damage assessment and disaster recovery. It uses any available method for detection of other networks, including radio-based detection (including use of software defined radio and cell phone–based methods), searching for nodes using the IP address range for the affected geographic area, and even AI (Artificial Intelligence)-based visual detection of some appliances and PCs (after visual detection, the seed still needs to find a network contact for a node to be invited).

The oppnet "calls to arms" the optimal subset of detected and contacted citizens, inviting all devices, clusters, and entire networks, which are able to help in communicating, computing, sensing, etc. In emergency situations, entities with any sensing capabilities (whether members of sensornets or not), such as cell phones with global positioning system (GPS) or desktops equipped with surveillance cameras, can be especially valuable for the oppnet.

Let us suppose that the oppnet is able to contact three independent sensornets in the disaster area, deployed for weather monitoring, water infrastructure control, and public space surveillance. They become helper candidates and are ordered (this is a life-or-death emergency) to immediately abandon their normal daily functions and start assisting in performing disaster recovery actions. For example, the weather monitoring sensornet can be called upon to sense fires and flooding, the water infrastructure sensornet with multisensor capabilities (and positioned under road surfaces)—to sense vehicular movement and traffic jams, and the public space surveillance sensornet—to automatically search public spaces for images of human victims.

17.2.4.3.2   Malevolent Oppnet Scenario: Bad Guys Gang Up

Suppose that foreign information warriors use agents or people unaware of their goals to create an apparently harmless weather monitoring sensornet. Only they know that the original sensornet becomes a seed of a malevolent oppnet when activated. The sensornet starts recruiting helpers.

The seed does reveal its true goals to any of its helpers. Instead, it uses a cover of a beneficial application, proclaiming to pursue weather monitoring for research. Actually, this opportunistic sensornet monitors weather but for malicious reasons: it analyzes wind patterns that can contribute to a faster spread of poisonous chemicals. Once the critical mass in terms of geographical spread and sensing capabilities is reached, the collected data can be used to make a decision on starting a chemical attack.

17.3 Related Technologies

This section includes an overview of related technologies followed by a qualitative comparison with oppnets.

17.3.1 Overview of Related Technologies

In this section, we review networks designed primarily for resource-sharing, such as peer-to-peer (P2P) and grid; those designed for monitoring and control via wireless sensor networks and the more conventional networks developed to achieve and maintain connectivity through MANETS, mesh, ambient networks (AN), and their like. We also briefly overview the opportunistic network paradigm that facilitates more pervasive networking scenarios.

17.3.1.1 Peer-to-Peer (P2P) Networks

P2P is a subclass of distributed networks, where resource (data, bandwidth, or computing power) sharing is achieved by direct exchange between the peers, rather than depending solely on centralized servers. On the basis of the level of intermediation of central servers, P2P networks can be broadly divided into pure P2P networks—those which have no central server or routers; and hybrid P2P networks—those which depend partially on central servers and routers to manage the network. In pure P2P networks, the role of clients and servers is completely merged making peers with equal standing which can act as a server, client, or both at the same time. In hybrid P2P networks, central servers—called supernodes or strong nodes—are used to manage information about data on peers and respond to requests for that information.

Oppnets follow similar decentralized pattern, where oppnet-enabled devices can act as a service provider, or as a user, or as both depending on its position in a particular oppnet hierarchy. What distinguishes oppnets from P2P systems is the implicit heterogeneity of oppnets due to which all the peers cannot be of equal standing. Devices with limited resources and capacity (lites) such as sensing devices can perform very limited functions and can only be ordered to perform their typical task. Seeds, on the other hand, having more communicative and computing powers, may be providing entirely different range of functionalities than helpers or lites. In addition, in P2P, exchange of resources between peers is the primary goal, whereas commonality of the goal for an entire oppnet demands much deeper coordination and collaboration among the nodes. To grow, oppnets may follow similar methods as P2P networks, that is, grow by joining; however, the growth of a P2P network is measured by the number of nodes joining the network, whereas the growth of an oppnet is determined by the enhancement in the potential of the oppnet to solve the problem at hand.

17.3.1.2 Grid Computing

In the beginning, grid computing was an effort to integrate various super computers around the world but now this term has much wider implications. A computational grid basically unifies distributed computers to a single computing resource, providing users with a transparent access to the entire set of resources [15]. Users have access to the complete resource pool but not to the individual peers. IBM defines grid computing as the ability, using a set of open standards and protocols, to gain access to applications and data, processing power, storage capacity, and a vast array of other computing resources over the Internet [16]. "A grid is a type of parallel and distributed system that enables the sharing, selection, and aggregation of resources distributed across 'multiple' administrative domains based on their (resources) availability, capacity, performance, cost and users' quality-of-service requirements" [17]. The term grid as used here is indicative of an analogy with electrical grids, which provide dependable and transparent access to electricity irrespective of its origin.

Oppnets share the following characteristics with grids [15]:

- Multiple administrative domains and autonomy: Resources in an oppnet, similar to grids, may be owned by different administrative domains or different organizations. Even if they agree to be the part of an oppnet, the autonomy of these resource owners must be honored and their local resource management and usage policies must be taken care of.
- Scalability: As the size of grid (and oppnet) network grows, the problem of potential performance degradation may arise. Consequently, applications that require a large number of geographically located resources must be designed to deal with latency and bandwidth problems.
- Dynamicity or adaptability: Distributed networks like grid or oppnet depend heavily on foreign resources so probability of resource failure may be high. As such, applications must tailor their behavior dynamically and use the available resources and services efficiently and effectively.
- CPU-scavenging: It is the process of creation of grid from the unused resources in a network of participants in a grid network. To reduce burden on the volunteer helpers, oppnets must follow similar resource utilization policy.

Although oppnets share many features with grids, the priorities and context of application for the two networks differ. While Grids are designed to focus primarily on computationally intensive operations and work in homogeneous environment, oppnets are being developed to predominantly deal with real-life situations in a physical world and collaborate in heterogeneous environment. As a direct implication of this, while grids allow remote sites to join or leave the environment whenever they choose, oppnets may force candidate helpers to join the network in emergency situations.

17.3.1.3 Wireless Sensor Networks (Sensornets)

Wireless sensor networks usually consist of hundreds or thousands of tiny sensor nodes deployed across a geographical area to collectively monitor physical or environmental conditions. The nodes are small devices having one or more sensing capabilities but with very small memory and processing powers. They are also equipped with communication capabilities, usually a radio transceiver, and a power source, usually a battery. They collectively and collaboratively collect and process information, and forward it to base stations. Sensor networks are now increasingly used in environmental, military, health, and home applications [13].

In typical situation, nodes are scattered in random fashion over the area to be monitored. Although this allows for the deployments in inaccessible terrains, it requires that the protocols and algorithms have to be designed to provide self-organizing capabilities. Owing to the limitation of power sources, which may be irreplaceable, algorithms also tend to focus on high power conversion power management, to prolong network lifetime. Apart from these constraints, other important factors affecting the design of sensor networks are fault tolerance, scalability, operating environment, transmission media, etc. [13].

Although lot of current research is going on in this area, little has been done to combine sensor networks with other existing networks. Oppnets have interfaces to sensor networks, which promise to integrate sensing capabilities to the applications when needed.

17.3.1.4 MANETs-Mobile Ad Hoc Networks

MANETs consist of mobile nodes that also support routing (to cooperatively make up for the absence of fixed routing infrastructures). Transient connections are established among nodes within range and may be broken down without any prior notice or consent of all parties. Because of the dynamicity and unpredictably introduced by rapidly changing topology of the network, coupled with absence of any centralized authority, distributed operation and continuous self-configuration become an essential characteristic of MANETs [18]. Because of a similar distributed nature and helper participation, continuous self-configuration is also essential for oppnets. In fact, the nature of self-management in oppnets may be more challenging as the goals of the nodes are much more than just facilitate routing. Yet another feature where oppnets differ from traditional ad hoc networks is the level of heterogeneity. Though these networks enable heterogeneous devices or networks to communicate with each other via the common largest network, the Internet, they do not enable devices in heterogeneous communication media to communicate with each other.

17.3.1.5 Mesh Networks

Mesh networks [19,20] are similar to ad hoc networks but these are supported by an infrastructure backbone provided by stationary but wireless routers. As in the case of MANETs, mesh networks are also self-configuring and self-healing. Multiple paths ensure that mesh networks do not have to depend on any single communication link. Even if some of the connections are broken, a mesh network can still operate forming a reliable networking system. As a result, a very reliable network is formed [21].

Mesh networks can either employ full mesh topology, where each node has direct connections with all other nodes; or partial mesh topology, where every node is connected only to the nodes with which they exchange most of their data.

Typically, these networks have two kinds of nodes: mesh routers and mesh clients. Although clients can also perform routing, separation of clients and routers, added by immobility of routers, make the design of protocols much simpler and cost-effective as compared with MANETs. However, building a large-scale routing backbone is often a challenge due to scalability problems. In spite of the scalability issues, additional reliability provided by mesh networking makes it a good candidate for utilization in emergency oppnet applications.

17.3.1.6 Ambient Networks

ANs is a European Commission sponsored project, which aims to develop a software-driven network infrastructure that will run on top of all current or future physical network infrastructures to provide a way for devices to connect to each other, and through each other to the outside world [22]. As opposed to the networking technologies mentioned above, ambient technologies do not deal with the communication link and interfaces between individual nodes but with the interfaces at the underlying network technology boundaries. The moment network boundaries are encountered, interfaces are realized by instant negotiation of agreements based on preconfigured policies.

The way ANs aim to provide end-to-end communication capabilities in heterogeneous internetworking environments may resemble how oppnets try to integrate devices, networks, or systems and manage oppnet resources through a distributed command center. However, the following features distinguish the two efforts [22–25]:

- AN is a global, universal network intended as a replacement for the Internet (beyond 3G) and all communication networks, whereas the oppnet is a local/wide area network meant to serve specific applications.
- AN requires heavyweight primitives whereas oppnet requires no or only lightweight primitives.
- AN is completely predesigned, AN is aware of the location of all sub-ANs, all its facilities are built-in or add-on, and only networks that have the needed primitives can be composed into ANs, whereas oppnet is mostly ad hoc system that has to discover helper devices.
- ANs contact each other so that any sub-AN can initiate connections, whereas oppnets have a mechanism where the seed oppnet nodes initiate discovery of devices.

17.3.1.7   Delay Tolerant Networks

Delay tolerant networking (aka. disruption tolerant networking) aims to improve connectivity between regional networks when connectivity is not continuous and prone to disruptions leading to large delays. Delay Tolerant Networks (DTN) started as an effort to deal with delays in the interplanetary Internet (floating nodes in space), where large distances cause much larger delays than in the Earth-bound Internet, and links are disrupted for minutes or hours. The idea has been extended to other networks having similar characteristics, for example, terrestrial mobile networks, military battlefield networks, etc.

DTNs overcome the problems associated with large delays by adapting store-and-forward message switching [26]. DTNs also enable interoperability between different regional networks having different characteristics by providing interfaces. The storage places are capable of holding the messages for indefinite periods as opposed to holding messages for just a few milliseconds in Internet routers. In oppnets, similar delays are expected due to two reasons. Firstly, due to high heterogeneity of the oppnet, communication capabilities of different devices may be at diverse levels. Secondly, helpers may get disconnected due to their own constraints and workloads.

17.3.1.8   Class 1 Opportunistic Networks

Class 1 opportunistic networks (e.g., Ref. [27]) can be viewed as a generalization of the MANET paradigm, in which the assumption of complete paths between data senders and receivers is relaxed [28]. This enables stations to communicate in disconnected environments, in which islands of connected stations appear, disappear, and reconfigure dynamically [28].

In class 1 opportunistic networks, there is no notion of utilizing resource of the nodes in the network to perform a network task. In contrast, in class 2 opportunistic networks, the network not only provides a communication backbone but also can provide computing, sensing, actuating, storage, or other resources or services. Also, oppnets can grow dynamically by admitting needed helpers, which facilitates execution of more challenging tasks. Such tasks are either beyond capabilities of traditional networks, or are much more difficult to achieve even in class 1 opportunistic networks. As shown in Section 17.2.3, oppnets provide relatively high-level primitives to facilitate building of complex applications.

Class 1 opportunistic networks are a proper subset of class 2 opportunistic networks or oppnets. In turn, class 1 opportunistic networks can be viewed as encompassing DTNs.

17.3.1.9 Spontaneous Networks

Spontaneous networking is a relatively new area of research focusing on a small subspace of ad hoc networks. The aim of the network is not just providing connectivity but also supporting the collaborative activity of a group of devices supported by wireless communication [29]. They resemble oppnet in some of their features, including heterogeneity of nodes and collaboration among the participants. However, hierarchy of nodes in oppnets and administrative capabilities of seed nodes is not found in spontaneous networks. Also, while spontaneous networks are based on the physical proximity of a restricted number of nodes located nearby each other, oppnets may grow considerably according to the needs of their tasks.

17.3.2 Qualitative Comparison of Related Technologies with Oppnets

In this section, we present Tables 17.5 and 17.6 (some technologies previously compared in Ref. [7]) that present a comparison of features of the discussed related technologies with oppnets.

17.4 Design and Implementation of the MicroOppnet

In this section, we present and discuss the design and implementation of MicroOppnet v.2.2, referred to below simply as the MicroOppnet. The MicroOppnet was developed as a proof of concept and a test bed for oppnets. It is a small-scale prototype that integrates UMA spectrums of BT, wireless Internet, and wireless sensor networks. In the following sections, we discuss the workings of the MicroOppnet and its design and implementation.

17.4.1 Overview of the MicroOppnet

The current version of the MicroOppnet, is a small-scale proof of concept and test bed for class 2 opportunistic networks, because it not only allows opportunistic communications but also opportunistically accesses sensornet nodes (SNNs) to perform sensing. It is, though, rudimentary in its class 2 opportunism, hence the prefix "micro" in its name.

The MicroOppnet is a platform on which functional parameters, such as oppnet components (including OVM primitives), protocols, and architectures are or will be implemented, tested, and fine-tuned. Nonfunctional parameters including quality of service (QoS) parameters, such as throughput, delay, reliability, accuracy, scalability, etc., can also be investigated on the MicroOppnet.

The seed oppnet in the MicroOppnet (Figure 17.8) consists of workstation A with a BT adapter and a serial port connection to sensornet base station (SBS) BS_1. The seed searches for BT devices and initiates a connection with them. Alternatively, a BT-enabled device—a cell phone labeled victim in our example—can find the seed and initiate a connection Once a connection has been established, the victim cell phone can send a message to the seed, for example, the help message. This message is then forwarded via base station BS_1, and then through the sensor network. In

Table 17.5 Comparison of Features of Selected Networks

Feature	Subfeatures (if any)	P2P Systems [8,14,24,41]	Computational Grids	Sensornets [22,30]	MANETS [7,31,32]	Mesh Networks [13,33]
Distinguishing features (w.r.t. other ad hoc nets)		Domination of peer nodes (over clients or servers), resource aggregation	Lightweight nodes with sensors, energy-constrained, densely deployed	Virtual pool of resources, users have access or knowledge of the pool and not of nodes	Rapidly changing topologies, lack of centralized entity	Some hosts are also routers, lack of centralized entity, very reliable
Deployment	Rapid	yes	Yes	Yes	Yes	Yes
	Incremental	in principle, yes	Yes	Possible	Yes	Yes
Configuration	Limits on minimal required configuration	Starts with a seed	Coordination among nodes required	No	No	No
	Nodes join/leave	Often (even w/o warning)		No	Often (even w/o warning)	Infrequently (once established)
Operation	Self-organizing	Yes	Yes	Yes	Yes	Yes (routers)
	Standalone/connected to Internet	Yes/possible	Possible/yes	No (requires remote task manager)/yes	Yes/possible	Yes/possible
	Centralized entities	Some (e.g., DNS)	In the form of administrative hierarchy	Sinks or base stations, gateways	none	none
	Resource aggregation	Yes	Yes	No	No	No

(continued)

Table 17.5 (continued) Comparison of Features of Selected Networks

Feature	Subfeatures (if any)	P2P Systems [8,14,24]	Computational Grids	Sensornets [22,30]	MANETS [7,31,32]	Mesh Networks [13,33]
Node types		Pure peer nodes, supernodes or super-peers and client peers	Base station or sink, sensor nodes (possibly with routing capabilities)	Computational devices creating pool of computational resource	Mobile, stationary (few)	Mesh routers, mesh clients, conventional clients
Node characteristics	Lightweight nodes	No	No	Mostly	Possible	As clients
	Software heterogeneity	Low	No	No	Low	Low
	Hardware heterogeneity	High	Possible	No	High	High
	Limited energy	No	No	Yes	Some	Possible
Node examples		Desktop, laptop	Mote, Internet gateway node	Super, cluster and ordinary computers/laptops, PDA	Pedestrians, soldiers, unmanned robots, vehicles, buildings	Desktop, laptop, PDA, Wi-Fi IP phone, RFID reader
Node mobility	Stationary nodes	Yes	Possible	Yes	Possible	Possible
	Mobility of combined hosts/routers	D.N.A.	No	No	High	D.N.A.

	Mobility of separate hosts (clients or peers)	Possible	Possible	D.N.A.	D.N.A.	Yes
	Mobility of separate routers	D.N.A.	Possible	D.N.A.	D.N.A.	Minimal
Communication	Wireless/wired	Yes/yes	Yes/yes	Yes/gateways to wired	Yes/possible	Yes/possible
	Limited bandwidth for some nodes	Yes	No	Yes	Yes	Yes
	Persists once established	No	Yes	Yes	No	Yes
	Connection to Internet	Typical	Mostly	Possible, via base station	Possible	Possible
	Limited node transmission radius	Yes if wireless	Yes	Yes	Yes	Yes (wireless mesh routers)
	Most communication between nearby nodes	Yes	No	Yes	Yes	Yes

(continued)

Table 17.5 (continued) Comparison of Features of Selected Networks

Feature	Subfeatures (if any)	P2P Systems [8,14,24]	Computational Grids	Sensornets [22,30]	MANETS [7,31,32]	Mesh Networks [13,33]
	Routing by	By underlying network (e.g., Internet)	By underlying network	Usually, by base stations or cluster heads; by sensor nodes also possible	All nodes	By mesh routers
	Interoperability	No	Possible	Usually via gateways	Possible	Possible
Topology	Arbitrary	Yes	Yes	Yes	Yes	Yes
	Time varying	Highly	Yes	Yes (mostly due to using up energy, jamming or noise, moving obstacles)	Highly	Limited
	Typical size (in nodes)	Even millions	$k \times 10$–$k \times 1000$	Even millions	$k \times 100$–$k \times 1000$	Arbitrary
	Typical area covered	MAN, countrywide	LAN-like or MAN-like	LAN-like or MAN-like	MAN, countrywide	LAN, MAN

Source: From L. Lilien, Proceedings of the 1st International Workshop on Mobile and Ubiquitous Context Aware Systems and Applications (MUBICA 2007), Philadelphia, Pennsylvania, August 2007.

Note: D.N.A, does not apply.

Table 17.6 Comparison of Features of Selected Networks

Feature	Subfeatures (if any)	Ambient Networks	DTN	Class 1 Opportunistic Networks	Spontaneous Networks [4]	Class 2 Opportunistic Networks (Oppnets) [16,34]
Distinguishing features (w.r.t. other ad hoc nets)		No administrative infrastructure, human interaction and proximity leveraged for configuration	Interoperation of heterogeneous networks major goal, deals with technology boundaries	No administrative infrastructure, human interaction and proximity leveraged for configuration	Tactical resource aggregation, high software heterogeneity, high interoperability	No administrative infrastructure, human interaction and proximity leveraged for configuration
Deployment	Rapid	D.N.A.	Yes	Yes	Yes	Yes
	Incremental	D.N.A	Yes	Yes (from the seed)	Yes	Yes (from the seed)
Configuration	Limits on minimal required configuration		Should support packet switching	Starts with a seed	No	Starts with a seed
	nodes join/leave	D.N.A	Yes	Mostly during growth or dismantling	Infrequently (except at session/subsession beginning/end)	Mostly during growth or dismantling
Operation	Self-organizing	D.N.A	—	Yes	Yes	Yes
	Standalone/connected to Internet	No/yes	Yes/no	Yes/possible	Yes/no	Yes/possible

(continued)

Table 17.6 (continued)   Comparison of Features of Selected Networks

Feature	Subfeatures (if any)	Ambient Networks	DTN	Class 1 Opportunistic Networks	Spontaneous Networks [4]	Class 2 Opportunistic Networks (Oppnets) [16,34]
	Centralized entities	Some (e.g., DNS)	None	None or in the seed	None	None or in the seed
	Resource aggregation	No	No	No	No	Yes
Node types		Peer nodes	Pure peer nodes, supernodes or super-peers and client peers	Peer nodes	Seed nodes, helpers	Peer nodes
Node characteristics	Lightweight nodes	Possible	No	Possible	No	Possible
	Software heterogeneity	High	No	High	No	High
	Hardware heterogeneity	High	Possible	High	Possible	High
	Limited energy	No	Possible	Infrequent	No	Infrequent
Node examples		Laptop, PDA, high-end mobile phone	Does not deal with nodes	Laptop, PDA, high-end mobile phone	Desktop, laptop, PDA, Wi-Fi IP phone, mote, RFID reader	Laptop, PDA, high-end mobile phone

Node mobility	Stationary nodes	Possible	No	Yes	Yes (when in use)	Yes
	Mobility of combined hosts/routers	Possible	No	Yes	No (when in use)	Yes
	Mobility of separate hosts (clients or peers)	Possible	Yes	Yes	D.N.A.	Yes
	Mobility of separate routers	Possible	Yes	Yes	D.N.A.	Yes
Communication	Wireless/wired	Yes/yes	Yes/no	Yes/yes	Yes (short-range for session establishment, then regular)/unlikely	Yes/yes
	Limited bandwidth for some nodes	D.N.A	Yes (connection in bits)	Yes	Yes (session setup)/no (session)	Yes
	Persists once established	D.N.A	No	Yes	Yes	Yes
	Connection to Internet	Typical	No	Possible	Unlikely	Possible

(continued)

Table 17.6 (continued) Comparison of Features of Selected Networks

Feature	Subfeatures (if any)	Ambient Networks	DTN	Class 1 Opportunistic Networks	Spontaneous Networks [4]	Class 2 Opportunistic Networks (Oppnets) [16,34]
	Limited node transmission radius	Yes if wireless	Yes	Yes if wireless	Yes	Yes if wireless
	Most communication between nearby nodes	No	Not necessary	Maybe	Yes	Maybe
	Routing by	By underlying network	All nodes	By most nodes or by routers	All nodes	By most nodes or by routers
	Interoperability	Yes	No	High	No	High
Topology	Arbitrary	Yes	Yes	Yes	Yes	Yes
	Time varying	Yes	Possibly	Mostly during growth or dismantling	Mostly at the beginning or end of session or subsession	Mostly during growth or dismantling
	Typical size (in nodes)	Even millions	Few	Arbitrary	Small to medium	Arbitrary
	Typical area covered	Planetary	Interplanetary	MAN	One room	MAN

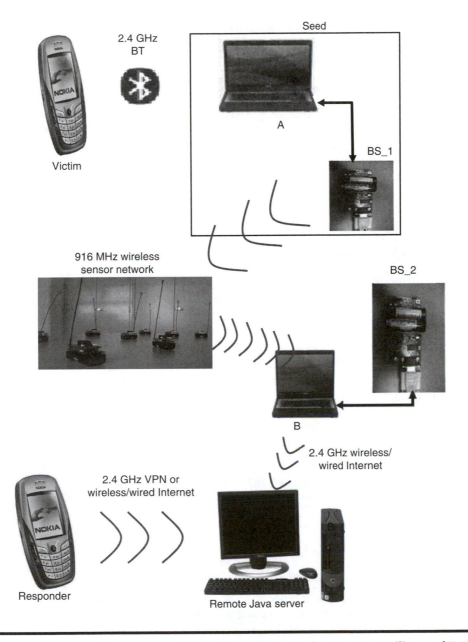

Figure 17.8 Structure of MicroOppnet v.2.2. (From Z. H. Kamal, A. Gupta, L. Lilien, and Z. Yang, *Conference: Proceedings The 3rd International Conference on Collaborative Computing: Networking Applications and Worksharing,* **White Plains, New York, November 12–15, 2007.)**

MicroOppnet the sensornet consists of ten Mica2 Motes and six Stargate sensornet gateways. Some of the gateways are also connected to Mica2 Motes.

Base station BS_2 at the other end of the sensornet is connected to laptop B. Once the help message is propagated via BS_2 to laptop B, a Java TCP/IP client socket connection is initiated with a remote Java server. The help message and the location of the device that sent it are logged on this server.

The Java server can be queried by remote users employing either traditional computing devices or Java-enabled devices. In our example, we employ cell phone with T-Mobile virtual private network (VPN) connection, labeled as responder.

The seed can broadcast to the sensornet a variety of messages in addition to help—for example, start_sensing, log_sensing, retrieve_log. The messages can be used, for instance, to start temperature sensing, to log temperature in the Electrically Erasable Programmable Read-Only Memory (EEPROM) of the sensor, or to retrieve the logged data from the sensor network. The retrieved temperature readings can be logged at the Java server. Then, they can either be queried by remote users via wireless Internet, or be broadcast by the seed on the BT channels.

The current version of the MicroOppnet integrates only three disparate communication media and frequency ranges, namely, BT (2.4 GHz), a sensor network (916/896/433MHz), and the wireless Internet 802.11b and 802.11g (2.4 GHz [34], the same frequency as BT).

17.4.2   Design of MicroOppnet

In this section, we present the flow of control for the MicroOppnet of Figure 17.8 in terms of the OVM primitives of Section 17.2.4. This flow of control, illustrated in Figure 17.9, can begin with (1) an active discovering of candidates—using SEED_discover; (2) with a passive wait—using SEED_listen, when candidates search for and initiate connection with the seed; or (3) with dispatching a task for the sensornet—using SEED_sendTask. In the MicroOppnet, communication for (1) and (2) is only over the BT medium.

Messages received from nodes wishing to use the MicroOppnet are processed, and tasks are delegated to the appropriate helpers. In the MicroOppnet, there are only two sets of helpers: the set of nodes in the sensornet, and the remote server.

Messages from a user such as victim in Figure 17.8 can be forwarded from the seed's SBS BS_1 to the helpers using the SEED_sendTask primitive. The nodes in the sensornet process the message using HLPR_processMsg and then perform the task (currently, only sensing or communication) using HLPR_runApplication. If the task is sensing, then the SNNs will start or stop sensing as required. Otherwise, they will forward either the received message or their temperature sensor readings as directed. When the message is received by another sensornet gateway or another base station (e.g., by BS_2), it is logged on a remote server. If the task was to retrieve sensor-measured temperature, then BS_2 aggregates sensornet readings and floods the result back through the sensornet to BS_1.

Devices such as responder (see Figure 17.8) can send the message retrieve_log to the remote helper server, which is in a listening mode with the HLPR_listen primitive. This allows the remote server's log to be queried for specific tasks and retrieve the appropriate messages. The server can process any TCP/IP socket connection.

Summarizing, the MicroOppnet supports only the following tasks: (1) communication tasks—flooding messages and retrieving sensor readings; (2) sensing tasks—starting and stopping sensing. All these tasks rely on opportunism. In more detail, the following is the exhaustive list of all tasks using resources opportunistically:

- Communication in the BT medium
- Communication in the sensornet medium
- Communication using TCP/IP in wired or wireless Internet
- Temperature sensing using SNNs

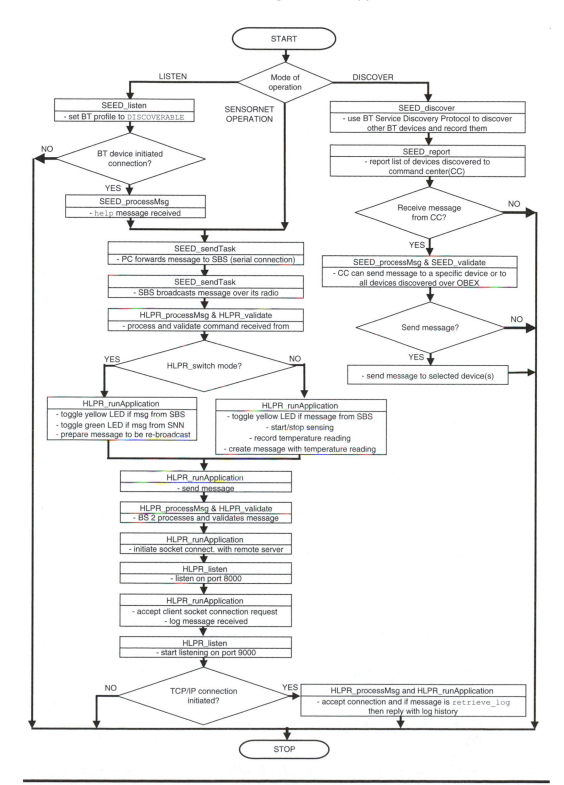

Figure 17.9 Flow of control in the MicroOppnet v.2.2. (From Z. H. Kamal, A. Gupta, L. Lilien, and Z. Yang, *Conference: Proceedings The 3rd International Conference on Collaborative Computing: Networking Applications and Worksharing,* **White Plains, New York, November 12–15, 2007.)**

The first three tasks use class 1 opportunism, and only the last task relies on class 2 opportunism—by leveraging the sensing resources of MicroOppnet helpers. Thanks to the last task, we can claim that MicroOppnet is a class 2 opportunistic network, albeit a rudimentary one (exploiting only one type of noncommunication resources).

17.4.3 Implementation Details for the MicroOppnet

A USB BT dongle equips the seed with a BT infrastructure. To exploit the BT Communication Framework, we use the BT Software Protocol stack provided by Atinav AveLink [15]. In this way, we can invoke the BT Service Discovery Protocol (SDP) using the API of the protocol stack to detect BT devices, and to either initiate connections with BT devices or to receive connections from BT devices.

The BT communication infrastructure consists of profiles that are built on top of layers/protocols to define further high-level functionality. There are numerous profiles that exist and, moreover, there are close dependencies between profiles. The lowest-level profile that most common BT profiles are dependent on is the generic access profile, which is used to establish a basic connection. After establishing an initial connection, we use generic object exchange (OBEX) profile, which uses the OBEX layer to exchange objects. Alternatively, we can use Logical Link Control and Adaptation Protocol (L2CAP) and RFCOMM Protocol (uses Serial Port Profile) for packet and stream data, respectively [34].

Our sensor network consists of Crossbow's Mica2 Motes and Stargate gateways [30]. The Mica2 Motes run UC Berkley's TinyOS [35] operating system, and are programmed with nesC [31]. The nesC code is compiled on a workstation and is flushed onto the Motes using Crossbow's programming boards.

The remote server, developed in Java using socket connections, runs on a Linux machine. Its flow of control is illustrated in Figure 17.10.

Cell phone programming is accomplished with Java MicroEdition (J2ME) and JSR-118 mobile information device profile (MIDP) 2.0 for resource-constrained devices, such as cell phones and PDAs. Java applications for such devices are called MIDlets. We use Java-enabled phones: Nokia 6600 (equipped with Symbian OS), Nokia 6103, and Motorola RAZR. Figure 17.11 illustrates the flow of control in the MIDlet of the victim and responder stations (see Figure 17.8). We found that Nokia 6600 is stronger than the other two models when it came to initiating BT connections with the seed or TCP/IP connections with the server.

Currently, the MicroOppnet v.2.2 uses MANET routing, in which every node in the oppnet is a router. We gained valuable insights on vulnerabilities of MANET routing in an opportunistic network. Consequently, we will scrutinize routing protocols developed for opportunistic networks and DTN, for example, the ones described in Refs. [3,8,9,33,36] to mention a few. We will evaluate them against the set of vulnerabilities and a number of very specific and low-level criteria that we have identified while implementing MicroOppnet and experimenting with it.

During the development of MicroOppnet, we learnt that Nokia 6600 was a much more robust cell phone (probably due to its Symbian OS), because MIDlets that created sockets or streams were not allowed on Nokia 6103 (locked by T-Mobile) and Motorola RAZR was not equipped to receive text messages over BT. We observed that T-Mobile WAP and GPRS connections do not allow unrestricted access to the Internet. Instead T-Mobile VPN was used to allow unrestricted MIDlet access to the Internet. Furthermore, we observed that the remote server should not be behind any firewall to allow MIDlet access to the server.

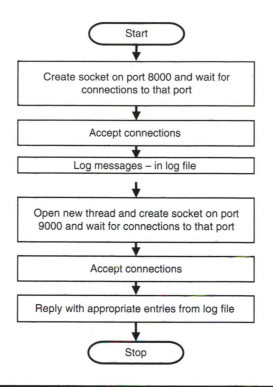

Figure 17.10 Flow of control for the remote Java server. (From Z. H. Kamal, A. Gupta, L. Lilien, and Z. Yang, *Conference: Proceedings The 3rd International Conference on Collaborative Computing: Networking Applications and Worksharing*, White Plains, New York, November 12–15, 2007.)

17.4.4 Sample Application Scenario for the MicroOppnet

To illustrate the use of MicroOppnet, let us consider an emergency scenario, for example, a fire in a large office building. Suppose that some workers were unable to evacuate. Most of them tried to use their cell phones to call for help. Many succeeded but many failed to get a connection as the cell phone infrastructure is overloaded with calls made by thousands of workers gathered outside of the building.

The firefighters can put a MicroOppnet (or, maybe, a MiniOppnet) to use. They deploy around the office building the MicroOppnet seed, consisting of laptops and networks connecting them. Now, the BT Class 1 connectivity (BT Class 1 has the range of ~100 m) becomes an essential communications capability, with the MicroOppnet using it to discover all kinds of BT-enabled helpers. An owner of any such helper, that is, an owner of a BT-equipped cell phone, PDA, laptop, etc., is now able to communicate with the firefighters via the extended MicroOppnet (consisting of the seed MicroOppnet plus all helpers that joined it).

We have so far used only class 1 opportunistic capabilities of the MicroOppnet. To show how class 2 opportunistic capabilities of the MicroOppnet can be used, suppose that the MicroOppnet is now commanded to contact and query for temperature readings from all sensing nodes within the building (they include a multitude of oppnet-enabled, in this case BT-enabled, smoke detectors with add-in multisensor capabilities). These temperature readings, aggregated at a Java server, are used to plot the heat profile for the building. The profile, together with location information gathered by

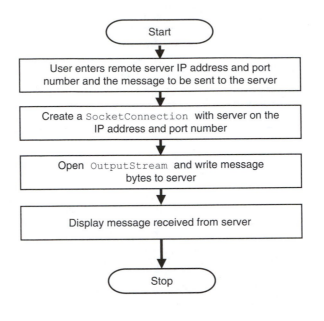

Figure 17.11 Flow of control for cell phone MIDlets. (From Z. H. Kamal, A. Gupta, L. Lilien, and Z. Yang, *Conference: Proceedings The 3rd International Conference on Collaborative Computing: Networking Applications and Worksharing*, White Plains, New York, November 12–15, 2007.)

BT-equipped helpers before, can be used by the firefighters to find the best routes for reaching the workers trapped in the building by fire.

Note that many other pervasive communications technologies could be used in parallel with BT (but our example should be clear and illustrative enough without discussing them).

17.5 Conclusion

We proposed categorization of opportunistic networks into class 1 opportunistic networks and even more opportunistic class 2 opportunistic networks or oppnets. We discussed a standard implementation framework for oppnets, OVM, and presented OVM primitives for control center, seed, oppnet helpers, and lites. We gave an overview of various oppnet applications and scenarios.

We detailed the design and implementation of a small-scale oppnet named MicroOppnet, which not only serves as a proof of concept but also is currently being extended as a test bed for designing, testing, and implementing: (1) oppnet primitives; (2) routing, privacy, and security protocols; and (3) architectures.

Our future work on MicroOppnet and its successors will include (1) extending the rudimentary class 2 opportunism of MicroOppnet to a more substantial class 2 opportunism—by implementing the opportunistic growth mechanisms of class 2 opportunistic networks fully and for multiple types of resources; (2) developing comprehensive privacy and security controls [2,15,22]; (3) incorporating opportunistic routing protocols [3,9,13,36–39]; (4) extending the prototype to a medium scale either by increasing the number of communication media used (e.g., adding worldwide interoperatility for microwave acces [WiMAX]), or by increasing the number of resource kinds that can be leveraged (e.g., including computation and storage); (5) stress-testing the oppnet components, protocols, and architectures on the testbed at the DETER Lab (funded by HSARPA and operated by the Information Sciences Institute at USC); and (6) developing a rapid application development (RAD) environment for oppnets and testing it and its products on our test bed [32].

In-progress and future MicroOppnet experiments include and will include the following:

- Studying the impact of interference from environmental factors on oppnet connectivity
- Studying the impact of link failures on connectivity and routing in oppnets
- Devising algorithms for efficient detection and localization of candidate nodes, clusters, or networks
- Identifying suspicious or inefficient oppnet nodes or clusters and removing them when necessary (even the members of the original seed oppnet can be fired)
- Devising controls for selected aspects of helper privacy and oppnet security

Some broad open issues and challenges in the oppnet paradigm can be categorized as follows [40]:

1. Optimizing the seed oppnet infrastructure. By developing measures and criteria for quantitative specification of oppnet features such as communication, computation, sensing, storage, energy resources, etc.
2. Developing methods for detecting helpers with useful resources and facilities. An integrated solution to detect devices by all diverse technologies, rather than using techniques such as tradition localization, GPS, ultrasound, etc.
   a. Designing integrated communication media for oppnets. Provide seamless communication across disjoint communication media.
3. Designing methods for inviting candidate helpers, and methods for controlling helpers. Develop primitives and protocols for inviting and admitting helpers.
4. Developing methods for deciding which tasks should be offloaded by oppnet to its helpers, and techniques for coordinating helper tasks by oppnets. Conjure protocols and primitives for collaborating and delegating tasks among oppnet nodes.
5. Proposing ways of managing oppnets, including control of privacy and security problems in oppnets. Algorithms for monitoring oppnet nodes to detect and identify suspicious or ineffective members of the oppnet.
6. Analyzing performance of oppnet algorithms and protocols, including the ones for localization, invitation, task offloading, and coordination. Develop measure and metrics for evaluating efficiency and effectiveness.

In conclusion, oppnets encompass the four fundamental paradigms of pervasive computing technology, and hence is a pervasive computing enabling technology. The goal of oppnets is bridging heterogeneous devices, networks, or systems under one umbrella, so that all their resources, such as computation, communication, sensing, actuation, storage, etc., can be integrated crossing boundaries imposed by technological discrepancies, such as programming language, hardware linguistics, and communication medium. It should be evident that oppnets are also a new paradigm for UMA technology.

17.6   Acknowledgments

This research was supported in part by the NSF under grants IIS-0242840 and IIS-0209059, and in part by the U.S. Department of Commerce under Grant BS123456. The authors would also like to acknowledge Computational Science Center and Information Technology and Image Analysis (ITIA) Center. Any opinions, findings, conclusions, or recommendation expressed in the chapter are those of the authors and do not necessarily reflect the views of the funding agencies or institutions.

APPENDIX OVM Primitives for Standard Oppnet Implementation Framework

This appendix includes OVM primitives for the CC, oppnet seed, helpers, and lites.

In the following description of primitives, the arguments of procedure calls are marked as IN, OUT, or INOUT. The meanings of these are

- Call uses but does not update an argument marked IN
- Call may update an argument marked OUT
- Call both uses and updates an argument marked INOUT

17.A.1 OVM Primitives for Control Center

Following is a list of procedures or primitives supported by oppnet control center:

1. **CTRL_initiate ()**
   CTRL_initiate broadcasts a Start-Up message to all oppnet seeds. Upon this command, the oppnet seeds will start to form a network.
2. **CTRL_terminate ()**
   CTRL_terminate broadcasts a message to all oppnet seeds. Upon this command, the oppnet seeds will start to release the oppnet helpers and terminate the network.
3. **CTRL_command (command, devices)**
   IN command: the command to be broadcasted (string)
   IN devices: the identification of a set of oppnet seeds that will receive the command (handles)

   CTRL_command broadcasts a command to all oppnet seeds specified in the set of devices.

17.A.2 OVM Primitives for Oppnet Seed

17.A.2.1 Primitives for Discovery and Identification of Oppnet-Enabled Devices

1. **SEED_discover (device, message)**
   IN device: the identification of the device who broadcasts this message (handle)
   IN message: information to be broadcasted to contact oppnet-enabled devices (string)

   SEED_discover broadcasts a message from an oppnet seed to all oppnet-enabled devices in range. The message has a predefined header that determines the type and length of the message. Each oppnet-enabled device is able to interpret the header.

2. **SEED_listen (buffer, count, time)**

OUT buffer: information received from oppnet-enabled devices (string)
OUT count: number of entries in buffer (integer)
IN time: Number of seconds that buffer is receiving (integer)

SEED_listen receives message from an oppnet-enabled device that responds to the seed discovering message. It is a nonblocking procedure, so the caller will not wait until the procedure returns. The buffer keeps receiving information until either the buffer is full, or time expires. The message received from an oppnet-enabled device has a predefined header that determines the type and length of the message.

17.A.2.2 Primitives for Managing Oppnet

1. **SEED_evalDevice (attributes, description)**

OUT attributes: whether the seed meets expectation (table)
IN description: the description of the device under evaluation (string)

SEED_evalDevice evaluates a device and return a table of the attributes.

2. **SEED_admitDevice (device, attributes, expect, oppnet, admitted)**

IN device: the identification of the device under evaluation (handle)
IN attributes: the attributes of this device
INOUT oppnet: table of devices that are admitted in the oppnet (table)
OUT admitted: whether this device got admitted (Boolean)

SEED_admitDevice admits a device into the oppnet. On the basis of the expectation of the device, SEED_admit add the device into a table that maps from an expectation to a set of devices that can fulfill the expectation.

3. **SEED_update (device, expectation, admitted)**

IN device: the identification of the device to be updated (handle)
IN expectation: the new expectation that the device can meet (string)
INOUT admitted: table of devices that are admitted in the oppnet (table)

SEED_update updates a device in the oppnet based on the new expectation of the device.

17.A.2.3 Primitives for Managing Workload

1. **SEED_sendTask (from, to, task, priority)**

IN from: ID of this seed who send task (handle)
IN to: ID of the device who will receive the task (handle)
IN task: the task (string)
IN priority: the priority of the task (handle)

SEED_sendTask informs a device of the task they need to perform.

2. **SEED_receive_task (from, to, task, priority)**

IN from: ID of this seed who send task (handle)
IN to: ID of the device who will receive the task (handle)
IN task: the task (string)
IN priority: the priority of the task (handle)

SEED_receiveTask informs a device of the task they need to perform.

17.A.2.4 Primitives for Releasing Helpers and Users

1. **SEED_release (devices, admitted)**

   IN devices: a set of devices to be released from oppnet (set)
   INOUT admitted: table of devices that are admitted in the oppnet (table)

   SEED_release releases a set of devices from the oppnet. The admitted table will be changed accordingly.

17.A.2.5 Primitives for Timer and Synchronization

1. **SEED_barrier (devices)**

   IN devices: the set of devices considered (set)

   SEED_barrier blocks the caller until all devices specified in the input parameter have called it.

2. **SEED_wait (seconds)**

   IN seconds: wait for certain amount of time (int)

   SEED_wait waits for certain amount of time before taking another action.

17.A.3 OVM Primitives for Oppnet Helpers

17.A.3.1 Primitives for Joining Oppnet

1. **HLPR_join (device, message)**

   IN device: identification of the device who asks helper to join oppnet (handle)
   IN message: information about the capability of this oppnet helper (string)

   HLPR_join responds to the request sent from device and express the willingness to join oppnet.

2. **HLPR_discover (device, message)**

   IN device: identification of this device (handle)
   IN message: information to be broadcasted to contact oppnet-enabled devices (string)

   HLPR_discover broadcasts a message from an oppnet helper to all oppnet-enabled devices in range. The message has a predefined header that determines the type and length of the message. Each oppnet-enabled device is able to interpret the header.

17.A.3.2 Primitives for Accepting or Assigning Oppnet Tasks

1. **HLPR_strongTask (device, task, respond)**

   IN device: identification of the device who assign tasks (handle)
   IN task: the tasks requested (string)
   OUT respond: whether such task will be performed (Boolean)

   HLPR_strongTask responds to the request sent from device and express the willingness to join oppnet. By accepting this task, the device will abort previous task.

2. **HLPR_weakTask (device, task, respond)**

IN device: identification of the device who assign tasks (handle)
IN task: the tasks requested (string)
OUT respond: whether such task will be performed (Boolean)

HLPR_weakTask responds to the request sent from device and express the willingness to join oppnet. By accepting this task, the device will put the task in a task queue.

3. **HLPR_assignStrongTask (device, task)**

IN device: identification of the device who assign tasks (handle)
IN task: the tasks requested (string)

HLPR_assignStrongTask assign tasks to a device. If accepted, the task will interrupt the previous task at the device.

4. **HLPR_assignWeakTask (device, task)**

IN device: identification of the device who assign tasks (handle)
IN task: the tasks requested (string)

HLPR_assignWeakTask assign tasks to a device. If accepted, the task will be queued.

17.A.3.3   Primitives for Internal Operation

1. **HLPR_evalTask (task, respond)**

IN task: the tasks requested (string)
OUT respond: whether such task will be performed (Boolean)

HLPR_evalTask evaluates whether the task requested from device will be performed. The task will not be declined if (a) the device who sent this task has a low security clearance level, (b) the helper is not able to perform such task by design (e.g. memory not enough), (c) the current resources prevent helper to perform the task (e.g. low battery), (d) in case of preemptive task, the security clearance level of this task is lower than the previous one, or (e) in case of nonpreemptive task, the task queue is full.

2. **HLPR_selectTask (task)**

OUT task: the task that is selected (string)

HLPR_selectTask select a task from the task queue to perform

17.A.3.4   Primitives for Quitting Oppnet

1. **HLPR_release (device)**

IN device: identification of the device to be released from oppnet (handle)

HLPR_release releases another oppnet helper or oppnet users from oppnet

2. **HLPR_leave(seedDevice, helpDevice)**

IN seedDevice: identification of a seed device whom to be informed of its release (handle)
IN helpDevice: identification of this helper device (handle)

HLPR_leave informs an oppnet seed that this helper will quit oppnet.

17.A.4 OVM Primitives for Oppnet Lites

17.A.4.1 Primitives for Joining Oppnet

1. **LITE_join (device, message)**

   IN device: identification of the device who asks helper to join oppnet (handle)

   IN message: information about the capability of this oppnet helper (string)

   LITE_join responds to the request sent from device and express the willingness to join oppnet.

17.A.4.2 Primitives for Accepting or Assigning Oppnet Tasks

1. **LITE_strongTask (device, task, respond)**

   IN device: identification of the device who assign tasks (handle)

   IN task: the tasks requested (string)

   OUT respond: whether such task will be performed (Boolean)

   LITE_strongTask responds to the request sent from device and express the willingness to join oppnet. By accepting this task, the device will abort previous task.

2. **LITE_weakTask (device, task, respond)**

   IN device: identification of the device who assign tasks (handle)

   IN task: the tasks requested (string)

   OUT respond: whether such task will be performed (Boolean)

   LITE_weakTask responds to the request sent from device and express the willingness to join oppnet. By accepting this task, the device will put the task in a task queue.

17.A.4.3 Primitives for Internal Operation

1. **LITE_evalTask (task, respond)**

   IN task: the tasks requested (string)

   OUT respond: whether such task will be performed (Boolean)

   LITE_evalTask evaluates whether the task requested from device will be performed. The task will not be declined if (1) the device who sent this task has a low security clearance level, (2) the helper is not able to perform such task by design (e.g., memory not enough), (3) the current resources prevent lite to perform the task (e.g., low battery), (4) in case of preemptive task, the security clearance level of this task is lower than the previous one, or (5) in case of nonpreemptive task, the task queue is full.

2. **LITE_selectTask (task)**

   OUT task: the task that is selected (string)

   LITE_selectTask select a task from the task queue to perform.

17.A.4.4 Primitives for Quitting Oppnet

1. **LITE_leave (seedDevice, helpDevice)**

   IN seedDevice: identification of a seed device whom to be informed of its release (handle)

   IN helpDevice: identification of this lite device (handle)

   LITE_leave informs an oppnet seed that this device will quit oppnet.

REFERENCES

1. U. Hansmann, L. Merk, M. S. Nicklous, and T. Stober, *Pervasive Computing*, 2nd Edition, Springer-Verlag, Berlin, Heidelberg, New York, 2003.
2. Wikipedia contributors, Generic access network, *Wikipedia, The Free Encyclopedia*, June 20, 2007, accessed June 28, 2007 http://en.wikipedia.org/w/index.php?title=Generic_Access_Network&oldid= 139404377.
3. L. Lilien, Z. H. Kamal, V. Bhuse, and A. Gupta, Opportunistic networks: The concept and research challenges in privacy and security, Proceedings of the International Workshop on Research Challenges in Security and Privacy for Mobile and Wireless Networks (WSPWN 2006), Miami, Florida, March 2006.
4. B. Bhargava, L. Lilien, A. Rosenthal, and M. Winslett, PervasiveTrust, *IEEE Intelligent Systems*, 19(5), 74–77, September/October 2004.
5. H. Karl and A. Willig, A short survey of wireless sensor networks, Technical Report TKN-03-018, Technical University Berlin, Berlin, October 2003.
6. L. Lilien, Developing specialized ad hoc networks: The case of opportunistic networks, Proceedings of Workshop on Distributed Systems and Networks at the WWIC 2006 Conference, Bern, Switzerland, May 2006.
7. L. Lilien, A taxonomy of specialized ad hoc networks and systems for emergency applications, Proceedings of the 1st International Workshop on Mobile and Ubiquitous Context Aware Systems and Applications (MUBICA 2007), Philadelphia, Pennsylvania, August 2007.
8. L. Lilien, A. Gupta, and Z. Yang, Opportunistic networks for emergency applications and their standard implementation framework, Proceedings of the 1st International Workshop on Next Generation Networks for First Responders and Critical Infrastructure (NetCri07), New Orleans, Louisiana, April 2007.
9. L. Lilien, Z. H. Kamal, V. Bhuse, and A. Gupta, Opportunistic networks: The concept and research challenges in privacy and security, in: K. Makki et al. (Ed.), *Mobile and Wireless Network Security and Privacy*, Springer Science+Business Media, Norwell, Massachusetts, 2007.
10. On Star Explained, OnStar Corp., 2007. Last accessed on June 23, 2007. Available at http://www.onstar.com/us_english/jsp/explore/index.jsp.
11. MANET Implementations, August 2004, http://www.comnets.uni-bremen.de/~koo/manet-impl. html.
12. National Strategy for Homeland Security, Office of Homeland Security, July 2002.
13. I. F. Akyildiz, W. Su, Y. Sankarasubramaniam, and E. Cayirci, Wireless sensor networks: A survey, *Computer Networks*, 38, 393–422, 2002.
14. M. Mutka, Personal communication, Department of Computer Science and Engineering, Michigan State University, East Lansing, Michigan, December 2006.
15. M. Baker, R. Buyya, and D. Laforenza, Grids and Grid technologies for wide-area distributed computing, Software-Practice & Experience, 32(15), 1437–1466, 2002.
16. IBM Solutions Grid for Business Partners Helping IBM Business Partners to Grid-enable applications for the next phase of e-business on demand.
17. Wikipedia contributors, Grid computing, *Wikipedia, The Free Encyclopedia*, June 28, 2007, accessed on June 30, 2007 http://en.wikipedia.org/w/index.php?title=Grid_computing&oldid=141246324.
18. P. Papadimitratos and Z. J. Haas, Securing mobile ad hoc networks, in: Ilyas, M. (Hrsg.), *Handbook of Ad Hoc Wireless Networks*. CRC Press, 551–567, 2002.
19. I.F. Akyildiz, X. Wang, and W. Wang. Wireless mesh networks: A survey, *Computer Networks*, 47(4), 445–487, March 2005.
20. R. Bruno, M. Conti, and E. Gregori, Mesh networks: Commodity multi-hop ad hoc networks, National Research Council (CNR), IEEE Communications, March 2005.
21. Wikipedia contributors, Mesh networking, *Wikipedia, The Free Encyclopedia*, June 21, 2007, accessed on June 30, 2007 http://en.wikipedia.org/w/index.php?title=Mesh_networking&oldid=139745759.

22. B. Ahlgren, L. Eggert, B. Ohlman, and A. Schieder, Ambient networks: Bridging heterogeneous network domains, Proceedings of the 16th Annual IEEE International Symposium on Personal Indoor and Mobile Radio Communications (PIMRC), September 2005.

23. Ambient Networks, Last accessed on June 30, 2007. Available at http://www.ambient-networks.org/.

24. N. Niebert, A. Schieder, H. Abramowicz, G. Malmgren, J. Sachs, U. Horn, C. Prehofer, and H. Karl, Ambient networks—An architecture for communication networks beyond 3G, *IEEE Wireless Communications* (Special Issue on 4G Mobile Communications—Towards Open Wireless Architecture), 11(2), 14–22, April 2004.

25. Wikipedia contributors, Ambient network, *Wikipedia, The Free Encyclopedia*, January 27, 2007, accessed June 30, 2007 http://en.wikipedia.org/w/index.php?title=Ambient_network&oldid=103581689.

26. V. Cerf, S. Burleigh, A. Hooke, L. Torgerson, R. Durst, K. Scott, K. Fall, and H. Weiss, Delay-tolerant network architecture, DTN Research Group Internet Draft, March 2003.

27. L. Pelusi, A. Passarella, and M. Conti, Opportunistic networking: Data forwarding in disconnected mobile ad hoc networks, *IEEE Communications*, 44(11), 134–141, November 2006.

28. Y. Wang, S. Jain, M. Martonosi, and K. Fall, Erasure-coding based routing for opportunistic networks, ACM Conference of the Special Interest Group on Data Communication (SIGCOMM 2005), Philadelphia, Pennsylvania, August 2005

29. L. M. Feeney, B. Ahlgren, and A. Westerlund, Spontaneous networking: An application-oriented approach to ad hoc networking, *IEEE Communications Magazine* 39 (6), 176–181, June 2001.

30. Crossbow Technology Inc., 2007. Accessed on June 30, 2007. Available at http://www.xbow.com/.

31. nesC: A programming language for deeply networked systems, UC Berkeley WEBS Project, 2004. Accessed on June 30, 2007. Available at http://nescc.sourceforge.net/.

32. RAPIDware: Component-based development of adaptable and dependable middleware, Network Systems (SENS) Laboratory, Michigan State University. Accessed on June 30, 2007. Available at http://www.cse.msu.edu/~mckinley/rapidware/.

33. Atinav, 2004–2006. Accessed on June 30, 2007. Available at http://www.atinav.com.

34. B. Hopkins and R. Anthony, Bluetooth for Java, Apress, 2003.

35. TinyOS, UC Berkeley, 2004. Accessed on June 30, 2007. Available at http://www.tinyos.net/.

36. D. S. Milojicic, V. Kalogeraki, R. Lukose, K. Nagaraja, J. Pruyne, B. Richard, S. Rollins, and Z. Xu, Peer-to-peer computing, Report HPL-2002-57, HP Laboratories, Palo Alto, California, March 2002.

37. E. M. Royer and C.-K. Toh, A review of current routing protocols for ad-hoc mobile wireless networks, *IEEE Personal Communications*, 6 (2), 46–55, April 1999.

38. P. Sistla, O. Wolfson, and B. Xu, Opportunistic data dissemination in mobile peer-to-peer networks, Proceedings of the 9th International Symposium on Advances in Spatial and Temporal Databases (SSTD 05), Angra dos Reis, Brazil, August 2005.

39. Z. Zhou, A survey on routing protocols in MANETs, Tech. Rep. MSU-CSE-03-8, Department of Computer Science, Michigan State University, East Lansing, Michigan, March 2003.

40. L. Lilien, Z. H. Kamal, and A. Gupta, Opportunistic networks: Research challenges in specializing the P2P paradigm, Proceedings of the 3rd International Workshop on P2P Data Management, Security and Trust (PDMST'06), Krakow, Poland, September 2006, pp. 722–726.

41. Peer-to-peer systems and applications, R. Steinmetz and K. Wehrle (Eds.), *Lecture Notes in Computer Science*, Vol. 3485, September 2005, http://www.peer-to-peer.info.

42. TMIO Intelligient Ovens, 23 April 2008, http://www.tmio.com/

43. Z. H. Kamal, A. Gupta, L. Lilien, and Z. Yang, The microoppnet tool for collaborative computing experiments with class 2 opportunistic networks, *Conference: Proceedings The 3rd International Conference on Collaborative Computing: Networking Applications and Worksharing*, White Plains, New York, November 12–15, 2007.

Index

A

Access categories (ACs)
    queues, 171
    user traffic priorities mapped to, 170–171
Access category index (ACI), 309
Access network convergence, 21
Access points
    AP detection mechanism, 91–92
    placement of, 64–65
    ultrathin, 67
Access services network (ASN), 124
ACK procedure, 173
ACK-Security-Block packet, 190
Adaptation Protocol (L2CAP), 382
Address Resolution Protocol spoofing, ARP spoofing
Ad hoc mesh network, 156
Ad hoc QoS multicast, 272
AES (advanced encryption standard), 12, 78, 215
Algorithmic attacks, 213
Ambient networks (ANs), 368–369
Analytic hierarchy process (AHP), 127
AP caching scheme, 194–195
AP List Request, 193
Application server level convergence, 22–23
AQM, *see* Ad hoc QoS multicast
ARF, *see* Auto rate fallback
ARP spoofing
    IP addresses and hardware addresses, 90–91
    protections against, 91
ASN gateway (ASN GW), 124
ASTM-E17.51, 300
Authentication and Key Agreement, 224
    third generation, mechanism of, 225
Authentication server (AS), 199
AUTHORIZATION-DATA, 220
AUTN (authentication vector), 225
Auto rate fallback, 59

B

Backoff algorithm, 236–237
BANs, *see* Body area networks

Base stations (BSs)
    average system throughput of, 239, 243
    backoff process, 236
    bandwidth allocation to SS, 180
    broadcasting process, 235–236
    BS_1, 370
    BS_2, 379
    data subcarriers, 235
    DL/UL throughput, 239–241
    listening process, 235
    OFDM subcarriers of, 234
        reconfiguration process, 236
    re-listening process, SSs, 236
    service flow between SS and, 181
    subcarriers distributions, 245–246
        and stations' starting order, 240–241
Base transceiver station, 224
Basic service set, *see* BSS
Basic Service Set Identifier (BSSID), 189
Beamforming and beam-steering antennas, 286
Benevolent oppnet, 364–365
BE scheduling, 179
B3G networks (Beyond 3G networks), 20
    multidimensional heterogeneity, 21
Bluetooth technology, 351
    cell phones, 370
    dongle, 382
    smoke detectors, 360, 383
    WPAN, 326
Body area networks, 353–354
Broadcasting algorithm, 235–236, 237
Brute force attack, 212
BS algorithms, 235–236
BSS, 65, 69–70, 164–165, 189, 208
BTS, *see* Base transceiver station
Bypass scanning, 193–194

C

Cache hit and cache miss, 195
Caching technique, bypass scanning, 193
Candidates, 353

393